AF729744

KEYS TO SUCCESS IN FORESTRY

NIPA GENX ELECTRONIC RESOURCES & SOLUTIONS P. LTD.
New Delhi-110 034

About the Authors

Dr. Manas Ranjan Nayak is a Scientist (Forestry) working in the Krishi Vigyan Kendra, Koraput, OUAT, Bhubaneswar, Odisha (India). He has more than 4 year of experience in research and extension in the field of forestry and agroforestry. He received his B. Sc. (Forestry) and M. Sc. (Forestry) from Orissa University of Agriculture and Technology, Bhubaneswar, India and Ph.D Forestry (Agroforestry) at Navsari Agricultureal University, Gujurat. He was the recipient of Merit Scholarscip during M.Sc., University Gold Medal from OUAT, Bhubaneswar, INSPIRE Fellowship, from DST, Govt. of India during course of study. He has received some professional's award like KVK Scientist Award, Young Scientist Award, Young Agroforestry scientist award-2020. He has qualified ICAR-SRF, ICAR-NET examination in agroforestry. He has more than 30 research papers published in various national and international journals and also published more than 40 popular articles on forestry, Agroforestry and environmental aspects. He was participated in several national and international conferences, seminars, workshop/training, summer school and presented oral and poster presentation in the thematic area of forestry, Agroforestry, Biodiversity and climate change with many best poster and oral awards of National and International level. He is also life member of 5 professional societies and also member of several committees of OUAT. He wrote 4 books: "Agrisilviculture in Odisha-Food crops and soil Fertility" (ISBN: 978-3-659-58141-0), "Forestry Solution" (ISBN: 818399098-3), Nursery Management (ISBN: 9789386200679) and Agroforestry Concepts & Principles (ISBN: 9789388892636) and 6 book chapter in different book.

Dr. Laxmikanta Behera is an Assistant Professor (Silviculture) working in the Department of Silviculture and Agroforestry, College of Forestry, Navsari Agricultural University, Navsari, Gujarat (India). He has more than 9 years of experience in teaching, guiding PG students and research in the field of Forestry. He did his graduation in Forestry from Department of Forestry, College of Agriculture, OUAT, Bhubaneswar whereas Master degree in Forestry with specialization in Agroforestry and Ecology along with Ph.D. in Forestry with specialization in Agroforestry degree from Navsari Agricultural University, He has qualified ICAR–NET examination in Forestry during 2009. He was awarded with Chancellor Gold Medal from NAU. Also received the ICAR Best Teacher award in the Forestry faculty of NAU during 2017 with many best poster and oral awards of National level Seminar/ Symposium. He has more than 50 research papers published in various national and international reputed peer reviewed journals, with book and book chapters and also published numerous popular articles on Forestry aspects. He has organized training cum workshops for students as well as for farmers. He has developed more than 10 technologies for farmer as well as for scientific community. Handling more than 5 external funded research projects as a principal investigator and co-principal investigator. He is also in the editorial board and reviewer panel of journals.

KEYS TO SUCCESS IN FORESTRY

Manas Ranjan Nayak
Scientist (Forestry)
Krishi Vigyan Kendra, Koraput,
OUAT, Bhubaneswar, Odisha, India

Laxmikanta Behera
Assistant Professor
Department of Silviculture and Agroforestry
College of Forestry
Navsari Agricultural University
Navsari, Gujarat, India

NIPA GENX ELECTRONIC RESOURCES & SOLUTIONS P. LTD.
New Delhi-110 034

NIPA GENX ELECTRONIC RESOURCES & SOLUTIONS P. LTD.

101,103, Vikas Surya Plaza, CU Block
L.S.C.Market, Pitam Pura, New Delhi-110 034
Ph : +91 11 27341616, 27341717, 27341718
E-mail:newindiapublishingagency@gmail.com
www: www.nipabooks.com
For customer assistance, please contact
Phone: + 91-11-27 34 17 17 Fax: + 91-11- 27 34 16 16
E-Mail: feedbacks@nipabooks.com

© 2022, Publisher

ISBN: 978-81-19235-40-7

All rights reserved. No part of this publication may be reproduced, stored in a retrieval system or transmitted in any form or by any means, including electronic, mechanical, photocopying recording or otherwise without the prior written permission of the publisher or the copyright holder.

This book contains information obtained from authentic and highly reliable sources. Reasonable efforts have been made to publish reliable data and information, but the author/s, editor/s and publisher cannot assume responsibility for the validity, accuracy or completeness of all materials or information published herein or the consequences of their use. The work is published with the understanding that the publisher and author/s are not attempting to render any professional services. The author/s, editor/s and publisher have attempted to trace and acknowledge the copyright holders of all material reproduced in this publication and apologize to copyright holders if permission and/or acknowledgements to publish in this form have not been taken. If any copyrighted material has not been acknowledged, please write to us and let us know so that we may rectify the error, in subsequent reprints.

Trademark Notice: NIPA, the NIPA logos and their presentations (the way they are written/ presented) in this book are the trademarks of the publisher and hence may not be used without written permission, if copied or used without authorization, the infringer will be prosecuted as per law.

NIPA also publishes books in a variety of electronic formats. Some content that appears in print may not be available in electronic books, and vice versa.

Composed and Designed by NIPA.

Preface

This book owes its origin to the sincere and hardworking analysis of questions being asked in examinations like ARS, NET, SRF, JRF, PG entrance examination at SAUs and various forestry related competitive examinations at state and national level. The compilation is an endeavour to present day relevant questions in user friendly or systematically in more reliable way. The questions presented in this book will give an idea about the model questions and will help to be focused.

The main objective of this book is to help the readers, especially students of forestry graduates, to quickly grasp the facts comprehensively and systematically from various subjects of forestry like Agroforestry, Silviculture, Forest mensuration, Forest Product Utilization, Tree improvement, Forest management, Forest protection, Tree physiology, Tree seed technology, Forest statistics, Rangeland management, Forest engineering, Wildlife management, Forest ecology and environment, Forest biotechnology, Forest soil, Remote sensing, Wood Science and Technology with other additional information related to forestry. This book consists of 21 chapters covering more than 4500 questions on various aspects of forestry. This book will meet out the growing need of forestry students to acquire the up to date information in the various subjects of forestry for getting success in forestry related competitive examinations.

We hope that the information provided and compiled in the form of questions would be useful for students of Forestry and Environmental faculties and it serves our purpose. Readers are welcome to point out errors and omissions, if any, in the book and also requested to send their valuable suggestions for further improvement of quality of this book.

Authors.

Contents

Glossary

A

Abatement: Decreasing the degree or intensity of, or eliminating, pollution.

Abiota: The soil, water, and air of an ecosystem or environment i.e. non living componenet is also known as Abiota.

Abiotic disease: An unhealthy state of being caused by a non-living factor or condition

Abiotic: Nonliving components of the environment such as air, rocks, soil particles, and plant litter.

Abrasion: General grinding or wearing of rock material by impact of objects in transport.

Absorption: The passage of one substance into or through another, either physically or chemically. It can also be defined as the uptake of water or other fluids or dissolved chemicals by a cell or an organism by capillary, osmotic, solvent or chemical actions.

Acid precipitation: A condition that occurs when moisture in the air combines with sulphur or nitrogen compounds to form acid. It is a serious threat to plant and animal life.

Acidification: Raising the acidity of a fluid by adding an acid or reduction in p^H, usually in soil. Acidification of soils and natural waters by acid rain or acid wastes can result in reduced biological productivity if the p^H is sufficiently reduced. Acidification can result in poorly structured or hard setting top soils that cannot support sufficient vegetation to prevent erosion.

Acre: A land area of 43,560 square feet. An acre can be any shape. If square, it would ensure approximately 209 feet per side.

Act: A congressional or legislative document contains the language of a proposed new law.

Adaption: Adjustment to a new or changing environment.

Adaptive behaviour: Changes in the habitat of animals that they have learned for the purpose of increasing their chances of survival.

Adaptive management**:** A process for ecosystem management that allows for continual change as more information becomes available on resource conditions and the effects of management actions on resources.

Adenosine triphosphate (ATP): A high energy molecule produced during photosynthesis; source of energy for muscle movement in animals.

Adhesion: A union of two surfaces those are normally separate.

Adsorb: Collecting gas or liquid molecules on the surface of another material.

Adsorption: One substance taking up another at its surface or the adhesion of molecules of gas, liquid, or dissolved solids to a surface.

Adventitious root: A root that grows downward from the main stem of a tree to provide added support to a tree growing in an aquatic environment; also known as a prop root.

Aerial fuel: Fuel located more than 6 feet above the ground in the mid to upper canopy of the forest.

Afforestation: The establishment of a forest through tree planting or seeding on land that lacked forest cover for a very long time or has never been forested.

Air exchange rate: The rate at which outside air replaces indoor air in a given space.

Air pollution: The presence of contaminations or pollutant substances in the air that interfere with human health or welfare, or produced other harmful environmental effects. The modification of the natural characteristics of the atmosphere by a chemical, particulate matter, or biological agent. Pollution can occur from natural changes or from human activities. The five major primary pollutants are carbon monoxide, hydrocarbons, nitrogen compounds, particulate matter, and sulphur dioxide.

Albedo: Reflectance; refers to the ratio of light from the sun that is reflected by the earth's surface, to the light received by it. Unreflecting light is converted to infrared radiation (Heat), which causes atmospheric warming. Thus, surface with a high albedo, like snow and ice, generally contribute to cooling, whereas surfaces with a low albedo, like forests, generally contribute to warming. Changes in land use that significantly alter the characteristics of land surfaces can alter the albedo.

Alfisols: Slightly acid soils with high mineral contents of Calcium, Magnesium, Sodium and Potassium.

Algae: Simple rootless plants that grow in sunlit waters in proportion to the amount of available nutrients. A general category photosynthesizing unicellular and multicellular organisms, primarily aquatic. They can affect water quality adversely by lowering the dissolve oxygen in the water. They are food for fish and small aquatic animals. Algae include many unicellular planktonic organisms, diatoms, and multicellular seaweeds.

Alkali: Any chemical substance which forms soluble soaps with fatty acids. Alkalis are also referred to as bases. Alkali may cause severe burns to the skin. Alkali turn litmus paper blue and have p^H value s from 8 to 14.

Alleles: Two genes that occupy the same position on homologous chromosomes and that cover the same trait (like 'flavors' of a trait).

Allelopathic effect: A response to competition where by live plants on the soil surface release small amount of poisonous chemicals into the soil around them as protection against invasion by other plants.

Alley Cropping: Growing annual crops between rows of trees or shrubs.

Alluvial pan: A geological formation of gravel, clay, sand and silt deposited by water; located where a stream slows down as it enters a plain or where a tributary joins a main stream.

Alluvial: Relating to and/or sand or soil deposited by flowing water.

Anaerobic: A life or process that not requires air or oxygen and is not destroyed by, the absence of oxygen. Used in reference to decomposition process that occur in the absence of oxygen.

Anaphase: the step during the cell division process known as mitosis during which the chromatids are pulled apart by the spindles.

Anatomy: The study of the structure of an organism.

Angiosperm: A class of tree that produces seeds inside an ovary or fruit; a flowering tree.

Annual ring: Bands which show tree growth for one year, as viewed on the cross section of a stem, branch, or root, or on a trunk core sample. Can be counted to determine a tree's age. Variation in width of rings records how the tree responded to growing conditions in different years.

Anther: the plant organ in which pollen grains develop and mature.

Anthropogenic: Man-made, not natural (or) Based on human activities; often used to refer to environmental changes caused by human activity.

Apical meristem: A rapidly dividing mass of cells located on the ends of stems and roots that cause the stems and roots to elongate and grow longer.

Aquatic: Adapted to water environment.

Aquifer: A stratum of earth or permeable rock that stores significant quantities of water. A confined aquifer is sealed above and underlain below by impermeable material.

Arable land: Lands that can be used for growing crops or for cultivation of crops.

Arboriculture: The scientific care of shrubs and trees in cities and towns.

Arid: A condition in which less than 10 inches of rainfalls each year and the level of evaporation is greater than the level of precipitation.

Artificial regeneration: Forest renewal that occurs when seeds or seedlings are planted at a harvest site.

Asexual reproduction: Propagation of plant or other organism from its part, such as seedling production from leaf, root or stem tissue.

Aspect: The compass direction which a slope faces.

Atmosphere: The air that surrounds the earth and is bounded by the earth's gravitational attraction. The gas layer surrounding a planet. Excluding the water vapor, the earth atmosphere consist of 78% nitrogen, 21% oxygen, 0.9 % argon, 0.035% carbon dioxide and minute quantity of other gases. Water vapor may constitute up to 3 per cent, depending up on the relative humidity and temperature of the atmosphere.

Autecology: Ecology dealing with biological relations between a species, or individual organism, and its environment.

Autotroph: An organism that produces complex organic compounds from simple inorganic molecules using energy from light or inorganic chemical reactions.

Adaptive management: A process for ecosystem management that allows for continual change as more information becomes available on resource conditions and the effects of management actions on resources.

Advanced reproduction: Young trees established before a regeneration cutting.

Aerial logging: Removal of logs from a harvest area by helicopter.

Age class: All trees in a stand within a given age interval, usually 10 or 20 years.

Allowable cut : The maximum volume of wood that can be harvested during a given period, usually equal to the forest's net growth.

Annual growth: The yearly increase in wood volume, usually expressed in terms of board feet or cubic feet per acre.

AUM (Animal unit month) : The amount of forage required by a mature cow(or other grazing animal) and her calf for one month.

Autecology: Ecology dealing with biological relations between a species, or individual organism, and its environment.

Adventitious: Organs that arise in unexpected positions e.g. adventitious roots growing from a stem or a leaf, or adventitious buds (and shoots) growing from a root segment.

Aseptic culture: Procedures in tissue culture used to prevent the introduction of fungi, bacteria,vira, mycoplasma or other microorganisms into cell, tissue and organ culture.

See Micropropagation, Sterile, Tissue culture.

Acquired character: A modification developed during the lifetime of an organism as the result ofenvironmental influences. Such modifications are not inherited.

Adaptability: (1) The process of evolutionary (genetic) adjustments fitting individuals or groups to their environment. Also the changed structure or function itself. (2) A tree's performance over a full rotation in a new environment.

Adapted: Refers to how well trees are physiologically suited for high survival, good growth, and resistance to pests and adverse environments. For exotics it refers to how well the trees will perform in their new environment.

Additive genes: Genes conveying effects additively. Additivity is the basis for most quantitative inheritance theory and for most breeding efforts. For ex., the expectation of hybrid performance isthe average of the parental values, i.e., mid-parental.

Advanced-generation selection: Selection of a tree by genetic test of crosses among parents from previous generation. Some form of family and within-family selection is usually used to choose advanced-generation selections.

Ageing (aging): A progression of physiological, cytological and biochemical events which ultimately leads to the death of an organ or an individual.

Air layering: Procedure to develop roots on undetached aerial parts of a plant by girdling or wounding the area where roots are desired and surrounding the area with a rooting medium and possibly applying rooting hormone.

Albinism: Complete or nearly complete absence of pigment in a plant or parts of a plant. The affected parts are white or nearly so. An albino plant usually soon dies.

Allele: One of an array of genes possible at a certain position (locus) on a given chromosome. Alternative (Mendelian) effects on the same character are produced by different alleles, e.g., as met in green or albino seedlings. If the array contains more than two genes, the genes are called multiple alleles. Multiple alleles arise by repeated mutations of a gene, each with different effects. No more than two alleles can be present in a given diploid organism.

Allopatric: Genetically isolated species or races inhabiting separate geographic areas or adjacentbut radically different habitats.

Allozyme (alloenzyme): An isozyme which differs from other variants of the enzyme as a result ofallelic difference, i.e different enzymes coded by the same gene locus.

Aneuploidy: A condition in which not all the chromosomes are present in equal numbers and hence the total number is not an exact multiple of the haploid set. It occurs when chromosomes failto separate during meiosis, so a gamete may either lack a chromosome or have an additionalcopy.

Anther: The pollen-bearing part of the stamen.

Anthesis: The period or stage of expansion of a flower; full bloom specified as diurnal, nocturnal, extended, etc. Often anthesis refers to the bursting of the pollen sacs with subsequent release of pollen.

Apex: Tip, topmost part, pointed end of anything. In shoot and root, the tip containing the apical meristem.

Apical meristem: A group of meristematic cells at the apex of root or shoot that by cell division produce the precur-sors of the primary tissues of root and shoot; may be vegetative, initiating vegetative tissues and organs, or reproductive initiating reproductive tissues and organs.

Apomixis: A general term that includes all forms of asexual reproduction that tend to substitute for sexual regeneration. Often used to mean the production of seeds without fertilization occurring. Depending on the type of apomixis, an apomict usually is genetically identical with its source plant(ortet).

Asexual reproduction: Reproduction without fertilization. New individuals may develop from vegetative parts such as tubers, bulbs, or rooted stems, or from sexual parts such as unfertilized eggs or other cells in the ovule.

Auxins: Class of plant hormones with main effect on elongation of shoots and roots. Most common natural occurring auxin is indole 3-acetic acid. Synthetic analogues, e.g. IBA and NAA, are widely used for promoting root formation on cuttings.

B

Back fire: A fire started along the inside edge of a firebreak to burn the fuel supply back to the wildfire.

Bacteria: Microscopic li8ving organism that can aid in pollution control by metabolizing organic matter in sewage, oil spills or other pollutants. However, bacteria in soil, water or in air can also cause plant, animal and human health problems.

Bare root stock: Seedling that have been removed from the soil in preparation for storage and shipping.

Bark beetle: An insect that bores into and beneath tree bark to lay its eggs. The feeding trails (galleries) girdle the tree and may kill it. Specific types of bark beetles attack different tree species. Most bark beetles can kill weakened trees.

Bark: The tissue covering the stems, branches, and roots of a plant, extending from the cambium layer to the outer surface.

Basal area: A key descriptive measure of trees and stands that indicates site occupancy. For a tree, it is the cross-sectional area of the trunk at breast height (4.5 feet above ground). The basal area of a tree 14 inches in diameter at breast height (DBH) is approximately 1 square foot. Basal area per acre is the sum of basal areas of the individual trees on an acre. A stand of 100 14-inch DBH trees would contain about 100 square feet of basal area per acre.

Baseline: A survey line that runs east-west.

Beam: A timber that is equal to or greater than 8 by 8 inches in dimension.

Beautification: Planting trees for ornamental purposes.

Biltomore stick: A measuring instrument used to determine the diameter of a tree at breast height.

Biodegradable: Capable of being broken down by bacteria or other living organisms.

Biodegradable: Materials that can be broken down into basic elements or natural substances by biological processes through micro organisms under natural conditions such as bacteria and fungi; includes most organic waste.

Biodiversity: Refers to the degree of species richness and natural genetic variation. Diversity can be defined as the variety of life in all its forms. Levels and combinations; includes ecosystem diversity, species diversity, and genetic diversity. For biological diversity, these items are organized at many levels, ranging from complete ecosystem to the biochemical structures that are the molecular basis of heredity. Thus the term encompasses different ecosystems, species, and genes.

Bioenergy: It encompasses biomass, the biological material used as biofuel, as well as the social, economic, scientific and technical fields associated with using biological sources for energy.

Biogas: Methane gas produced by animal and human dung, crop residues, and other organic matter; can be used as a fuel or fertilizer. It is also called biomass gas.

Biogeochemical cycle: A pathway by which a chemical elements or molecule moves through both biotic (bio) and abiotic (geo) parts of an ecosystem. Elements involved in biogeochemical cycles include carbon, nitrogen, sulfur, and phosphorus.

Bioindicator: A living organism, such as lichen, that provides an early warning of unhealthy conditions when it disappears from an area where it was once found in abundance.

Biological amplification: The accumulation of higher levels of pollutants in organisms higher up in the food chain.

Biological control: Controlling plants, diseases, and animal pests by the use of natural enemies; or inhibiting the reproduction of pests by methods that result in the laying of infertile eggs, etc.

Biological diversity: Richness and abundance of species, and variety of natural communities, in a forest environment. Both the number of species and the number of individuals of each species are important in considering the extent of biological diversity in an area. Also referred to as biodiversity.

Biological magnification: The phenomenon where by contaminant concentrations increase in organisms at successively higher trophic levels along a food chain. Refers to the process where by certain substances such as pesticides or heavy metals move up the food chain, work their way into rivers or lakes, and are eaten by aquatic organisms such as fish, which in turn are eaten by large birds, animals or humans. The substances become concentrated in tissue or internal organs as they move up the chain.

Biological oxidation: Decomposition of complex organic materials by microorganisms. Occurs in self-purification of water bodies and in activated sludge waste water treatment.

Biological oxygen demand: (BOD) A measure of the amount of oxygen consumed in the biological processes that break down organic matter in water. An indication of the extent to which water is polluted by sewage or other organic waste. It is measure of the dissolved oxygen consumed by micro organisms as they breakdown the organic matter in a simple of water; the greater the quantity of organic matter, the greater the oxygen required for its decomposition. It is expressed in parts per million of dissolved oxygen consumed. The greater the BOD, the greater the degree of pollution. Also called Biological Oxygen Demand.

Biological succession: Changes that occur as living organisms replace other lower-order organisms in an environment.

Biological value: The relative worth of the life forms that populate an area, taking into account economic values and effect on climate, watersheds, water temperature, soil erosion, wildlife, and so on.

Biomass: Vegetation and waste materials containing large amount of vegetative matter that is rich in cellulose.

Biome: A climatic and geographically defined area of ecologically similar communities of plants, animals, and soil organisms, often referred to as ecosystems.

Biome: A group of ecosystem within a region that have similar type of vegetation and climatic condition.

Biosphere: The natural system of the earth and the atmosphere that supports life. The parts of the earth, including air, land, surface rocks, and water, within which life occurs, and which biotic processes in turn alter or transform. Is the interface between the hydrosphere, geosphere and atmosphere.

Biota: The animal and plant life of a given region, the plant and animal life of an ecosystem; often referred to as flora and fauna.

Biotic disease: An unhealthy condition caused by living agent of infections such as bacteria, fungi, viruses, micro plasmas, parasites, nematodes and so on.

Bipinately compound: Multiple leaflets attached to side branches that grow opposite one another on a central leafstalk.

Blackwood: A species of acacia or wattle (*Acacia melanoxylon*), that is widespread in eastern Australia, southern south Australia and Tasmaina and produces a high quality hard wood timber.

Blade : The flat part of leaf; usually green in colour.

Blow up: A crown fire that suddenly erupts into an extremely intense fire.

Blue gum: Generally refers to the forest trees, Tasmania blue gum (*Eucalyptus globulus*), which is grown widely for wood and hardwood chips production.

Board foot(BF): A unit of measurement for lumber equal to a piece of wood 1 foot long, 12 inches wide and 1 inch thick.

Board: A cut of wood less than 2 inches thick and greater than 4 inches wide.

Boreal forest: A forest located in the northen zone of the north American continent characterized by a wet, cold climate.

Boreal forests: Boreal forest consists mainly of coniferous evergreen trees. Found in regions that have long winters and moderate to high annual rainfall.

Boundary Plantings: Trees planted along boundaries or property lines to mark them well.

Brown rot: A type of wood rot that occurs in trees as fungi break down the cellulose in the cell walls, especially in heart wood.

Bucking: A process by which a felled tree is cut into desirable lengths.

Bulk density: The mass per unit volume of a solid material. Usually expressed as lb/ ft^3 or g/cm^3.

Bunching: The process of assembling the logs in an area into small piles.

Butt rot: Decay or rotting of the interior wood near the base of a tree.

BLOW-DOWN : Uprooted trees as a result of high wind.

Blue stain : A fungal discoloration, carried by bark beetles, common in pines.

Board foot : A unit for measuring wood volumes equaling 144 cubic inches, commonly used to measure and express the amount of wood in a tree, sawlog, or individual piece of lumber. For example, a piece of wood measuring 1 foot x 1 foot x 1 inch, or a piece measuring 1 foot x 2inches x 6 inches, would contain 1 board foot of wood.

Bole : The main tree trunk.

Bolt : A short log or a squared timber cut from a log, less than 8 feet long.

Brush : Commonly refers to undesirable shrubs and small trees.

Buck : To cut a log into specified lengths.

Budcap : A piece of paper or other suitable material covering the terminal budof a young seedling to prevent deer browse.

Buffer : A protective strip of land or timber adjacent to an area requiring attention or protection. For example, a protective strip of un harvested timber along a stream.

Broadcast burn : A controlled fire (prescribed fire) used to burn slash without piling it.

Prescribed burn : The application of fire to a designated area under conditions ofweather, soil moisture, and time of day that will accomplish specific silvicultural, wildlife, grazing, or fire-hazard-reduction purposes.

Butt : The base of a tree or log.

Backcrossing: Crossing an offspring (usually hybrid) to either one of its parental types. Repeated backcrossing is particularly effective in transferring simply inherited characters, e.g., disease resistance, from a donor parent to a deficient but otherwise, valuable recurrent parent.

Background pollination: Pollination by long distance wind borne pollen. Some tiny pollen grains may be suspended in the air currents and carried hundreds of kilometres. Such background pollination cannot be totally excluded in e.g. seed orchards.

Base population (Selection population)**:** The population of trees from which the tree breeder chooses to make selections for the next generation of breeding. The base population can be a natural stand (first generation) or a genetic test.

Biological diversity (Biodiversity)**:** The variability among living organisms from all sources and the ecological complexes of which they are part; this includes:

- *Ecosystem diversity*: the variety and frequency of different ecosystems
- *Species diversity*: the frequency and diversity of different species
- *Genetic diversity*: the frequency and diversity of different genes and/or genomes. It includes the variation within a population and between populations atmosphere.

Biotechnology: The application of scientific and engineering principles to the processing of materials by biological agents to provide goods and services. Biotechnology comprises any technique that uses living organisms to make or modify a product, to improve plants or animals or to develop microorganisms for specific purposes.

Bisexual (Hermaphroditic or Monoclinous)**:** Having functional male and female reproductive organs in the same flower.

Block: A subunit of the test area consisting of plots. In field design usually equivalent to a replicate.

Breeding arboretum: A collection of selected trees or species established for breeding. If the collection is preserved vegetatively, it is sometimes known as a clone bank. Spacing, culture and protection are designed to stimulate early and prolific flowering for controlled pollination and seed harvest.

Breeding plan: A detailed schedule of operations to be undertaken in order to implement the breeding strategy.

Breeding population: A subset of individuals from a base population that is selected for their desirable qualities to serve as parents for the next generation of breeding.

Breeding strategy: An overview of the management of the genetic improvement of a tree species used in man-made forest.

Breeding value: The value of a parent judged by the performance of its progeny from several crosses. The breeding value of an individual is defined as twice its general combining ability (GCA).

Breeding zone: A subdivision of a species' geographical distribution in which individuals interbreed. Delineation of breeding zones may be done by means of climatic or edaphic data.

Breed true: A population of homozygous individuals derived by inbreeding which produce a new generation of homozygous individuals when inter-mating.

Budding (graft budding)**:** Grafting by inserting a bud, with a small amount of tissue, into a slit or hole made in the bark of a stock plant. After a union has formed, the portion of the stock plant above the bud is removed. A number of budding methods exist, the figure illustrates the most common ones.

Buffer area (zone): A zone surrounding a seed production area in a plantation in order to limit contamination by external pollen sources. The buffer is composed of the same species as the SPA and therefore contributes pollen to, or cross pollinate with the stand. The buffer zone is rogued according to the same criteria as the SPA but no flower/fruit promoting operation is undertaken since seeds are not collected in the buffer zone.

C

C3 and C4 plants: C4 plants comprise about 5% of all plants, are most abundant in hot and arid conditions, and include crops like sugarcane and soybeans. During photosynthesis they form molecules with 4-carbon atoms and structure at the given levels of co_2. C3 plants, the other 95%, photosynthesis to form 3 carbon molecules and increase photosynthesis with as co_2 levels increase.

Callus tissue: Plant tissue that is not differentiated into leaf, root, stem, or other specialized tissue.

Calvin cycle: A series of chemical reactions that converts carbon dioxide to simple sugars during the process of photosynthesis.

Cambium: A layer of meristem tissue located between the wood and the bark of a tree from which new wood and bark are produced.

Canker: A fungus infection in a tree that kills its soft tissues, such as cambium and the bark.

Canopy: The top part, or crown of a tree, consisting of branches and foliage.

Capillary action: Movement of water through very small spaces due to molecular forces called capillary forces.

Capillary fringe: The porous material just above the water table which may hold water by capillarity in smaller void spaces. Also referred to the zone above the water table within which the porous medium is saturated by water under less than atmospheric pressure.

Carbon credit: A market driven way of reducing the impact of green house gas emissions; it allows an agent to benefit financially from an emission reduction. There are two forms of carbon credit, those that are part of national and international trade and those that are purchased by individuals. Internationally, to achieve Kyoto Protocol objectives, caps (limits) on participating country's emission are established. To meet these limits countries in turn, set 'caps' (allowances or credits: 1 convertible and transferable credit = 1 metric tone of co_2 emissions) for operators. Operators that meet the agreed 'caps'. Operators can then choose the most cost effective way of reducing emissions. Individual carbon credits would operate in a similar way of carbon offset..

Carbon cycle: The biogeochemical cycle by which carbon is exchanged between the biosphere, geosphere, hydrosphere, and atmosphere of the earth. The cycling pathway through which carbon circulates throughout nature.

Carbon pool: A storage reservoir of carbon.

Carbon sink: Any carbon storage system that causes a net removal of greenhouse gases from the atmosphere e.g., oceans and rain forests.

Carbon stocks: The quantity of carbon held within a carbon pool at a specified time.

Carbonization process: A process also known as destructive distillation, by which wood is heated in the absence of oxygen to produce charcoal and volatile gases.

Carnivore: An animals or other organisms whose diet consists of meat.

Catchment area: All of the land surface from which water eventually flows into a particular stream. This area is also called a watershed.

Catchment: An area of land drained by a major rives and its tributaries.

Cell membrane: The outer flexible membrane that surrounds the contents of cell.

Cell wall: The rigid outer layer or covering of a plant cell.

Cell: A small structure that consists cytoplasm and a nucleus; the basic unit of life.

Cellulose xanthate: A viscous liquid obtained from the cellulose component of wood that is used to manufacture rayon, cellophane, photographic films, and so on.

Cellulose: The substance of which the cell walls of plants are composed; wood fibre used in paper manufacturing.

Centriole: A cell structure found at earlier end of a cell that anchors the fibres, drawing chromosomes apart during cell division.

Centromere: the point of attachment for a pair of chromatids.

Chain saw: A gasoline powered saw used to fell and trim trees as one of the procedure in a logging operation.

Chain: The unit of forest measurement of distance; one chain is equal to the 66-foot length of a surveyor's chain.

Chemical control: The use of chemicals to control weeds and insects.

Chemical factor: A form of stress to forests that originates from gases, toxic metals and acids.

Chemical pulping: A processing method that use chemicals to dissolve the lignin component of wood in the manufacturing process for high strength paper.

Chipping: Debarked logs, often of lower quality, are chipped or flaked into small pieces for use in particle board, paper manufacture or fuel wood.

Chlorophyll: A green substance found in plant cells that aids the plant in capturing energy from sunlight and converting its to sugars and starches through the process of photosynthesis.

Chloroplast: A plant cell structure containing chlorophyll that converts raw materials to sugars and starches using energy obtained from sunlight.

Chlorosis: Discoloration of normally green plant parts caused by disease, lack of nutrients, or various air pollutants.

Choker: A cable fastened around a log for the purpose of transporting it to the yard or landing.

Chromatid: Half of a replicated chromosome.

Cleaning operation: Removal of forest vegetation that competes with young trees during the seedling to sapling stage of maturity.

Clear felling: The complete removal of a crop of trees from a site in single operation.

Clear wood: Wood which is free from knots; also known as clear timber.

Clear-cutting: A harvest method in which all of the trees in the stand are cut at the same time.

Climax community: The populations of plants that occupies an environment when succession of species is completed and plant populations have become stable.

Climax forest: Consists of the trees that occupy an environment when plant populations have become stable and succession of species is complete.

Cloning: A process of producing plant, which all have the same genes as the original plant from which the clone was made.

Clumping: A spatial pattern of tree establishment where small groups, or clumps of trees are distributed throughout a stand, as opposed to random or even distribution of single trees.

Coastal plain: A land area, particularly along the Atlantic or golf coasts, consisting of a series of a terraces that generally run parallel to the coastline.

Co-dominant: A crown class that includes trees with crowns that are located at or near the same height as most of the trees next to it. It access full sunlight from above, but very little sunlight from the sides.

Cohort: Aggregation of trees that starts as the result of a single disturbance. If age of trees within a cohort is very narrow then it is a single age class that is even-aged. (Smith et al 1997)

Coke: A highly combustible wood product obtained by heating wood to temperatures in excess of its combustion temperature in large ovens from which oxygen is excluded.

Collenchyma: Thick walled plant cells that add strength to stems and stalks.

Commensalism: A living relationship between two organism in which one organism is benefited and the other does not benefit, but is not harmed.

Commercial thinning: A partial cut in stands where the timber removed is sold; conducted in even-aged stands, provides an interim timber harvest during a stand rotation.

Compartment: A small area of forest from which trees are harvested.

Compressed Natural Gas (CNG): An alternative fuel for motor vehicles; considered one of the cleanest because of low hydrocarbon emissions and its vapor are relatively non-ozone producing. However, vehicles fueled with CNG do emit a significant quantity of nitrogen oxides.

Condensation: The process where in water vapor in the atmosphere is cooled forming small water droplets. As water droplets increase, clouds are formed.

Conifers: Cone-bearing trees, most of which are evergreen and have needle like leaves.

Conk: A growth that arises on a tree trunk due to a fungal infection inside the tree, and form which reproductive spores are released into the environment.

Conservation easement: A permanent agreement between a landowner and a land trust or government agency that limits the use of land to enhance conservation and prevent development of the property.

Conservation tillage: The practice of reducing or eliminating tillage operations and leaving crop residues on the soil to prevent erosion.

Contact pesticide: A chemical that kills pests when it touches them, instead of by in gestation. Also, soils that contains the minutes skeletons of certain algae that scratch and dehydrate waxy-coated insects.

Containerized seedling: Very young trees produced from seeds in individual containers under green house condition.

Contour strip farming: A kind of contour farming in which row crops are planted in strips, between alternating strips of close-growing, erosion-resistant forage crops.

Controlled burn: The use of fire to burn trash materials from the soil surface in preparation for planting or as a fire prevention measure in established forests; sometimes known as a prescribed burn.

Coppice method: A form of sexual reproduction of a forest in which all of the trees are cut and new forest growth is generated from the stumps of the harvested trees; also known as sprout method.

Cork cambium: Meristem tissue that produces the cork on the outside of a root.

Cork: A protective outer covering of root and stem tissues consisting of dead plant cells containing a waxy material.

Cortex: loosely arranged parenchyma cells located in the interior of a root in which sugars and starches are stored.

Cover crop: Plants used to hold the soil during the fallow season. A crop that provides temporary protection for delicate seedling and/or provides a cover canopy for seasonal soil protection and improvement between normal crop production periods.

Cross-banding: Orientation of the grain of the wood in each layer of plywood to align the long wood fibers across one another.

Crown class: A category of similar trees based upon the position or elevation of their crowns compared to the crowns of other nearby trees.

Crown fire: A surface fire that builds in intensity until it ignites the aerial fuels, including the upper foliage of trees.

Crown: Consists of the foliage, particularly in the upper branches of a trees.

Cruise: The process of estimating the timber yield in a forest.

Cuticle: A protective coating of waxy material that forms on the outside of plant tissues such as stems and leaves.

Cutting cycle: The length of time in years that elapses between timber harvests.

Cuttings: Plant parts obtained from leaves, stems, roots, or buds that are treated with a hormone to stimulate rooting and growth of a new plant.

Cyclone: Intensive low pressure weather systems; mid-latitude cyclones are atmospheric circulation that rotate clock wise in the southern hemisphere and antic lock wise in the northern hemisphere and are generally associated with stronger winds, unsettled conditions, cloudiness and rainfall. Tropical cyclones causes strom surges in coastal areas.

Cytoplasm: All of the structures and substances within a cell except the nucleus.

Cambium : The growing layer of cells beneath the bark of a tree.

Candidate species : Plants and animals being considered for listing as federally endangered or threatened species in the United States.

Canopy: The uppermost layer in a forest, formed collectively by tree crowns.

Canopy layers : Forests with varying age classes may have several height classes. For example, an overstory canopy layer of trees overtopping a lower canopy of other trees or shrubs.

Carriage : In logging, a load-carrying device with cable systems from which logs are suspended for yarding or loading.

Catface : A wound on a tree or log usually caused by sideswiping with equipment or by wildfire.

Calks : Steel pegs in soles of heavy boots to give loggers secure footing. Also called corks.

Certified forester : A designation by the Society of American Foresters that confirms that the forester possesses knowledge and is competent to practice forestry in a credible manner.

Chain : A unit for measuring distance. A chain is 66 feet. An acre contains an area equal to 10 square chains. A section of land is 80 by 80 chains square, or 640 acres.

Check scale : An estimate of the usable wood volume in a log as measured by anindependent third party.

Chips : Small pieces of wood used for pulp, fuel, or pressed board.

Clearcut harvest : A harvest and regeneration technique removing all the trees (regardless of size) on an area in one operation. Clearcutting is commonly used with shade-intolerant species such as Douglas-fir or lodgepole pine, which require full sunlight to reproduce and grow well. Clear cutting produces an evenaged stand.

Climax community : A relatively stable plant community which has changed through stages and has dominant plant populations suited to the environment.

Clinometer : An instrument for measuring vertical angles or slopes; usually hand-held and pocket-sized.

Codominant trees : Trees whose crowns form the general level of the stand, receiving full light from above but comparatively little from the sides.

Cold deck ; A pile of logs left for later transportation.

Community : An assemblage of different organisms living and functioning togetherin a particular area. Usually named for the dominant plants, animals,or major physical components of the area.

Competition : In a forest, the struggle for water and light among neighboring treeshaving similar requirements.

Cone : The female reproductive part of conifers. This structure has overlapping scales that cover the seeds.

Conservation : The protection, improvement, and wise use of the natural environment (such as the forests, soils, and water systems) to prevent destruction and exploitation while still producing goods and services for the human population.

Conservation biology : The science of diversity, scarcity, and survival of species. Deals with active management to protect and maintain genetic variety within species. Deals also with the concepts of biological diversity, sustainability and relationships between biotic and abiotic resources.

Cord : A stack of wood that has a gross volume of 128 cubic feet. A standard cord measures 4 feet x 4 feet x 8 feet, and should contain approximately80 cubic feet of solid wood.

Critical habitat : Specific areas within the geographic range occupied by a species listed by the federal government as endangered or threatened. The physical and biological features considered necessary for the survival and recovery of the species.

Crook : An abrupt bend in a log—a defect.

Crop tree: A tree identified to be grown to maturity for the final harvest cut. Usually selected on the basis of its species, its location relative toot her trees, and its quality.

Crotch: The fork of a tree or branch.

Crown class : A relative designation of tree crowns.

- *Dominant trees* are those with crowns above the general level of the canopy.
- *Codominant trees* are those with crowns forming the general level of the canopy.
- *Intermediate trees* are those with crowns below the general level of the canopy.
- *Suppressed trees* are those much shorter than the general level of the canopy.

Crown closure : The point when, in a young stand, the crowns of the trees begin to touch each other.

Crown differentiation : The process whereby some trees grow faster and develop large, full crowns, while others fall behind in height and have smaller, sparser crowns.

Crown ratio: A measure of the length of a tree's live crown relative to total tree height.

Cruise: A survey of forest land to estimate timber quantity.

Cubic foot : A cube 12 inches on a side. One cubic foot of wood in a log usually produces from 3 to 7 board feet of lumber because of the cylindrical log shape and sawing losses.

Cull : A tree or log of merchantable size which, because of a defect, is useless for its intended purpose.

Cunit : A measurement equal to 100 cubic feet of solid wood, often used in relation to pulpwood.

Cutting area: A portion of the forest where timber is being cut or will be cut.

Cutting cycle: The planned time interval between harvesting operations in the same uneven-aged stand. For example, a cutting cycle of 10 years in an uneven-aged stand means that a harvest cutting of trees is made every 10years.

Callus: 1) The mass of parenchyma cells that develops from and around wounded plant tissues in order to diminish evaporation from the wound and initiate healing. In grafting it occurs at the junction of a graft union, arriving from living cells of both scion and stock. The production and interlocking of these parenchyma (or callus) cells constitute one of the important steps in the healing process of a successful graft. 2) In micropropagation used to describe an undifferentiated mass of cells.

Candidate tree: A tree that has been selected for grading because of its desirable phenotypic qualities but has not yet been graded or tested.

Canopy: the cover of branches and foliage formed by the crowns of trees.

Canopy density: the relative completeness of canopy usually expressed as a decimal coefficient, taking closed canopy as unit.

Compensatory afforestation: mandated afforestation to be done by the user agency as a compensation for forest land diverted for non-forestry purpose.

Cultural services: the non-material benefits people obtain from ecosystems through spiritual enrichment, cognitive development, reflection and aesthetic experience.

Catkin: Pendulous inflorescence modified for wind pollination. It is a spike of usually unisexual and reduced flowers, as in Betula and Populus. The term should not be applied to the male or staminate strobilus of conifers.

Cambium: A thin tissue of the plant located between the bark and the wood. Its cells are meristematic, i.e. they are capable of dividing and forming new cells.

Cell: The fundamental unit of a living organism and the basis of its structure and physiology. A plant cell consists of a cell wall and protoplast(cytoplasm, nucleus, cell membranes and organelles (e.g. cloroplasts and mitochondrion).

Certified tree seed: Seed collected from trees of proven genetic superiority, as defined by acertifying agency, and produced under conditions that assure genetic identity. These could come from trees in a seed orchard, or from superior ('plus') trees in natural stands with controlled pollination.

Character (trait): A distinctive but not necessarily invariable feature exhibited by all individuals of a group and capable of being described or measured; e.g., size, straightness, form, and performance. A character of a given individual will have a certain phenotype (state) as determined by the individual's genotype and environment. Selection is based on characters. Desired characters for wood producing species can be fast growth and straightness, while un-desired characters may be crookedness, low branching and susceptibility to diseases.

Chimera (Chimaera): Plants composed of two or more genetically different tissues growing separately but adjacent to each other in one plant. Chimeras may be sexually reproduced if they are in the lineage of reproductive cells.

Chromosome number: The number of chromosomes characteristic of a species. The number of chromosome sets associated with this number must also be specified; thus in Pinus the chromosome number may be expressed as n = 12 or as 2n = 24, depending on whether reference is to sex cells or normal diploid vegetative cells.

Chromosome: A microscopic, usually rod-like body carrying the genes. Number, size, and form of chromosomes are usually constant for individual species. The chromosome consists of chromatide which is made up of DNA, packaged with protein and coiled.

Clonal plantation: A plantation (wood producing population) established from vegetative propagated materials, e.g. cuttings or tissue cultured plants. The clonal plantation can be composed of several clones.

Clonal test: Comparison and evaluation of a number of clones in replicated tests. Such tests provide estimates of respective genotypes, but do not necessarily provide information on breeding behavior.

Clone: A population of genetically identical cells or individuals. Such a population is obtained by mitotic division or by asexual reproduction.

Clone (or clonal) bank: A collection of selected trees or species established for breeding. Thecollection is established and preserved by vegetative propagation.

Co-dominance: The condition in heterozygotes where both members of an allelic pair are expressed and contribute to the phenotype, which is then a mixture of the phenotypic characters produced in either homozygous condition. A hypothetic allele pair producing red and white flowers respectively in homozygotes would in a co-dominance situation produce both red and white flowers. In incomplete dominance the flowers would be pink. **Comparison or check tree:** Trees that are located in the same stand, are of nearly the same age, are growing on the same or better site as the select tree and against which the select tree is graded. Trees chosen as comparison trees are the best in the stand, with characteristics similar to "crop" trees that would be chosen in a silvicultural operation.

Conelet: An immature cone. The term is applied to the young cone from the time of female 'flower' scale closure after pollination until the initiation of rapid development of the cone a few months before maturity.

Controlled pollination: Transfer or permitted transfer of pollen from a known source to receptive flower parts of known seed parents. All other pollen, including own, are excluded (e.g. by covering flowers with isolation bags, and/or emasculation prior to pollination).

Coppice shoots: Shoots appearing from the stump after the tree has been cut. For some species coppice shoots can also be induced by girdling. Coppice shoots are often used as scions or cuttings in vegetative propagation as they are juvenile and usually have an orthotropic growth.

Cross: Same as Hybrid. Also an abbreviation of Cross-pollinate.

Crossability: Relative success in obtaining viable offspring from cross-pollinations, especiallyinterspecific ones.

Crossing over: Exchange of cromatide segments between homologous chromosomes during meiosis, resulting in gene recombination.

Cross-pollination: Pollination by a genetically different plant. An outcross is a cross to an unrelated individual.

Cryopreservation: Maintaining tissues or seeds for the purpose of long-term storage at ultra low temperatures, typically between -150°C and -190°C.

Cutting : Detached portion of stem or other plant part which when rooted, produces a whole plant.

Cyclophysis: The process of maturation of the apical meristem on a vegetative propagule.

Cytogenetics: Cytological aspects of genetics; i.e., chromosome behavior as related to sterility, or chromosome damage as related to mutations following irradiation.

Cytology: The study of the cell, i.e. its structure, function, development, and reproduction in relationto growth, differentiation, and heredity.

Cytoplasm: The living matter (protoplasm) within a cell, excluding the nucleus.

Cytoplasmic inheritance (extrachromosomal inheritance)**:** The determination of charac-teristicby genes in the cytoplasm (plasmagenes) rather than in the chromosomes. They are normally transmitted through the female gamete, which contributes most of the cytoplasm to the zygote. Genes inherited in this manner are found on DNA present in small quantities in the chloroplasts, mitochondria and sometimes in the cytosol itself.

Cytosol: The soluble part of the cytoplasm when all the cell organelles and other particles have been removed by centrifugation.

D

Dead wood: A standing tree that has died before it is harvested , such as a tree that has been killed by fire or insects.

Debarking: The process of removing the bark from a log by applying pressure and frictions to its outer surfaces.

Decibel (db): A measurement of noise intensity or the loudness of a sound.

Deciduous: Shedding leaves annually or at a particular stage of growth.

Deforestation: A forest losing 40 per cent or more of the trees (or) the conversion of forested areas to non-forest land for agriculture, urban use, development, or waste land.

Deforestation: The cutting or burning of forests to clear the land of trees.

Degradation: Any damage to the natural resources that living things need to survive.

Dehydration synthesis: A process by which a molecule of water is removed each time a single molecule, such as glucose, is added to a larger molecule, such as starch.

Dendrology: The scientific study of trees.

Denitrification: The biological reduction of nitrate to nitrogen gas by denitrifying bacteria in soil.

Denitrification: The process by which some kind of bacteria break down nitrate compounds, releasing nitrogen gas into the atmosphere.

Depredation hunt: A control measure for reducing excessive populations of large game animals by allowing sportsmen to hunt them legally.

Destructive distillation: A processes by which wood is heated in the absence of oxygen to produce charcoal and volatile gases.

Diameter at Breast Height (dbh): A forest measurement used to estimate the maturity of trees. Usually diameter taken at 1.37 m from the ground level.

Dieback: A term used to describe the gradual death of trees. Dieback can be caused by a complex interaction of factors including fungal infection, insect attack, water logging and soil salinity.

Diploid: A cell having twice as many chromosomes as it would have during a non-reproductive stage.

Direct seeding: Planting tree seeds to generate new forest growth.

Discount rate: a rate used to determine the present value of future benefits.

Direct-use value (of ecosystems): the benefits derived from the services provided by an ecosystem that are used directly by an economic agent. These include consumptive uses (e.g. harvesting goods) and non-consumptive uses (e.g. enjoyment of scenic beauty).

Double counting of services: erroneously including the same service more than once in an analysis. Ecosystem services**:** the direct and indirect contributions of ecosystems to human wellbeing.

Disease: An unhealthy condition that can be traced to a specific cause , generally with consistent symptoms.

Disk refiner: A machine that separate wood fibers between two mechanical disks as the disks rotate in opposite direction.

Dispersed Trees: Trees planted alone or in small numbers on pastures or otherwise treeless

Dissolved oxygen: Amount of oxygen gas contained in water or sewage, usually given in parts per million at a specified temperature and atmospheric pressure. It is a measure of the ability of water to support aquatic organisms. Water with very low dissolved oxygen content (less than 5 ppm), which is usually caused by too much or improperly treated organic wastes, does not support fish and similar organisms.

Distillation: The act of purifying liquids through boiling, so that the steam or gaseous vapors condense to a pure liquid. Pollutants and contaminants may remain in a concentrated residue.

Dominant: A crown class that includes trees extending above nearby trees with full access to sunlight from above and partial sunlight from the sides.

Dominant: The allele of a gene that masks or suppresses the expression of an alternate allele; the trait appears in the heterozygous condition.

Dot-grid method: A technique for determining the land area on a map by counting the number of dots that fall within the boundary lines.

Doyle's rule: A table used by the timber industry to estimate the number of board feet in a log.

Draft: A supply of fresh air drawn by the convection column into the base of a fire.

Duff: Ground fuel made up of decaying plant material located on or just beneath the surface of the forest floor.

Deciduous tree : A tree that loses its leaves or needles during the fall and winter.

Deck : A pile of logs ready for loading and transporting.

Decomposition : The breaking down of dead organic matter by decay organisms.

Defect : That portion of a tree or log that makes it unusable for the intended solid wood product. Defects include rot, crookedness, cavities, and cracks. Severe defects cause the log to be classified as a cull.

Defoliators : Insects that feed on foliage.

Dendrology : The study of tree identification.

Density : A measure of site occupancy. The quantity of trees per unit of area.

Usually expressed as trees/acre.

Desertification : The process of desert expansion caused by loss of soil and vegetative cover through climate change, overgrazing, or other influences.

Diameter limit : The smallest (but occasionally the largest) size to which trees or logs are to be measured, cut, or used. Usually refers to the stump, breast height, or top diameter.

Direct seeding : Sowing tree seed directly to forest soil to regenerate a forest.

Dormancy : A biological process in which a plant ceases most growth activities and simply maintains existing tissue.

Duff : Various stages of decaying organic matter found on the soil surface.

Dehiscence: The splitting open of an anther or fruiting structure. Dehiscent fruits are fruit types that open or disintegrate while still attached to the tree e.g. folicles, capsules and cones.

Diffuse flowering: Flowering throughout the year or flowering outside the main flowering season.

Dichogamy: The maturation of male and female organs on the same plant at separate periods, so that pollen presentation and pollen receptance do not coincide. Dichogamy fosters natural crosspollination.

Dimorphism: Existence of two distinct forms of a structure, for instance sun leaves - shade leaves,juvenile leaves - adult leaves, males - females etc.

Dioecious: A species or other taxon in which individuals are unisexual (male or female), e.g.rattan and some casuarinas.

Diploid: A cell is diploid when it contains two sets of chromosomes (2n), one set from each parent. Most higher organisms are diploid except for their sex cells and associated tissue.

Distal: The distal end of either the root or the shoot is that furthest from the stem-root junction of the plant and nearest to the tip of the root or the shoot.

Diversity: The condition of being different, showing differences. In a tree breeding context diversity refers to the variation of genotypes in the population.

DNA (Deoxy-ribo-Nucleic Acid): A doubled-stranded organic molecule consisting of sugars and bases which carries the genetic information or 'code'. On the figure G,C,A and T indicate the4 bases Guanine, Cytocine, Adenine and Thymine.

Dormancy: 1) Seed dormancy**:** Viable seeds which fail to germinate though exposed to conditions normally appropriate for germination. Seed dormancy may be classified as innate, enforced or induced. Dormancy can be due to e.g. physical constraints to water absorption, chemical or hormonal influence etc. In order to make a dormant seed germinating, dormancy has to be broken by pretreatment. 2) Bud dormancy**:** Buds, on the stem or branches, which fail to develop into shoot sunless the dormancy is broken.

Double fertilization: The process, unique to the angiosperms, in which one of the two sperm nuclei from the pollen tube units with the egg nucleus to form the diploid zygote and the other sperm nucleus unites with two of the nuclei of the embryo sac to form the triploid endosperm.

E

Earth day: An event established in 1970 by Gaylord Nelson and held every year on April 22 to raise/promote environmental awareness.

Earth summit: The united nations Conference on Environment and Development, which took place at Rio de Janerio, Brazil in June, 1992.

Earthworks: Constructions made of earth, usually to conserve or control water.

Ecological sustainability: Allowing ecosystem to function healthily by balancing the needs of nature with the needs of humans and not using resources faster than they can be replaced.

Ecologically sustainable development: Ecologically sustainable development meets the needs of the present generation without compromising the needs of future generations.

Ecology: The branch of biology that describes relationship between living organisms and the environments in which they live.

Economic rotation age: The point in time (stand age) during the growth of a stand when the rate of increase of the net value of the stand begins to decline; the economic rotation age generally occurs before the physical rotation age (culmination of mean annual increment).

Ecosphere: The 'bio-bubble' that contains life on earth, in surface waters, and in the air. In ecology this term refers to the global ecosystem of the planet.

Ecosystem: A community of living organisms in addition to all of the nonliving features of the environment such as water, air, sunlight and soil.

Ecosystem: It is a natural dynamic complex of plant, animal and micro organism communities (biotic factor) and their non-living (abiotic) environment all interacting as a functional unit.

Ecosystem: The interaction between a community of living organism in a particular area and the non living environment.

Ecotone: A habitat created by the juxtaposition of distinctly different habitats, an edge habitat; or an ecological zone or transition area between two adjacent ecological communities.

Elemental cycle: The recurring circular flow of elements from living organisms to non-living materials and back again.

Eluviation: The loss of soil components from a soil horizon due to downward leaching and movement of water through the soil profile.

Embryo-sac: A female gamete in plants consisting of the cell mass located in the ovule that develops into the embryo and the endosperm of a seed after fertilization.

Emergent: A crown class that includes trees with crown completely above the crown of other trees. These trees have access to full sunlight from above and form the sides.

Emulsifier: A chemicals that aids in suspending one liquid in another. Usually an organic chemical in an aqueous solution.

Endangered species: Animals, birds, fish, plants, or other living organisms threatened with extinction by anthropogenic (man-caused) or other natural changes in their environment. Requirements for declaring a species endangered are contained in the endangered species Act.

Endemic: an endemic species is one that is confined to a specific country, region, or location.

Endodermis: The innermost layer of cells in the cortex of a root, surround by a waxy waterproof material that restricts the flow of water and some dissolved materials.

Energetic: It is the study of how energy flows within an ecosystem; the routes it takes, rates of flow, where it is stored and how it is used.

Energy efficiency: The amount of fuel needed to sustain a particular level of production or consumption. It is typically defined as annual primary energy consumption per dollar of gross domestic product. Using less energy to provide the same level of energy service.

Enhanced greenhouse effect: The increase in the natural greenhouse effect resulting from increases in atmospheric concentrations of greenhouse gases due to emission from human activities.

Environment impact statement: A science based study of the harvest area that describes the expected effects of human activities on the environment and the wildlife in the area.

Environment: A general description of the surrounding of an organism encompassing air, water, and land which support its life and receive its waste products.

Epidemic: Spreading rapidly and effecting a large portion of a population at one time. Also, an outbreak of disease with such extensive characteristics.

Epidermis: The outer layer of cells that protect plant leaves, stems, flowers, seeds and roots.

Ergonomics: Study of human characteristics for the appropriate design of living and work environments.

Erosion: The losses of top soil due to the forces of wind or flowing water.

Erosion: The process of wearing away of a surface by physical means such as running water, wind, glacial ice, and gravity. The term is usually applied to the loss of soil through the agency of wind/ water.

Eutrophication: The enrichment of water bodies with nutrients, primarily nitrogen and phosphorus, which stimulates the growth of aquatic organisms. These typically promote extensive growths of algae. As the algae die and decompose, high levels of organic matter and the decomposing organisms deplete the water of available oxygen, causing the death of other organisms, such as fish. Eutrophication is a natural, slow aging process during which a lake, estuary, or bays evolves into a bog or marsh and eventually disappears. During the later stages of eutrophication the water bodies is chocked by abundant plant life due to higher levels of nutritive compounds such as nitrogen and phosphorus. Human activities can accelerate the process.

Evaporation: A change in the form of substance, such as water, forms a liquid to a gas.

Even-aged stand: A population of trees in which most of the trees are approximately the same age.

Evergreen: A tree that bears green leaves in all seasons.

***Ex situ*:** The term *ex situ* or "moved from its original place," means excavated or removed.

Exotic species: A plant or animal species introduced from another country.

Extractive: One of several wood products that is dissolved and removed from wood through the use of solvents.

Ecosystem management : A type of natural resource management that considers the relationships between all organisms, including humans, and their environment.

Ecology : The science that studies the interaction of plants and animals with their environment.

Ecosystem : Ecological system. An interacting system of living organisms and their environment. The dynamic relationships of living (biotic) and nonliving(abiotic) components of a region, forest, stand, or smaller area, as well as the forces, such as weather and wildfire, that affect them.

Ecosystem health: A measure of the overall capacity of an ecosystem to maintain biological diversity, normal productivity, sustainability, and resilience to disturbance.

Ecotone : A transitional area between two forest or rangeland communities containing the characteristic species of each as well as characteristics of its own. A point of abrupt change, such as a prairie-forest junction or a land-water interface.

Edaphic : Of the soil or influenced by the soil rather than by climatic factors, especially pertaining to living organisms.

Edge effect : The tendency toward greater species variety and greater density of animal and plant life in the margin where two ecological communities meet.

Emphasis Area Concept: An alternative to reserve areas and buffer zones that integrates neighboring and uses and protects the emphasized use through disturbance management.

Endangered species: A plant or animal vulnerable to extinction in all or a significant portion of its range. Identified by the Secretary of the Interior in accordance withthe Endangered Species Act (1973).

Endemic population level : A normal (insect) population level. Commonly used with insect populations. Environment the external conditions, both physical and biological, in which an organism lives. Includes climate, soil, topography, food supply, and all other influences affecting development.

Environmental impact statement (EIS): A document evaluating the probable consequences of a proposed project that might significantly alter the environment. Required by the National Environmental Policy Act (1969) for any such project a U.S. Government agency plans to undertake, regulate, or fund. An EIS is released in draft form (DEIS) to other agencies and the public for comment and review.

Ephemeral stream: A stream that flows only sporadically, such as after storms.

Epidemic; Widespread insect or disease incidence beyond normal proportions.

Even-aged management: Stand management designed to remove (harvest) all trees at one time, or over a short period, to produce even-aged stands. Most trees are within20 years of the same age.

Evergreen tree: A tree that retains some or most of its leaves, or needles, throughout the year.

Exotic species: A nonnative plant or animal species introduced by humans, either deliberately or accidentally.

Extinct: Said of a species or other taxonomic group having no living members.

Ecological niche: A localized environment that favours the permanent survival of some particular populations. Such habitat may be discontinuous or may be part of a gradient. Unique niches may favour hybrids or mutants that are at a disadvantage in other environments.

Ecosystem: A dynamic complex of plant, animal and micro-organism communities and their nonliving environmental interaction as a functional unit.

Ecotype: A race genetically adapted to the selective action of a particular environment. Most differences among ecotypes show up only when different ecotypes are tested in a uniform environment. Ecotypes are described, e.g., as climatic or edaphic.

Ecotyic differentiation: The process of producing distinct individuals within a species, sub species,or variety in a given environment that are different morphologically and physiologically from others of the same species in another environment.

Electrophoresis: The process by which charged particles move, typically through a gel in response to an electrical field. Due to different speed of movement in the field the particles, e.g. proteins, are separated by the process. Similar particles tend to move with a similar speed. Hence it is possible to verify a biochemical relationship by comparing the electrophoresis pattern and to identify particles by plotting them against a known reference.

Elite tree: A tree verified by appropriate testing as highly superior or desirable for a specified environment and breeding program. The superiority of crossbred parents is determined by progeny tests, and that of clones by clonal tests.

Emasculation: Removal of immature anthers to prevent self-pollination or unwanted cross-pollination of surrounding plants.

Embryo: The young plant individual after fertilization (or parthenogenesis). The embryo is usually differentiated in the seed into radicle (root), hypocotyl (stem) and cotyledons (leaves).

Endogamy: Crossing of related individuals with loss of heterozygosity.

Endosperm: Typically triploid tissue in the seeds of many angiosperms, furnishing food for the embryo. Its hereditary characters may be determined independently from those of the embryo. The so-called endosperm of conifers is haploid female tissue.

Environmental variation: Variation in trials or performance of individuals caused by environmental differences at the growth site e.g. soil condition and exposure.

Enzyme: Proteins that promote chemical processes in living organisms without themselves being changed.

Epitasis: The masking of phenotypic effect of either or both alleles by a gene of a different locus.

Escape: An apparently pest-resistant individual that appears so because it was never attacked. Also, an introduced tree found wild.

Eugenic: Tending to improve the genetic qualities of future generations. Applied especially to man-made practices such as reserving the best phenotypes for seed trees.

Exotic (non-native): A plant grown outside its natural range of distribution.

Experimental design: The layout of a test or trial, e.g. provenance or progeny test. The test area is divided into a number of blocks or replications, which are further subdivided into plots. Examples of experimental designs of field trials are completely random design, Randomized complete block design, Latin square design, lattice design or split block design.

Explants: The basic plant part removed from the mother plant when used for micro-propagation/tissue culture. Corresponds to cutting, air-layer or scion in macro propagation.

***Ex-situ*:** Material moved from its place of origin to another place.

Existence value: the value that individuals place on knowing that a resource exists, even if they never use that resource (also sometimes known as conservation value or passive use value).

F

Farm forestry: The practice of producing wood and agricultural products on the same land also known as agroforestry.

Felling: The cutting down or harvesting of forest trees.

Fermentation: A process by which bacteria, yeasts, or enzymes are used to change the chemical makeup of a product, such as producing alcohol from wood.

Fertigate: Apply fertilizer through an irrigation system.

Fertilization: The fusion of male and female gametes; also the addition of plant nutrients to the soil.

Fiberboard: A reconstituted board made of wood fibers that become cross-banded in the panel due to the random arrangement of the fibers in the mat from which the board is formed.

Filament: The stalk of the stamen in a flower.

Filter strip: A narrow strip of land on both side of a running stream on which permanent vegetation is maintained. It is also called a riperine zone.

Fire break: A zone located in the pathway of a fire from which fuel has been removed to stop the fire from spreading; also called a fire line.

Fire point: The lowest temperature at which a liquid vaporizes at a sufficient rate to support continuous combustion.

Fire storm: Erratic behavior by a crown fire, such as hurling burning debris beyond the leading edge of the fire and jumping across fire lines and rivers.

Fire suppression: All of the activities aimed at discovering and extinguishing a fire.

Fire line: A zone located in the pathway of a fire from which fuel has been removed to stop the fire from spreading; also called a fire break.

Flitch: A thick piece of high quality hardwood that is processed into veneer.

Food and Agriculture Organization (FAO): A United nations agency based in Rome and given the charges to improve efficiency of production and distribution of agricultural products.

Food chain: A series of step through which energy from the sun is transferred to living organisms; members of the food chain feed on lower ranking members of the community.

Food chain: It describe the feeding relationships between species within a ecosystem. A hierarchy of organisms, each feeding on the lower one.

Food pyramid: An arrangement of organisms in a ranking order according to there dominance in the food web.

Food web: A group of interwoven food chains.

Food web: The feeding relationships by which energy and nutrients are transferred from one species to another. The inter connected food chains in an ecosystem. Plants, herbivores, and carnivores all form parts of the food web.

Forest cover: all lands, more than one hectare in area, with a tree canopy density of more than 10 percent irrespective of ownership and legal status. Such lands may not necessarily be a recorded forest area. It also includes orchards, bamboo and palm.

Forest Inventory: the measurement of certain parameters of forests to assess the growing stand and stock and other characteristics of forests.

Forest estate model: A representation of the growth and natural dynamics of a predefined area of forest; used to simulate the effects of harvesting and silviculture on the long-term availability of timber supply from defined forest areas.

Forest floor: The layer of decaying plant materials on the soil surface acting as a mulch to preserve moisture.

Forest growth model: A computer program that prepares estimates of forest yields by applying all of the factors known to affect forest growth.

Forest growth: The increase in the volume of wood in a forest that occurs over a specific period of time.

Forest inventory: A record of the estimated volume of the different varieties of timber present in the forest at a given time.

Forest litter: Dead leaves, branches and other materials shed by trees and other plants form a layer of mulch on the forest floor.

Forest management zone(FMZ): A distinct area or unit for which a forest management plan is developed and implemented.

Forest type map: A map of the locations of different type of trees prepared from aerial photographs of the forest with the aid of instruments called a stereoscope.

Forest: A large tract or land area where mixture of interrelated community of trees, herbs, shrubs and/or animals. Among the most dominant living organisms are trees.

Forester: A person trained in forest management, who usually has a university degree in forestry.

Forestry: The managing and protecting of forest land for timber production and other values.

Forestry: The science of planting and managing forest for specific purposes, such as timber production, conservation and so on.

Forwarder: A machine that is used to move logs from the cutting area to the yard or landing area.

Freshwater biome: A set of similar ecological communities found in or near water that is not salty.

Frugivore: an animal that lives/feeds on fruit. Frugivores are typically primates that cannot digest cellulose in leaves.

Frutting body: A reproductive structure of fungi that grows out of the trunk of an infected tree, where it releases reproductive spores into the environment.

Fungi: Threadlike plants lacking chlorophyll that obtain their nourishment from other organic materials.

Firebreak : An existing barrier, or one constructed before a fire occurs, from which all or most flammable materials have been removed.

Fire cycle: The average time between fires in a given area.

Fire scar: An injury or wound on the stem of a tree caused or accentuated by fire.

Forage value: The relative importance of plants for grazing purposes.

Forest; A plant community dominated by trees and other woody plants.

Forester: A professional who has been educated in forestry at a college or university.

Forest management: The application of scientific, economic, and social principles to managinga forest property for specific objectives.

Forest practice; A specialized law that is intended to foster the protection of public resources, such as air, water, and wildlife, from adverse affects of forest management activities.

Forest Plan : A document that guides all natural resource management activity and establishes management standards and guidelines for a National Forest, embodying the provisions of the National Forest Management Act (1976).

Forestry ; The science, art, and practice of managing and protecting tree and forest resources for human benefit.

Forest type : A group of tree species which, because of their environmental requirements, commonly grow together. Examples afforest types are the Douglas-fir / hemlock type or the spruce/fir type. Also, a descriptive term used to group stands with similar composition and development characteristics.

Form class : A measure of bole taper derived by dividing diameter inside bark at a given Height (usually 17.3 or 33.6 feet) by DBH, times 100. These values often are entered when using tree-volume tables.

Frost crack ; A longitudinal tree bark split caused by freezing sap. Usually found on Thin-barked trees.

Fuel ladder ; Lower limbs along the bole of a tree that allow ground fire to "climb" into the tree crown.

F1: The first (filial) generation of a cross. If each parent is true-breeding (homozygous), the F1individuals resemble each other.

F2: The second generation, produced by intercrossing or selfing the F1 individuals. Characteristically the individuals within an F2 generation vary greatly.

Fertilization: Union of nucleus and other cellular constituents of a male gamete (sperm) with those of a female gamete (egg) to form a zygote. In some species, fertilization may occur months after pollination.

Fixation index: A measure of the departure of genotype frequencies from those predicted under Hardy-Weinberg conditions.

Flower: Angiosperm reproductive structure bearing pistils, stamens, or both, and usually also sepals and petals. So called flowers of conifers are the male and female strobilus before and during pollination.

Fluting: Non-circular cross-section of the stem often caused by buttress roots, i.e. large plank like up growth of the upper side of the roots, providing support for the tree. Buttressing and fluting is heritable to varying degree. It is normally a non-desired character in forest tree breeding.

Forest tree breeding: Practices applying knowledge of genetics to develop improved trees. Conventional tree breeding is based on selection of individuals for certain desired characters and mating these individuals to produce an improved popupation. Forest tree breeding may can note breeding systems varying from harvesting seed from only the best sources (mass selection) to sophisticated multi-phase, multi-generation programmes or controlled pollination.

Forest tree improvement: Usually synonymous with forest tree breeding, but may refer to tree breeding in combination with silvicultural practices.

G

Gamete: A haploid reproductive cell.

Gang saw: A multiple bladed lumber processing saw that is capable of making several cuts at the same time

Gene: A genetic structure that controls the traits and characteristics expressed in an organism.

Gene: A unit of heredity; a section of DNA sequence encoding a single protein.

Genetic diversity: The genetic variability within a group of related organisms. One of the three levels of biodiversity that refers to the total number of genetic characteristics.

Genome: The entire set of genes in an organism.

Genome: The total genetic composition of an organism. The genome includes both the genes and the non-coding sequence of the DNA. In modern molecular biology the genome of an organism is its heredity information encoded in DNA.

Genotype: The genetic constitution of an organism or cell, as distinct from its expressed features.

Genotype: The genetic makeup of an organisms

Geographic Information System (GIS): Satellite technology used to make observations and photographic images of the earth surface features and conditions.

Germination: The process by which a seed sprouts and begins to grow.

Girdling: To cut or remove a ring of bark completely around a tree; a condition often caused by gnawing rodents.

Global Environment Monitoring System: (GEMS) A part of the united nations Environment program's Earth watch, based in Nairobi, Kenya; purposes of GEMS are to coordinate environmental monitoring and assessment activities world wide and it provide an environmental data exchange service.

Global Positioning System (GPS): The use of satellite technology to accurately and consistently identify exact locations.

Global warming: The observed increase in the average temperature of earth's atmosphere and oceans in recent decades, as a result of human activities.

Goal programming: A LP minimization problem where the objective function sums deviations from state dout put goals; if all goals can be met exactly, then the objective function value is zero. Goals may be weighted equally or they may have different weights; goals with higher weights are favored over those with lower weightings. (Remsoft, 2004)

Grassed waterway: Natural or constructed watercourse or outlet that is shaped or graded and established in suitable vegetation for the disposal of runoff water without erosion.

Green chain: Parallel moving chains that carry boards to new locations in a sawmill.

Greenhouse gases: Gases in the atmosphere that stop heat escaping from the earth and contribute to global warming. The major natural greenhouse gases are water vapor, carbon dioxide, and ozone.

Gross National product (GNP): A measurement of the value of a national economy.

Ground fire: A fire that burns the decaying fuels on and beneath the surface of the ground.

Ground fuel: Combustible materials located beneath the surface that has begun to decay.

Ground water: Water found below the surface, usually in porous rocks, or soils, or in underground aquifers, that often supplies wells and springs. Ground water often serves as principal source of drinking water, there is growing concern over contamination from leaching agricultural or industrial pollutants or leaking underground storage tanks.

Ground water: Water found in or obtained from an aquifer.

Growing stock: the sum (by number or volume) of all the trees growing/living in the forest or a specific part of it.

Growth habit: The natural shape of a mature tree or other plant.

Growth impact: A calculation of the extent of insect damage, taking into account the timber losses due to reduced growth rates and deaths of trees.

Growth rings: The rings formed in the trunk of a tree as wood of different densities is produced at different times during the growing season.

Guide meridian: A survey line that is established running north-south for each interval of slightly les than 24 miles on either side of the principal meridian.

Gum naval stores: Wood products refined from a liquid known as oleoresin, obtained by injuring live pine trees and collecting the liquid as it flows from the wounds.

Gymnosperm: A class of tree such as pine, spruce, or cedar that bears seeds in cones.

Girdling : A cutting or insect gallery that completely encircles the tree trunk. Girdling usually results in the death of the tree.

GIS : Geographic Information System. A computer database designed to store and retrieve geographic information linked to topographic or land features.

Grading : Evaluating and sorting trees, logs, or lumber according to quality.

Grow-mat : Specialized fabric material placed around newly planted seedlings to inhibit weed competition and promote initial seedling growth. Also known as "fabric-mulch."

Growing stock : All the trees growing in a stand, generally expressed in terms of number, basal area, or volume.

Gametes: Mature sex cells, either sperm or eggs.

Gene: The smallest transmissible unit of genetic material consistently associated with a single primary genetic effect. The genes are ultra-microscopic and act as if linearly arranged at fixed places (loci) on a chromosome. The gene consists of a segment of DNA chain with a definite nucleotide sequence that contains the instruction for a complete protein. Each gene, by interacting with other genes on the same or other chromosomes and the environment, governs a certain physiological effect in the cell and is expressed as one or more characters.

Gene bank: A collection of superior and sometimes special trees maintained for present and future use in research and tree breeding.

Gene centre: Centre of diversity. A geographical area that shows considerable genetic diversity of certain plants.

Gene conservation: Protecting and maintaining the genetic variety of a species in order to keep a genetic resource for future research and improvement.

- *In situ gene conservation*: Protection of a species and its genetic variation at its native site, e.g. by establishing forest reserves.

- *Ex situ gene conservation*: Protection of a species and its genetic variation outside its

native site i.e. established plantations.

Gene expression: The process of turning the genetic code into visible or identifiable traits through protein synthesis.

Gene flow: Passage of genes from one population to another (also called gene migration).

Gene frequency: The proportion of an allele in a population compared with other alleles of that gene. Thus if a gene A has two alleles, a and A, and the frequency of a=0.2 (20%), then the frequency of A must be 0.8 (80%) since A+a=1 (100%).

Gene pool: The total sum of all the genes and their alleles present in a breeding population or species at one time.

Genetic code: The exact sequence of amino acids laid down in a protein, determined by the sequence of nucleotides in messenger RNA which in turn depends on the sequence of the nucleotides in the DNA.

Genetic drift: Random change in allele frequencies in a population from one generation to the next because of small population size (also called random drift).

Genetic engineering (recombinant DNA technology): The isolation of useful genes from a donor organism or tissue and their incorporation into an organism that does not normally possess them.

Genetic erosion: The loss of genetic variation and the consequent narrowing of the genetic base of cultivated plants through the introduction of improved varieties that can be grown over wide areas.

Genetic gain: Average improvement in a progeny over the mean of the parents. Gain is achieved by selection in the parental generation; the amount depends on selection intensity, parental variation, and heritability.

Genetic makeup: Total genetic content of an individual or species (also called genome).

Genetic marker: Any allele used as an experimental probe to mark a nucleus, chromosome or agene. Genetic markers can be classified as Biochemical markers (e.g. isozymes) and Molecular markers (e.g. DNA markers).

Genetic response: The change in genotype frequency following selection. After one generation of selection the genetic response, R, is expressed by R = h2 x D, where h2 is the heritability of the character under consideration, and D is the selection differential.

Genetic resource: A unit of heritable variability of actual or potential value.

Genetics: Genetics is the basic science dealing with causes of resemblances and differences among organisms related by descent. It takes into account the effects of genes and the environment. When the basic knowledge of genetics is applied to breeding trees, the effort is preferably referred to as forest tree breeding or forest tree improvement.

Genotype: (1) An individual's hereditary constitution, with or without phenotypic expression of the one or more characters it underlies. Also the gene classification of this constitution expressed in a formula. The genotype is determined chiefly from performance of progeny and other relatives. It interacts with the environment to produce the phenotype. (2) Individual(s) characterized by a certain genetic constitution.

Germination: The physiological processes in the first stages of growth of seed and pollen grain. Pollen germination may be restricted by Incompatibility, seed germination by Dormancy.

Germination test (viability test): Standard test carried out to determine the quality of a seed lot, i.e. the percentage of viable seeds. The germination test is carried out under prescribed standard conditions under the, for the particular species, optimal germination conditions of temperature, humidity and light. Dormant seeds must be pretreated in order to break dormancy before the germination test.

Germplasm: (1) Within an individual or group, the collective hereditary material that is the physical basis for inheritance; i.e. the genotype, with particular reference to its transmission to the next generation.

Girdling: Removal of a piece of bark of the tree in its full or partial circumference. Girdling impedes normal passage of water and nutrients. Sometimes used as an imposed stress factor to promote flowering or for rejuvenation in connection with harvesting of cuttings or scions.

Grafting: Uniting parts of separate individuals by matching their tissue so that union and growth can occur. Commonly, an aerial portion (scion) is joined with a rootstock. Both the site of the union and the composite individual are termed grafts. Because the scion retains its original genetic constitution, grafting is valuable for preserving, testing, and propagating.

Gregarious flowering: Profuse flowering during a short period of time.

H

Habitat: An ecological or environmental area that is inhabited by a particular species or in other words the natural abide of an organism including its total environment.

Habitat: The environment in which a plant or animals lives. Each living organism has its preferred habitat. Some organism are much more sensitive than other to changes in the environment, and are therefore more valuable.

Habitat: The living environment of an organism.

Haploid: A cell that contains a single chromosome from each homologous chromosome pair.

Hard water: Water that contains sufficient concentration of metal ions, primarily those of calcium and magnesium to cause the formation of precipitates with soaps and detergents.

Hardboard: A reconstituted board similar to fiberboard except that it is press-bonded between heated steel plates, resulting in a density range for this product of 0.5 to 1.3 grams per cubic centimeter.

Hardiness: A measurement of the tolerance of a particular species of tree or other plant to restrictive environmental factors such as climate, altitude, temperature, and availability of moisture.

Hardwood: Wood from a broad leaved tree.

Hardwoods: Woods produced by species of trees that bear flowers and have broad leaves.

Harvest constraints: Limits imposed on a condition or activity within a forest estate; defined as part of an overall forest estate model; are usually defined by forest-level objectives.

Harvest scheduling flexibility: The relative variability permitted in the order and timing of stand harvest activities in a forest estate model; defined by the interaction between forest-level objectives, harvest constraints, and the current and future conditions of the forest estate.

Harvesting: The cutting down of a crop of forest trees for wood.

Head rig: A large saw used to cut logs into marketable dimensions of lumber; also known as head saw.

Head saw: A large saw used to cut logs into marketable dimensions of lumber; also known as head rig.

Head water system: A small catchment area or watershed located at the top end of a stream or river system.

Heart rot: A disease caused by fungi that result in the decay of wood; that accumulates at the center of the trunk.

Heart wood: Old, dark-colored, non-functional xylem located in the center of a tree stem or trunk that has become saturated with tannins, gums, and resins, thus eliminating then flow of water and dissolved materials.

Heart wood: The central core of the dark hard wood in the tree trunks, consisting of old cells blocked resins, tannins, and oils.

Heat of vaporization: The amount of heat that is required to convert water from a liquid to gases.

Heavy metals: Metallic elements with high atomic weights usually greater than 50 and relative density greater than 7; (e.g. iron, zinc, mercury, chromium, cadamium and lead); can damage living things at low concentrations and tend to accumulate in the food chain.

Hectare: A metric measurement for land area that is equal to 10,000 square meters; equivalent to 2.471 acres.

Hemicellulose: A polymer that makes up 27 to 29 % of the material found in wood.

Herb layer: The bottom layer of vegetation in a forest, consisting of ferns, grasses, and other low plants that grow on the forestry floor beneath the shrub layer.

Herbicide: A chemical used to kill plant.

Herbivore: An animals or other organism that eats plants.

Heterozygous: Having two different genes for a particular characteristic.

High forest: A forest that has been propagated from seeds.

Homologous chromosome: Each chromosome of an identical pair of chromosomes.

Homozygous: Having identical genes (one from each parent) for a particular characteristic.

Horizon: One of several different layers in a soil profile.

Humus: Organic matter in the soil; a source of nutrients for soil organisms and plants.

Hydrapulper: A machine that is used to reduce recycled paper to pulp.

Hydrologic cycle: The movement of water in the form of vapor forms the oceans to be clouds to the earth as precipitation, and back to the oceans through rivers and streams. It also called as water cycle.

Hydrology: The science dealing with the properties, distribution, and circulation of water.

Hydrosphere: Encompasses all of the water and ice that is found on the surface of the earth, and that found in the atmosphere in the form of water vapor and clouds.

Hydrosphere: The water portion of the planet, this would include water found in the sea, streams, lakes and other water bodies, the soil, ground water, and in the air, approximately 75 per cent of earth's surface is covered by water.

Hypsometer: A measuring device that uses trigonometry or geometry to calculate the height of a tree without taking a direct measurement.

Habitat: The local environment in which a plant or animal naturally lives and develops.

Habitat type: Classification of a land area according to dominant plant forms(usually trees and shrubs) and physical characteristics. Can help to indicate the biological potential of a site.

Hardwood: A term describing broadleaf trees, usually deciduous, such as oaks, maples, cottonwood, ashes, alders, and elms.

Harvest; Removing trees on an area to obtain an income or usable product.

Haul back line : In cable logging, the line used to pull chokers or the carriage from the landing out to the felling area.

Hazard reduction or Slash reduction: The burning, crushing, or scattering of slash to reduce the risk offorest fires or forest fire damage to an area.

Heart rot : A decay in trees, characteristically confined to the heartwood. It usually originates in the living tree.

Heartwood : The inner core of a woody stem, wholly composed of nonliving cells and usually differentiated from the outer enveloping layer (sapwood) by its darker color.

Heel-in ; To store seedlings before planting by placing them in a trench and covering the roots with soil.

Height :

- *Merchantable height.* Tree height (or length of the tree trunk)up to which a particular product may be obtained. For example, if 6-inch minimum diameter saw logs were being cut from the tree, its merchantable height would be its height up from the stump to a diameter of 6 inches. The products being cut determine merchantable height.
- *Total height.* Tree height from ground level to top.

Haploid: Having one complete set of chromosomes per cell. These are the n chromosomes normal in sex cells as compared to the 2n in vegetative cells.

Hardy-Weinberg law: The law stating that provided certain conditions are met, the geno-type frequency in a population of organisms will remain constant and distributed as p2, 2pq and q2 for the genotypes A1A1, A1A2 and A2A2 respectively, where p= the frequency of the allele A1 and q = the frequency of the allele A2 such that p+q=1. Narrow-sense heritability is the fraction of total variation that contributed by additive effects of genes; i.e., it is the ratio of additive genotypic variance to phenotypic variance. Broad-sense heritability, applicable to vegetative propagated species, includes also non-additive effects. High heritability indicates that individual phenotypes are indicative of their genotypes. If calculated from parent-progeny data, it estimates the degree of resemblance between parent and progeny.

Heterogeneous: Exhibiting dissimilarity among members of a group. Not to be confused with heterozygous.

Heterosis: Hybrid vigor exhibited when mean F1 hybrid phenotype falls outside the range of the parents, e.g. more vigorous growth, greater yield or increased disease resistance.

Heterozygous: Having one or more sets of unlike alleles, e.g. the dominant with the recessive gene. Thus, an cell or plant is heterozygous whereas the AA's and aa's are homozygous. Refers also to differences in the arrangement of genes on the chromosomes. A heterozygous organism (heterozygote) does not generally breed true and is known as a hybrid with respect to the genes in question.

Homogeneous: Exhibiting similarity among members of a group; not to be confused with homozygous.

Homologous chromosomes: A pair of similar chromosomes, one of maternal, and one of paternal origin, present in diploid cells. The members of each pair are similar in size, shape, gene content, and arrangement. They pair physically at meiosis.

Homozygous: Having one or more sets of similar alleles, e.g., both dominant (AA), or both recessive (aa). The homozygote breeds true when mated with the same genotype.

Hormone (growth substance): Substance that has a marked and specific effect on plant growth and that produces this effect when present in very low concentrations. Certain hormones stimulate plant growth by promoting cell division or cell elongation while others inhibit certain developmental processes. The term covers substances produced within the plant as well as artificial, often structurally related, chemicals that have similar effect. Examples of hormones are auxin, gibberellins and acetylene. Synthetic analogous of auxins are widely used as rooting hormone. Others are used for instance for flower induction.

Human well-being: concept prominently used in the Millennium Ecosystem Assessment –it describes elements largely agreed to constitute 'a good life', including basic material goods, freedom and choice, health and bodily well-being, good social relations, security, peace of mind, and spiritual experience.

Hybrid: The offspring of genetically distinctly different parents. The term is applied to the progeny from mating within species (intraspecific) as well as to those between species (interspecific).

Hybridization: The formation of a hybrid. In the context of DNA technology the formation of a double stranded molecule by complementary base pairing between two single stranded molecules.

Hybrid sterility: The reduced ability of some hybrids to produce viable gametes. This is caused by the absence of homologous pairs of chromosomes, so that bivalents cannot form during meiosis. The resulting gametes are thus aneuploid.

I

Illuviation: A buildup of trans located soil components that accumulate in the B horizon of a soil; a process by which subsoil is formed.

Impaired water: Waters that exceed the total maximum daily load of one or more specific pollutants.

Improved Fallows: Areas left to grow up in selected trees in trees-crop rotation systems.

***In situ*:** In its original place; unmoved unexcavated; remaining at the site or in the subsurface.

***In vitro*:** Testing or action outside an organism (e.g., inside a test tube)

***In vivo*:** Testing or action inside an organism.

Incendiarism: The willful setting of destructive fires.

Incendiary fire: A fire that has been deliberately started by a person.

Increment borer: An instrument used to extract a core of wood from the cross-section of a living tree to determine its age or the condition of its health.

Indigenous: A species which occurs naturally in an area and has not been introduced from elsewhere.

Individual Trees: Trees occurring alone, whether spontaneously emerging or planted.

Infiltration: Movement of water into the surface of the soil through small openings between soil particles.

Ingrowth: Growth that occurs in a forest that is in excess of the growth accounted for in individual trees; growth due to new trees that were not present or were too small to be tallied when the original timber yield measurements were taken.

Initial point: A permanent physical feature that is prominent in the landscape that serves as a beginning point for a rectangular land survey.

Insecticide: A chemical used to kill insects.

Incentives (disincentives), economic: a material reward (or punishment) in return for acting in a particular way which is beneficial (or harmful) to a set goal.

Indirect-use value (of ecosystems): the benefits derived from the goods and services provided by an ecosystem that are used indirectly by an economic agent. For example, the purification of drinking water filtered by soils.

Integrated Pest Management (IPM): A pest control strategy that uses an arrey of complementary methods: natural predators and parasites, pest resistant varieties, cultural practices, biological controls, various is physical techniques, and the strategic use of pesticides.

Integrated Pest Management (IPM): The use of natural insect enemies and limited chemical applications to control harmful insects while providing some protection for useful insects.

Intermediate cutting: A silvicultural practice used to improve the forest by removing some of the trees to provide enough space for the remaining trees to grow; also known as thinning.

Intermediate: A crown class that includes trees having crowns that extended upward into the lower levels of the canopy formed by dominant and co-dominant trees.

International log rule: A table used to estimate the number of board feet in a log by calculating the volume of each 4-foot log section adding that volume together.

Inter phase: A resting or non reproductive stage in the life span of a cell.

Isotope: A variation of an element that has the same atomic number of protons but a different weight because of the number of neutrons. Various isotopes of the same element may have different radioactive behaviors, some are highly unstable.

Its projects and activities play a part to stop the degradation of earth's natural environment, and to built a future in which humans live in a harmony with nature.

Increment borer: A hollow, auger like instrument used to bore into a tree trunk to remove a cylindrical cross section (core sample) of wood. It is used to expose annual growth rings.

Indicator species; A plant or animal species whose presence in an area indicates that certain specific habitat conditions prevail throughout the area.

Integrated pest management (IPM): A method that evaluates alternatives for managing forest pest populations on the basis of pest-host and pest-environment relationships.

Integrated resource management: A term used to indicate the simultaneous consideration of ecological, physical, economic, and social aspects of lands, waters, and resources in developing and carrying out multiple-use, sustained-yield management. Intermediate cut removing immature trees between stand establishment and stand harvest to improve the quality or reduce competition among remaining trees. In contrast to a harvest cut, an intermediate cut may or may not generate.

Intolerant species: Tree or shrub species that are incapable of establishing or growing in the shade of other trees or shrubs.

Invasive species: An insect, plant, or mammal that has expanded its range, often to the Detriment of native species.

Inbred line: A population derived by inbreeding. It is relatively pure genetically (homozygous) and hence true-breeding. The approach to homozygosis is often expedited by recurrent selfing accompanied by selection.

Inbreeding: Production of relatively homozygous offspring by mating closely related organisms, usually by selfing. This procedure, especially if carried out for a number of generations, exposes undesirable recessive characters and is used with selection to fix desirable ones, i.e, to render them true-breeding. Inbreeding of normally crossbred organisms frequently results in an 'inbreeding depression' of vigor or fertility, but the vigor can usually be restored and may even be increased by crossbreeding again, cf. heterosis. Inbreeding is mathematically expressed by the inbreeding coefficient F, which ranges from 0 (no inbreeding) to 1.

Incompatibility: The failure of gametes from genetically similar material to fuse due to physiological or morphological mechanisms. For example, pollen tube growth may be deficient even though the egg and sperm cells are potentially functional. Also used synonymous with graft incompatibility.

Independent culling: Multi trait selection that involves setting of minimum values for each trait of interest. Individuals must meet these minimum criteria if they are to be maintained.

Indirect selection: Choosing individuals or populations on the basis of a trait (usually of no economic importance in itself) believed to be associated with a desired trait of economic importance, rather than on the basis of the desired trait.

Individual selection: From a family or population single plants are selected on their own merit as parents or ortets. Sometimes called phenotypic selection.

Inheritance: The acquisition of characters or qualities by transition from parent to offspring.

***In situ* conditions:** Conditions where genetic resources exist within ecosystems and natural habitats, and in the case of domesticated or cultivated species, in the surroundings where they have developed their distinctive properties.

Internodes: The part of a plant stem between two nodes. In conifer referring to the stem separating two distinct whorls of branches.

***In vitro*:** Describing experiments on biological processes that are carried out in laboratory apparatus.

***In vivo*:** Describing experiments investigating biological processes that are carried out in living organisms.

Isolation: A condition that prevents or limits breeding among populations, e.g., phonological difference in flowering time, chromosome difference, and distance. A knowledge of effective pollination distance is important to the relative isolation of seed orchards. For wind pollinated species back ground tree pollen in the air usually cannot be eliminated, but contamination is often minimized by dilution zones of border trees.

The term also refers to the condition of isolating flowers during controlled pollination.

Isozyme (Isoenzyme): Different forms of the same enzyme may be formed by different loci or different alleles at the same locus. In the latter case they can be called allonyms'.

J

Juvenile: Young or non-mature stage of a tree, a population or propagation material. Ant. matures. He juvenile stage of trees often refers to the stage prior to flowering. Also used in morphological context e.g. juvenile leaves.

Juvenile selection: Selection from non-mature stands in order to reduce generation interval. Juvenile selection can only be implied if there is a high juvenile-mature correlation.

K

Kiln drying: Steam seasoning of green timber to reduce the moisture in the wood and give it increased stability and strength.

Kyoto protocol: An international agreement adopted in December 1997 in Kyoto, Japan. The protocol sets binding emission targets for developed countries that would reduce their emissions on average 5.2 per cent below 1990 levels.

Kerf: Width of a cut made by a saw.

L

Lagrangian relaxation: Modeling techniques where hard constraints are transferred into the objective and penalties are imposed on the objective function if they are not satisfied (violated).

land up to 100 000 hectares in size delineated according to topographic or geographic features such as a watershed or series of adjacent watersheds. The district manager must establish objectives for a landscape unit, and may vary or cancel an objective.

Landing: A site to which logs are gathered in preparation for hauling them to a sawmill or processing plant; also known as yard.

Landscape unit: A landscape unit is defined by the district manager. It is generally an area of

Layering: A form of vegetative reproduction in which live tree branches are buried in the debris in the forest floor, where they generate roots and live stems to regenerate a forest.

Leaching: The process by which soluble constituents are dissolved and filtered through the soil or sludge deposits or other materials by a percolating fluid.

Liberation: A type of stand improvement in which undesirable older trees are removed to make sunlight available to young trees.

Light reaction: A chemical reaction that occurs during photosynthesis in which energy is captured from the sun.

Lignin: A major component making up 25% of the total materials in wood; functions to bind wood components together.

Limbing: The process of removing limbs from logs during harvesting.

Limnology: The study of the physical, chemical, hydrological, and biological aspects of fresh water bodies.

Linear programming model: A method for determining an optimal level of output from an activity – the optimal level of output is determined by the availability of conditions and inputs required to perform activities defined in the model; often used to define the timing and quantity of harvest levels that can be obtained from a forest estate while maintaining the forest in a condition that provides for all other defined forest values.

Linear programming: An optimization technique that will choose the best alternative from a set of possible or feasible alternatives. It is a method of allocating scarce or limited resources among competing activities in an optimal manner. LP models are composed of a linear objective function subject to linear constraints. (Remsoft, 2004)

Line-plot cruising: In systematic sampling, a processed used together timber yield data.

Lipid: Any one compound of a group of compounds consisting of facts or similar substances.

Litigation: A lawsuit or the process of conducting a lawsuit.

Litter: Ground fuel that is made up of decaying plant material located on or just beneath the surface of the forest floor.

Living Fences: Fences in which the posts are living trees, or in which the entire fence

Load: A measurement of the amount of a pollutant a water body receives in a specific period of time.

Loading jack: A mechanical armed raised between a set of rollers to lift lumber to another set of rollers on a higher level.

Lobed simple leaf: A leaf from a broadleaf tree with divisions along the leaf margins.

Locus: A fixed location on a strand of DNA where a gene or one of its alleles is located.

Logging: The process of harvesting trees.

Low forest: A forest that has been regenerated from the roots, stumps, or branches of other trees.

Lumber: Timber that has been sawed into standard dimensions, such as boards, planks, beams, and so on; wood that is cut to cross-section dimensions less than 5 by 5 inches.

Landscape: A large land area composed of interacting groups of ecosystems, including all the physical and biological aspects of such an area regardless of ownership.

Landscape ecology: The study of biological interactions across a large land area, or watershed.

Liberation cutting: Also known as release cutting. An operation to release young trees from overtopping older trees.

Litter: The uppermost layer of the soil, made up of freshly fallen or slightly decomposed organic materials. See also Duff.

Log: A piece of the woody stem of a tree, usually cut to a specified length.

Log rule: A table that expresses log volume based on log diameter and length. The Scribner Decimal C Rule is the legal rule in Washington State and is expressed in board feet.

Log scale : The lumber content of a log as determined by a log rule.

Lop : To chop or saw branches, tops, or small trees after fellings. These pieces, or debris, are usually left on the ground to decompose.

Latin square design: Experimental design in which the test area is divided into rows and columns, one plot of each seed lot is planted in each row and each column.

Lethal gene: A gene which, when expressed, is mortal, e.g. Albinism.

Linkage: The association of inherited characters because their genes are in proximity on the same chromosome. There are as many linkage groups as chromosome pairs. Linkage is rarely complete, i.e., new combinations of linked characters can occur due to crossing over between chromatids of homologous chromosomes during meiosis.

Local seed source: Source native to the locality in which the seedlings are to be grown, i.e. belonging to the indigenous geographic race. Its seed-collection zone is usually defined experimentally as being within a certain distance or elevation of the planting site.

Locus (loci): The position occupied by a gene on a chromosome. Alleles of a gene occupy the same locus on homologous chromosomes.

M

Magma: Molten rock that sometimes forms beneath the surface of the earth that often collects in a magma chamber and is ejected by volcanos.

Management unit plans: A forest management plan approved under a tree farm licence or a woodlot licence, usually in effect for a period of five years and specifies proposed management activities to establish, tend, protect and harvest timber resources, and to conserve other resource values.

Margin: The distinctively shaped outer edge of a leaf; a useful structure for tree identification process.

Mathematical programming model: A model based on mathematical algorithms or sets of equations designed to solve a problem based on a pre-defined set of assumptions and starting conditions; a linear programming model is a type of mathematical programming model.

Mature: A tree that is fully developed for a specific purpose; usually 10 to 24 inches in diameter.

Mechanical control: Pest management using machines or tools to destroy or removes weeds and harmful insects.

Mechanical pulping: Separation of wood fibers using abrasive machine action.

Megaspore mother cell: A diploid plant cell from which a female gamete is formed.

Megaspore: A haploid plant cell occurring during the formation of female gametes.

Meiosis: A cell division process through which the number of chromosomes is reduced by half during the formation of male and female gametes.

Melting point: Temperature at which a solid substances changes to a liquid state.

Meristem: Plant cells, located in growth zone, that are dividing rapidly to form new plant tissues such as stems and roots.

Mesotrophic: Reservoirs and lakes which contain moderate quantities of nutrients and are moderately productive in terms of aquatic animals and plant life.

Metaphase: An intermediate step in cell reproduction during which the chromosomes become aligned at the center of the cell.

Metes and bounds: A land survey system that uses natural physical features in the landscape as starting point for land measurements.

Microbe: A micro organism, an organism that can only be seen with the aid of a microscope, especially a bacterium producing disease or fermentation.

Micropyle: A small opening through which pollen enters the ovule of a flower during fertilization.

Microspore mother cell: A diploid plant cell from which male gametes are formed.

Microspore: A haploid plant cell that divides to form a pollen grain.

Midrib: A reinforced vein in a leaf extending from the leaf stem to the tip of the leaf.

Mitosis: A type of cell division that occurs in an animal or plant, resulting in growth.

Model: Advanced programming available as a software package that is used to simulate an actual system.

Monitoring: The routine counting, testing or measuring of environmental factors or biota to determine their status or condition.

Monoculture: A forest that is made up predominantly of only one kind of trees.

Monohybrid cross: A genetic cross involving a single pair of genes (one trait); parents differ by a single trait.

Monomer: A simple sugar molecule, such as glucose; also known as monosaccharide.

Monosaccharide: A simple sugar molecule, such as glucose; also known as monomer.

Monte Carlo simulation: A modelling technique that allows some activities to occur based on a probabilistic basis. Although the activity may occur with probability (P) in any given run, there is no guarantee that it will occur. Therefore, each simulation run can yield different results; in a regular inventory projection model where activities occur with certainty, the answer is always the same no matter how many times you run the model. (Remsoft, 2004)

Mortality: A category of timber losses due to the deaths of trees.

Mulch: A layer of material placed around plants to hold moisture, prevent weed growth, and enrich or sterilize the soil.

Multi-function forest: A forest area with a range of different uses, such as wood production, recreation, water catchment protection and conservation of native fauna and flora.

Multistage remote sensing: A strategy for landscape characterization that involves gathering and analyzing information at several geographic scales, raging from generalized levels of detail at the national level through high levels of detail at the local scale.

Mutualism: A living relationship between two organisms that is beneficial to both. The dependency of one or both organisms may be such that neither organism can survive without its mutual partner.

Mycorrhiza: An association between a fungus and the roots of a forest tree that usually benefits them both.

Macroclimate: The climate of a large region as a whole, considered apart from modifying irregularities of land and vegetation. See also Microclimate.

Mainline: In cable logging, the line used to retrieve logs. See also Skyline logging. Also the main access road to a forest area.

Management plan : A written plan for the organized handling and operation of a forest Property. It usually includes data and practices designed to provide optimum use of forest resources according to the landowner's objectives.

Mature forest : In an ecological sense**:** A mature forest possesses varied biological Diversity throughout the ecosystem.

Mean annual increment (MAI) : The annual increase in size (volume) of a tree. Or the increase in size(volume) of a stand at a certain age, divided by that age in years.

Menstruation : That phase of forestry dealing with the measurement of volume, growth, and development of individual trees and stands.

Microclimate : The climate of a small, specific area reflecting such local differences as soil surfaces, vegetation, and atmospheric characteristics.

Microsite : An environmental feature that is small in scale but unique in character. Micro sites often have a significant impact on natural regeneration.

Mortality: Death of forest trees as a result of competition, disease, insect damage, drought, wind, fire, air pollution, and other factors.

Multi-storied : Forest stands containing trees of different heights.

Macropropagation: Vegetative propagation from cuttings, air-layering, grafting or other large plant part, distinguished from micro propagation or tissue culture.

Marker**:** A character or gene which by virtue of linkage can be used to indicate the presence of another gene.

Mass propagation: Large scale multiplication of improved material (breeding population) to be used for plantation (wood producing population). Mass propagation of clonal material is usually carried out by cuttings.

Mass selection: Use of seed from individuals chosen for a certain common phenotype from a larger population, and then bulked. The larger population provides the male parentage, and selections are not intermitted prior to seed use. Thus, seed may be collected from selected trees of a wild stand and used in a commercial planting or for establishing a seedling seed orchard. The process may be repeated through successive generations. Note, however, that collecting seed from a rouged orchard is recurrent selection.

Masting: Ecological term used to describe the strategy of some plant species to produce large crops of seeds at long interval, e.g. dipterocarps.

Megaspores (macrospores): The female cells produced by sexual cell divisions (meiosis), to give rise to the embryo sac.

Meiosis: The process by which diploid cells divide to form haploid cells. The process consists of two consecutive specialized nuclear divisions ultimately leading to the formation of eggs or sperm.

Mendel's principles: Hereditary differences have particular bases called genes. Genes are present in pairs, each of which is an allele. One allele may dominate the other. During meiosis paired alleles separate. During fertilization alleles from male and female part unite. The resulting off spring exhibit segregation ratios for various character combinations according to laws of probability, because each segregating allele is distributed randomly to one gamete and independently of other genes (provided the genes are not linked), and because eggs and sperm unit randomly.

Meristem: Tissue primarily concerned with formation of new cells by division, e.g. apical meristem and cambium.

Meristem culture: The culture of excised meristem on suitable nutrient media under asepticconditions.

Micro propagation: Production of plants from very small plant parts, tissues, or cells grown aseptically in a test tube or other container where the environment and nutrition can be rigidly controlled. The term often used synonymous with tissue culture.

Microspores: The male cells produced by sexual cell divisions (meiosis) and maturing to pollen grains.

Mitosis: Normal division of a nucleus into two identical daughter nuclei by a process of duplication and separation of chromosomes.

Moderately Dense Forest (MDF): all lands with forest cover having a canopy density between 40 and 70%.

Monoecious: Of a population or species having functional male and female organs in separateplaces on the same plant.

Morphology: The external and internal form and structure of whole plants, organs, tissues, or cells; also the study of such form and structure, including life cycles of organisms. Internal morphology is often known as anatomy. Functional aspects of structure are an aspect of physiology rather than morphology.

Multiple alleles: A series of alleles of a particular gene.

Mutagen: Any agent that causes an increased frequency of mutation. Mutagens are typically short wave electromagnetic radiations (e.g. ultraviolet irradiation, x-rays and cosmic rays) ionizing radiations(e.g. alpha or beta particles) and chemicals (e.g. nitrous acid and proflavine) that react with nucleotides.

Mutation: A heritable change in the genetic constitution, often recognized as a sudden deviation from the ancestral phenotype. If a chromosome changes, the type of the mutation is preferably specified as a change in structure (aberration) or number. Although mutations are infrequent, and usually recessive and harmful, they are the material of evolution and sometimes breeding. A changed individual or part is also known as a mutation or mutant. Because the spontaneous mutation rate is low, external mutagens, such as X-rays or chemicals, have been used to induce mutations.

N

NADPH (Nicotine adenine dinucleotide phosphate-oxi-dase): A high energy molecule produced un plant cells during photosynthesis.

Native forest: Trees and vegetation that naturally grow in an area and are not introduced from elsewhere.

Natural regeneration: Reforestation of an area from seeds or vegetation without intervention by people.

Natural capital: an economic metaphor for the limited stocks of physical and biological resources found on earth, and of the limited capacity of ecosystems to provide ecosystem services.

Net Present Value (NPV): The NPV of a time series of cash flows, both incoming and outgoing, is defined as the sum of the present values (PVs) of the individual cash flows.

Non-use value: benefits which do not arise from direct or indirect use.

Natural regeneration: Regrowth of forests as seeds fall naturally into the ground or shoots sprout from the roots and stumps of fallen trees.

Naval stores: Products extracted from wood by using organic solvents, such as fatty acids, turpentine, and rosin.

Nectar Crop: Trees valuable as a source of nectar for honey bees.

Needle leaf: A uniquely shaped leaf that is narrow in width and relatively long in length.

Nitrogen cycle: The recurring circular flow of nitrogen from living plant animal tissues to nonliving atmospheric nitrogen and back again.

Nitrogen fixation: A process by which certain strains of bacteria convert nitrogen gas from the atmosphere to nitrogen compounds useful to plants.

Nitrogen-fixing bacteria: A bacterial strain capable of converting nitrogen gas into nitrates and other nitrogen compounds required by plants as nutrients.

Noise duration: The length of time that a person is exposed to a sound.

Noise intensity: the amount of energy in the sound waves from a noise source; loudness of a noise.

Nonrenewable resource: A resource formed solely that cannot be replaced once it is used up.

Nucleoplasm: The materials that makes up a cell nucleus.

Nucleus: A structure in a cell containing hereditary materials.

Nursery stock: Tree seedlings or sapling raised for the purpose of transplanting them to forests or other locations.

Nursery: A farm where young trees are raised before they are transplanted to new locations.

Net present value (NPV) : The maximum dollar amount that can be invested without reducing the rate of return below a specified interest rate. Reflects the magnitude of investment as well as the flow of benefits and costs over time. Sometimes called present net value (PNV).

Noncommercial cutting: Cutting trees to improve growth conditions of the stand. Trees cut in this operation do not yield a net income, usually because the trees cut are too Small to be marketable.

Nutrient cycling: The biological, geological, and chemical circulation of inorganic elements such as nitrogen, phosphorus, and potassium through the soil, living organisms, water, and air, thus providing nutrients to animals and vegetation in the process.

Natural selection: The force that causes some individuals in a population to contribute more descendants (and genes) than others to subsequent generations and so, over a period of time, leads to changes in the genetic composition of populations.

Nodal cutting: Cutting consisting of a small piece of stem with a single leaf and an auxiliary bud attached, i.e. a single node.

Node: A point of the stem from which one or more leaves arise. In the mature stem the nodes are usually well separated by internodes which elongate during growth. In conifers node refers to the branch whorls.

Nucleus: The cell part that contains the genetic material. The nucleus consists of chromosomes, nucleolus and nucleolus. It is separated from the cytoplasm by the nuclear membrane.

Nucleic acid: Two types viz. DNA and RNA. They are made up of long chains of molecules called nucleotides.

Nucleotide: The building blocks of DNA and RNA. Each nucleotide is composed of phosphate, sugar and one of the four bases, Adenine, Cystine, Guanine and Thymine (Urasil in RNA). The sequense of the bases is the basis of the genetic code, i.e the formation of proteins.

O

Oleoresion: A mixture of resin and oils obtained from the sap of pine trees.

Omnivore: An animals that eats both plants and other animals.

Omnivore: An animals that feeds in more than one tropic level. e.g. Humans.

Open forests: Forests in which the leaves shade 30 to 70 % of the forest floor. As well as trees, they have a lower layer of shrubs and a ground layer covered in grasses.

Open Forest (OF): all lands with forest cover having a canopy density between 10 and 40%.

Opportunity costs: foregone benefits of not using land/ecosystems in a different way, e.g. the potential income from agriculture when conserving a forest.

Option value: the value of preserving the option to use services in the future either by oneself (option value) or by others or heirs (bequest value). Quasi-option value represents the value of avoiding irreversible decisions until new information reveals whether certain ecosystem functions have values which society is not currently aware of.

Optimal Solution: An optimal solution is a feasible combination of decision variables that yields the highest objective function value for a maximization problem and the lowest objective function value for minimization problem. Although no other solution exists for a given problem that will yield a higher objective function value, other solutions may exist that yield the same objective function value. These are termed alternative optimal solutions. (Remsoft, 2004)

Optimal:A mathematical programming concept for a class of solutions to a problem definition in which the objective function value is at a global maximum or minimum. (Remsoft, 2004)

Organ: A group of several tissues, such as roots and stems, that functions as a single unit.

Osmosis: The passage of a liquid from a weak solution to a more concentrated solution across a semi permeable membrane that allows passage of the solvent but not the dissolved solids.

Ovary: The part of a flower that produces the ovule or egg cell.

Overmature: A tree for which harvest has been delayed until the wood has begun to decay or die.

Oversight: Responsibility for, or supervision of , the care of a natural resource such as trees.

Ovule: An immature female germ cell located inside the ovary of a flower.

Ozone layer: A layer of ozone in the earth's upper atmosphere, at about 12 to 30 miles in altitude, where the intensity of short wavelength ultraviolet light from the sun (between 130 and 200 nanometers) is sufficiently high to convert normal diatomic oxygen(O_2) to Ozone(O_3). The ozone thus formed shields much of he earth by absorbing ultraviolet radiation at wavelengths that are mutagenic.

Obligate species: A species restricted to a particular environment or a particular mode of life, such as a plant or animal found only in a narrowly defined habitat, e.g., a tree cavity.

Old growth: A forest ecosystem containing old trees, usually over 150 years old, and associated plants and wildlife characterized by diverse structure including gaps filled by shade-tolerant species.

Organic matter: Material produced by plants and animals, such as leaves, branches, bark, wood, hair, fur, and bones.

Overrun : The excess lumber sawn from logs over the estimated volume or log scale, usually expressed in percent of log scale.

Overstocked: A stand or forest condition, indicating more trees than desired.

Overstory : That portion of the trees in a stand forming the upper crown cover.

Ortet: The original plant from which a clone derives.

Orthotropic growth: The situation where the vegetative propagule assumes an upright or normal tree form.

Ovulate: Having seed-bearing organs only. May apply to individual flowers, or strobili, or to female plants of dioecious species in conifers and other gymnosperms.

P

Palmately compound : Multiple leaflets arranged on spines or veins that radiate from a single point; shaped like a hand with the fingers spread apart.

Parasite: An organism that derives its nutrition by feeding on another living organism i.e., derives nourishment at the expense of its host, usually without killing it.

Parasite: An organism which lives on or in and obtains its food from other living organisms.

Parasite: An organism, such as fungus, that lives in or on plants, harming its host by feeding on fluids or damaging tissues of the host organism.

Parasitism: A living relationship between two organisms that benefits one organism and harms the other.

Parenchyma: Thin-walled cells loosely packed together to form spongy tissue with air spaces between them; cells that make up much of the material in plant leaves, roots, stems, and fruits.

Parent material: Rocks (Not including bedrock) located in the C horizon of a soil.

Parthenogenesis: A type of reproduction inn certain types of insects, such as aphids, in which mature females produce female offspring without mating with a male.

Particle board: Dried wood particles glued and squashed together in layers to forms boards.

Particleboard: A reconstituted board containing a high percentage of wood shaving in the central core with layers of wood flakes on either side of the core and fine saw dust near the surface.

Particulate matter: Tiny particles of dust and waste materials that are suspended in the air.

Peat: plant materials found in deep deposits in wet areas, such as marshes and bogs, where decay is minimal.

Pericycle: The outer layer of the vascular cylinder of a root; lateral roots develop from this tissue, growing out through the cortex and epidermis of the primary root.

Permeability: The rate at which water infiltrates or seeps into the soil.

Permeable: The capacity of a membrane to allow fluids to pass through it.

Pest: An animal or plant species that occurs where it is not wanted .

Pesticide residue: Chemicals that remain in food, soil, and water after weed killers and other sprays have been used.

Petal: the inner floral leaves of a flower that are often brightly colored.

Petiole: A plant leafstalk forming the attachment to the main stem.

P^H: A measure of the relative concentration of hydrogen ions in a solution; this values indicates the acidity or alkalinity of the solution in a scale of 0 to 14, , 7 indicates a neutral solution; p^H value greater than 7 are basic, and those below 7 are acidic. It is calculated as the negative logarithim to the base ten of the hydrogen ion concentration in moles per litter.

Phenotype: The physical appearance of an organism (Genotype + environment)

Phloem: A vascular plant tissue through which sugars and plant foods, manufactured in leaves, flow to the stem and roots.

Photogrammetry: A method for measuring land area by producing aerial photographs of a known scale from which the total area within the boundary lines is determined.

Photosynthesis: A process that uses chlorophyll to capture energy from the sun and combining it with carbon dioxide, water, and nutrients to form plant tissues.

Photosynthesis: The biochemical process by which green plant makes food from carbon dioxide and water using light energy from the sun.

Photosynthesis: The series of chemical reactions by which plant cells transform light energy into chemical energy mediated by chlorophyll through producing simple sugars and oxygen, from carbon dioxide and water.

Physiology: The branch of biology concerned with the life functions and processing of living organisms.

Phytoplankton: The portion of the plankton community comprised of tiny plants capable of photosynthesis, such as unicellular algae, diatoms.

Pine cone: The woody fruits of a conifer tree, made up of overlapping scale.

Pinnately compound: Multiple leaflets arranged along opposite side s of a central leafstalk.

Pioneer species: the first plant to grow naturally in an area that has been cleared or burned.

Pistil: A female reproductive structure found in flower.

Pith: A structure located in the center of a stem consisting of parenchyma cells, which have the primary function of storing plant food.

Planer: A machine that reduces rough lumber to standard lumber dimensions by shaving or planning off the outer surfaces.

Planimeter: An instrument that accurately measures the land area of a map of known scale by tracing the perimeter boundary.

Plank: A piece of wood that measures 1.5 to 4 inches in thickness and more than 11 inches in width.

Plantation forest: A population of trees established by planting seeds or seedlings and intensively managed for maximum production.

Plantation: Trees planted for the production of timber or other values. planted with trees.

Plywood: A laminated wood products made of several sheets of veneer and lesser quality wood bonded together by adhesives.

Pneumatic power: The use of compressed air to deliver power to a location remote from the power source.

Poaching: Hunting wildlife illegally.

Polar nuclei: haploid nuclei that develop during the formation of female gametes.

Pole: A young tree that ranges from 4 to 10 inches in diameter.

Pollen grain: A male sex cell in plants that fertilizes female flower parts to produce fruits and seeds.

Polymer: A natural or synthetic chemical structure where two or more like molecules are joined to form a more complex molecular structure.

Polymer: Any of a large group of compounding consisting of two or more simple sugars such as starch, carbohydrates, or cellulose; also known as polysaccharide.

Polysaccharide: Any of a large group of compounding consisting of two or more simple sugars such as starch, carbohydrates, or cellulose; also known as polymer.

Pores: Specialized vessel, visible in wood, through which dissolved nutrients passed when the trees was living.

Porosity: Degree to which soil, gravel, sediment, or rock is permeated with pores or cavities through which water or air can move.

Precautionary Principle: If an action has a suspected risk of causing harm to the environment, in the absence of scientific consensus that the action is harmful, the burden of proof that it is not harmful falls on those taking an act.

Provisioning services: the products obtained from ecosystems, including, for example, genetic resources, food and fiber and fresh water.

Public goods: a good or service in which the benefit received by any one party does not diminish the availability of the benefits to others, and where access to the good cannot be restricted.

Predator: an animal that kills and eats other animals.

Prehauler: A large machine used to load and haul logs to the landing.

Prescribed burn: A controlled forest fire purposely started to burn the trash on the forest floor, thus reducing the risk of a catastrophic fire due to the presence of too much fuel in the forest; also known as a controlled burning.

Preservation: Impregnating logs or sawn timber with chemical preservatives.

Primary consumer: An animal that eats plants.

Primary growth: Elongation of plant cells accounting for the length wise growth of roots and stems.

Primary succession: The development of an ecological community in an area where living organisms were not previously found, as on a newly formed volcanic island.

Primary succession: The sequential development f communities on an area that was not previously occupied by organisms, such as the surface of a recent volcano or deglaciated land or abandoned strip mine.

Primary tissue: The first specialized tissues that forms in the area of maturation, such as the epidermis, cortex, and the vascular cylinder tissues in the root.

Principal meridian: A survey line that runs north-west.

Producer: A plant that converts solar energy and other plant nutrients to starches and sugars.

Producer: Another term for autotrophy. Plants perform photosynthesis and provide food to consumers. Also called primary producer.

Propagation: A process by which an organism reproduces.

Prophase: The first stage of active cell reproduction.

Protoplasm: All of the structures and substances located within a cell.

Protoplast: A membrane bound cell from which the outer wall has been partially or completely removed. The term often is applied to plant cells.

Pruning: Cutting branches from the trunk or crown of trees, often to promote the production of clear wood.

Pruning: The removal of unwanted branches from a trees; shaping a tree by controlling the growth of the limbs and branches.

Partial cut: Any cutting scheme that removes less than the total tree stand at any one time.

Photosynthesis : The chlorophyll process in trees and other "green" plants changing carbon dioxide and water into carbohydrate using the sun's energy.

Pile and burn: A controlled burn where slash is concentrated, usually by machinery, before burning.

Pioneers; Shade intolerant species that are the first trees to invade freshly disturbed sites, such as red alder or western larch.

Pistol butt: Bases of trees curving down slope and then upward. This may indicate unstable or moving soil.

Pitch; Resin occurring in the wood of certain conifers.

Pitch-out: A defense mechanism of vigorous trees against insects, whereby a flow of pitch pushes bark beetles back out their entry holes through the bark.

Plant association; A vegetation community, in which the dominant species forms the highest level in the hierarchy of plant species.

Plantation: A reforested area established by planting trees.

Planting stock; Seedling trees ready for planting.

Plug seedling: A seedling tree grown in a small container, under carefully controlled. Environmental conditions, such as in a greenhouse. Seedlings are removed from containers for planting. Also called containerized seedling.

Pole saw: A saw attached to a long pole for pruning without using a ladder.

Pole stand: A stand of trees whose diameters range from approximately 5 to 9 inches.

Population: Organisms (trees, shrubs, herbs, animals, insects, etc.) of common ancestry that occupy a particular forest area.

Provenance: The geographical source or place of tree seed origin.

Pruning : Removing live or dead branches from standing trees to improve wood quality.

Pulpwood: Wood cut primarily for the manufacture of paper, fiberboard, or other wood fiber products. Dead trees often are cut for this purpose.

Punky : A soft, weak, often spongy condition in wood; caused by decay.

Parthenocarpy: Development of fruit without viable seed. May be induced artificially as by foreign pollen or hormones.

Parthenogenesis: Reproduction from an unfertilized egg; a type of apomixis. The resultant embryo may be either haploid or diploid.

Phenology: The study of relations between plant development and seasonal climatic changes, such as in temperature or day length, especially as such changes affect periodic phenomena like leafing, flowering, and dormancy.

Phenotype: The plant or character as we see it; state, description, or degree of expression of character; the product of the interaction of the genes of an organism (genotype) with the environment. When the total character expressions of an individual are considered, the phenotype describes the individual. Similar phenotypes do not necessarily breed alike.

Physiological character: A character related to the life processes of an organism; a function in contrast to form or structure. Often it is not discernible except by performance, e.g. rate of growth or resistance to certain pests.

Pistillate: Having female organs only. May apply to individual flowers or inflorescences, or toplants of a dioecious species in angiosperms. Collectively the pistils are called the gynoecium.

Plagiotropic growth: The situation where the vegetative propagule does not assume a normal tree form, but continues to grow like a branch. Ant. orthotropic growth.

Plant breeding: The improvement of a plant production through selection and controlled propagation, e.g. crossing.

Plus stand: A stand containing a preponderance of good phenotypes, not necessarily plus trees.

Polarity: The condition that plant parts maintain their spatial orientation when separated from the mother plant. The phenomenon is important in cuttings and grafting.

Pollen: The microspores of seed plants, produced in vast numbers in the male part of the flower(anther).

Pollen dilution zone: A zone surrounding seed orchards and seed stands where trees of the species occurring in the seed orchard or stand and possibly hybridizing species have been cut in order to protect the stand from contamination by outside pollen.

Pollen tube: An outgrowth of a germinating pollen grain through the stigma, through which the sperm passes to fertilize the egg.

Pollen vector: The carrier, such as wind, insects or birds, carrying pollen from one plant to another.

Pollination: Deposition of pollen on the receptive part of the female flower. In angiosperms this is the stigmatic surface, in gymnosperms the ovule tip.

Polyembryony: The formation of more than one embryo in an ovule.

Polygenes: Several non-allelic genes all affecting the same character and approximately additivein their effect (polymeric systems).

Polyploid: Having more than twice the basic number n of chromosomes of the ancestral species inits vegetative cells. A cell, tissue, or organism having three sets (3n) is called triploid, and four sets(4n) tetraploid

Progeny test (trial): Evaluation of parents by comparing the performance of their offspring. Accuracy is usually gained because several to many offspring per parent are evaluated under more controlled conditions than exist for the parent. Preferably established on two or more environmentsto study family x environment interaction.

Propagule: A plant part, such as a bud, tuber, root, or shoot, used to propagate an individual vegetatively.

Protandry: Termination of the shedding of pollen of a plant or flower prior to receptivity on the same plant or flower.

Protein synthesis: The biochemical process by which the genetic material of the cells is expressed as the amino acid sequence of a protein. The proteins consist of polypeptide chains which are built up from 20 different amino acids. The number, sequence and type are unique for a particular protein. The sequence of amino acids is determined by the sequence of bases of the DNA via the process illustrated below: (a) The base sequence is transcript to a complementary base sequence of themRNA; (b) the mRNA moves to the ribosome's where the synthesis occurs; (c) the tRNA binds specific amino acids at one end and a triplet of bases (anticodon) at the other; (d) the anticodon binds to a triplet (codon) of the mRNA; (e) the individual amino acids bind together in a polypeptidechain; (f) the tRNA is released to bind to another free amino acid. The matching of codons and anticodons and the attachment of amino acids to the polypeptide chain is called translation.

Protogyny: Termination of receptivity prior to the maturation of pollen on the same plant or flower.

Provenance: 1) The original geographic source of seed, pollen, or propagules. In forest tree breeding the term usually refers to the original native source of a population. When a population is removed from its sources and has grown elsewhere for a number of generations, it is referred to as a 'land race'. 2) The place in which any stand of trees are growing. The stand may be indigenous or non-indigenous.

Provenance resource stand: A forest plantation of known provenance and broad genetic base the boundaries of which are marked in the field and recorded on official maps, and which may be used for plus tree selection, seed collection, provenance conservation etc.

Provenance trial (test): A well-designed comparison of population samples from a range (preferably the whole range) of a species. Should be established in two or more environments to enable assessment of provenance x environment interaction.

Proximal: The proximal end of either the root or the shoot is that nearest to the stem-root junction of the plant and farthest from the tip of the root or the shoot.

Pure line (pure strain): A succession of generations recognized by their ability to produce genotypicalidentical offspring when selfed or crossed between themselves. The members of a pureline are said to be 'breed true'. By inference such individuals are deemed to be homozygous.

Q

Quarter section: A tract of land measuring approximately 160 acres; one fourth of a section.

Quantitative genetics: The part of genetic science which deals with the hereditary transmission of differences among individuals that are quantitative, e.g. growth rate, wood density etc. Most of these characters are determined by additive genes.

Quantitative inheritance: Inheritance of characters, such as size and form, which vary continuously (quantitatively). From one to hundreds of genes may control such characters. These genes(multiple factors or polygenes) act cumulatively and cannot be detected by their individual effect. Quantitative characters are usually subject to considerable modification by the environment and hence require refined statistical techniques in their study.

R

Radial growth: Growth resulting in the increased diameter in a tree.

Rain forests: Forest with a very dense canopy (70 to 100% cover) which allow little sunlight to penetrate to their lower levels.

Range: The east-west location of a township from a principal meridian.

Rangeland: An area that provides vegetation for grazing animals.

Receptacle: The base of a flower.

Recessive: An allele that is masked by a dominant allele; does not appear in the heterozygous condition, only in homozygous.

Recharge area: The area within a catchment in which surface water (from rainfall, irrigations or streams) soaks into the soil and increase s ground water levels.

Recombinant DNA Technology: The transfer of a desired gene, such as disease-resistant gene, to the genetic material of a new plant or animal, resulting in the expression of the new trait.

Rectangular survey: A land survey system that uses an initial point as the beginning point for each survey and that establishes east-west baselines, north-south meridian, and intermediate survey lines.

Recycling: Reusing materials that would otherwise be disposed of as waste.

Red gum: A native tree, also known as river red gum (*Eucalyptus camaldulensis*), that was once widely cut for its timber. Often found in Native forest reserves, generally near watercourses.

Regulating services: the benefits obtained from the regulation of ecosystem processes, including, for example, the regulation of climate, water and some human diseases.

Resilience (of ecosystems): their ability to function and provide critical ecosystem services under changing conditions.

Reforestation: The return of a population of forest plants to an area where they are previously been destroyed or from which they were removed.

Regeneration: Regrowth of plants and trees, either naturally or resulting from management practices such as sowing seeds or planting tube stock.

Regeneration: The ability of a plant to reproduce itself from seeds or vegetative plant parts; development of new plant tissues to replace missing parts.

Remote sensing: The collection or interpretation of information about an object without physical contact with the object; e.g., satellite imaging, aerial photography, and open path measurements.

Renewable source: A resource that is not used up but continuous to produce more of its kind at regular intervals.

Reproductive organ: A plant organ in which sexual reproduction occurs; a flower.

Resin duct: A natural channel in a live tree through which pitch flows to different locations in the tree.

Resin: Any of a group of thick, sticky liquid consisting of plant juices such as pitch or gum; a wood product used to manufacture varnishes, lacquers, naval stores, and so on.

Resolution: A written document that express the intentions of a legislative body.

Respiration: A process in which energy and carbon dioxide are released due to the breakdown of plant tissues during periods of darkness.

Respiration: The process by which living organism take in oxygen from the atmosphere and give out carbon dioxide.

Rhizomorphs: Thin strand of fungal tissue that enter into the surfaces of tree roots, where they infect the tree with new colonies of fungi.

Rill erosion: The removing of thin layers of soil by surface runoff; can be easily smoothed out or obliterated by normal tillage.

Riparian zone: A narrow strip of land on both sides of a running stream on which permanent vegetation is maintained. It is also called a filter strip.

Rodent: Small gnawing animals identified by the four large incisor teeth located in the front of their mouths.

Rodenticide: A chemical used to kill rodents.

Root cap: A group of specialized cells that develops at the root tip from which a slimy material is produced, helping the root to pass through the soil.

Root graft: A connection between the roots of different trees where they have grown together at the point of connection.

Root hair: Epidermal cells that have developed into long, thin, threadlike projections to facilitate absorption of water and dissolved nutrients from the soil.

Rotation age: The actual or expected age of a stand when it is harvested.

Rotation age: The age or stage of maturity at which a stand of trees is harvested.

Rotation length: The number of years between planting and harvesting a forest. The time varies with species, site conditions and growth rates.

Run-off: The portion of precipitation, snow melt, or irrigation water that runs off the land into streams or other surface-water. In broad terms, the water available for human use. It can carry pollutants from the land into receiving waters.

Rust: A class of plant diseases that results in spotted red or brown discoloration of the stems and leaves of a tree.

Rangeland: A large open land area on which livestock wander and graze. The native vegetation is mainly grasses, forbs, and shrubs.

Reforestation: Reestablishing a forest on an area where forest vegetation has been removed.

Refuge: To save, rather than manage, portions of an ecosystem or forest.

Regeneration cutting: A harvesting technique that provides for stand regeneration, such as clearcutting, seed-tree, selection, and shelter woodcutting methods.

Release cutting: Improving the species composition in young stands by cutting inferior species. This operation releases the better trees from competition.

Relief culvert: A structure that relieves surface runoff from road sided itches. Prevents volume and velocity of runoff from reaching excessive buildup.

Resilience: The ability of a forest ecosystem to recover from a disturbance.

Riparian: Pertaining to the area along the banks of a river, stream, or lake.

Riparian ecosystem: An ecosystem that is transitional between land and water ecosystems. The soils, plants, animals, and other organisms found in such an area.

Root collar: On a tree, the transition zone between stem and root. Usually recognizable in trees and seedlings by a slight swelling.

Roots: That portion of a tree, generally underground, which anchors the tree and absorbs water and nutrients from the soil.

Rot : Wood in a state of decay.

Rotation age: The number of years required to establish and grow trees to a specified size, product, or condition of maturity. Rotation ages vary from 7–10years for intensively managed plantations to over 100 years for extensively managed stands.

Round wood: Wood products which are round, such as pulpwood, posts, pilings, utility poles, and fencing materials.

Ramet: An individual member of a clone.

Recessive (allele): Describing an allele that is only expressed when homozygous. They are in most cases considered to be the mutant form of dominant alleles. Recessive alleles are in crossings referred to with small letters, dominant alleles with capitals.

Reciprocal cross: The repetition of a cross where the sexual function of the parents is reversed, i.e. female B x male A is the reciprocal of female A x male B. Used in full dialed cross.

Recurrent selection: Selecting individuals or families and intermitting them or allowing them to interpollinateto produce the next generation, e.g. a seed-production area or a clonal seed orchard. The new generations are generally used as breeding populations in which to start repeated cycles of selection and breeding.

RNA: Ribo-Nucleic-Acid. A single stranded organic molecule consisting of sugars and bases which acts during the protein synthesis by translating and transcribing the genetic code. There are three types of RNA viz. messenger RNA (mRNA), transport RNA (tRNA), and ribosomal RNA (rRNA)See DNA, Protein synthesis, Transcription, Translation.

Rogueing (Culling): Systematic removal of individuals not desired for the perpetuation of apopulation.

Rooting hormone: Hormone (growth substance) which specifically promotes root formation androot growth. The active compound is usually β-indole-butyric acid (IBA) or α-napthalene-aceticacid (NAA). Sold under various trade names e.g. RooTone or Seradix.

Rootstock (Stock): The root-bearing plant or plant part, usually stem or root, onto which another plant is grafted.

Root suckers: Shoots appearing from adventitious buds on superficial lateral roots. Roots suckers are genetically identical to the mother plant and used in vegetative propagation.

S

Saccharification: A process by which the polymers that make up cellulose in wood are converted to simple sugars.

Salinization: The accumulation of salts in the soil, waterways or the water table to a level that causes damage to plants.

Salvage cutting: A harvest of trees that have been damaged by fire, wind, insects, and so on.

Sanitation cutting: A harvest of trees that are infected with diseases or insects for the purpose of preventing the problem from spreading to healthy trees.

Sapling: A young that is more than 3 feet in height and up to 4 inches in diameter.

Saprophyte: A fungus that obtains nutrition from dead organic materials.

Saprophyte: An organism that lives on dead or decaying organic matter that help natural decomposition of organic matter in water.

Saprophytes: Plants that live and feed on dead organic material.

Sapwood: light colored wood through which water and dissolved plant nutrients flow.

Sapwood: The pale outer part of the wood of a tree which contains living cells.

Savanna: A savanna is a rolling grassland scattered with shrubs and isolated trees, which can be found between a tropical rainforest and desert biome. This receives more rainfall but not enough to support forests. Savannas are also known as tropical grasslands. They are found in a band on either side of the equator on the edges of tropical rainforests.

Saw head: A large blade mounted on a mechanical tree feller for the purpose of sawing through the tree at its base after it has been grasped by the mechanical arm.

Sawyer: A person who operates the head saw or head rig in a saw mill.

Scale leaf: A leaf that consists of a number of tiny thin and flat overlapping leaf structures.

Scaling: Measurement of forest products to determine the quantity or amount.

Sclerenchyma: A specialized type of plant cell that strengthens tissues by adding fiber to them; cells that form the shells of nuts.

Sclerophyll: Sclerophyll means 'Hard-leaved' and refers to plant species that have developed drought resistant leaves with thick cell walls and a heavy cuticle.

Secondary consumer: A carnivorous animal that obtains its nutrition by eating primary consumers and other carnivores.

Secondary growth: Growth in diameter in roots, stems, and branches.

Secondary succession: It occurs on substrate that previously supported vegetation before a disturbance destroyed the plant life. It is a response to a disturbance.

Secondary succession: The gradual change in species of plants that live in an area during the time that a damaged ecosystem is returing to its original stage of ecological development.

Secondary tissue: The tissue responsible for the increase in the diameter of roots, stems, and branches as they grow.

Section: A tract of land approximately 1 mile square with a surface area of 640 acres.

Sediment: Soil particle such as silt, clay, and sand. It is considered to be the most significant pollutant to come from forested area.

Seedling: A young plant grown from seed.

Seedling: A young tree in the very early stages of development.

Seed-tree method: A timber harvest method in which mature tree of the desired species are protected from cutting in locatins scattered throughout the forest for the purpose of producing seeds.

Selection cutting: A harvest method used to remove a sustained yield of wood from the forest at regular intervals by cutting only the most matured trees.

Semichemical pulping: A process of producing wood pulp in which wood is exposed to a mild chemical treatment to partially separate the wood fibers, followed by processing through a mechanical risk refiner.

Senescent: An advanced stage of maturity in which a tree shows symptoms of heart rot, decay, and other defect due to age.

Sensitivity analysis: A method used to determine the relationships between measures and assumptions used in an analysis, and to determine the effects of changing the value of measures and assumptions on the results of an analysis; is usually conducted by recalculating an analysis a number of times each with change in one measure or assumption at a time.

Sepal: A protective leaf like structure that closes over a flower during hours of darkness.

Sexual reproduction: The production of male and female gametes and the process by which they join together to produce offspring.

Shade intolerance: A characteristic of plants that describes their propensity to establish and grow in director full sunlight conditions.

Shade tolerance: A characteristic of plants that describes their propensity to establish and grow under indirect or reduced light conditions most often in the shade of other plants growing around or above them.

Sheet erosion: The loss of thin layers of soil from unprotected land surfaces, commonly agricultural lands; accomplished primary by wind and water.

Shelterwood system: A timber harvest method in which mature trees are left in the harvested area in sufficient numbers to provide shade and protection for seedlings. The mature trees are harvested once the seedling have become established.

Short duration fire cycle: The use of prescribed fires on a regular basis to eliminate fuels from the forests.

Short rotation intensive culture (SRIC): Wood crops harvested every 3 to 7 years.

Shrub layer: Vegetation consisting of short woody plants that occupy the stratum between the herb layer and the under story of the forest.

Shrub: A short woody plant.

Sieve element: A specialized plant structure that makes up sieves tubes and phloem tissue; important in the flow of plant foods to stems and roots.

Sieve tube: A specialized plant structure made up of sieve elements; an important component in phloem tissue, facilitating the flow of plant foods to stems and roots.

Silt load: The amount of eroded soil carried in the flowing waters of streams and rivers.

Silt: The tiny soil particles that are easily eroded by becoming suspended in flowing water or blown as dust in the wind.

Siltation: The process by which tiny soil particles are suspended in flowing water and deposited in the beds of streams, rivers, lakes, and oceans.

Silvics: The life history and general characteristics of forest.

Silviculture prescription: A site-specific plan that describes the forest management objectives for an area; must be consistent with any higher level plan that encompasses the area to which the prescription applies; identifies the method for harvesting the existing forest stands and a series of silviculture treatments that will be carried out to establish a free growing crop of trees in a manner that accommodates other resource values as identified; subsequent documents, including cutting authorities and logging plans, must follow the intent and meet the standards specified in the silviculture prescription.

Silviculture regime: A series of carefully sequenced and implemented activities at specific time periods and intensities to achieve desired objectives.

Silviculture: It is the art and science of controlling the establishment, growth, composition, health, and quality of forest trees or crops to meet diverse needs and values of the many landowners, societies and cultures over the parts of the globe that are covered by dry land.

Silviculture: Management of forests and their environments to establish and cultivate populations of trees and to promote the growth and to harvesting of trees for commercial purposes.

Silviculture: The management of forests for the production of timber products.

Simulation model: An integrated set of equations and numerical techniques that mimic the relationships and development of actual systems (e.g., trees, stands, forests).

Skidder: A large machine used to drag logs from the harvest site to the landing.

Slab: The outer shell of a log that is removed as the log is sawed into lumber.

Slash: The leaves and branches left on the ground after harvesting operations.

Smalian's formula: A dependable log scaling method that is used to estimate the amount of solid wood in a log by using a formula that takes into account the log length and the measurements from both ends of the log.

Smog: A term coined by Dr. H. A. Des Voeux in 1905 of England to describe the conditions of fog that had soot or smoke in it. Smog is a combination of various gases with water vapor and dust.

Smog: High concentration of chemical pollutants and particulate matter suspended in the atmosphere; atmospheric haze caused by the action of ultraviolet light on atmospheric pollutants.

Soft wood: Wood obtained from a conifer or needle bearing tree; immature plant materials obtained from new growth trees.

Softwood: The open-grained wood of conifer species such as pines, firs, and cedars.

Soil conservation: Prevention and control of soil erosion and degradation by careful land management.

Soil conservation: Soil management practices that prevent soil loses and maintain soil fertility.

Soil profile: A description of the texture and content of each of the different horizons or layer in a soil.

Soil texture: The proportion of mineral particles of different sizes that are found in a sample.

Species diversity: The total number of different species and their relative abundance with given area.

Species: A population or group of individual plants or animals which are able to interbreed to produce fertile offspring and/or have common characteristics derived from a common gene pool.

Spiking: The illegal act of driving steel spikes into trees with the intention of damaging saws and milling equipment; an act of protest by anti-logging demonstrators.

Spindle: A bundle of fibers that functions to separate chromosome pairs during cell division.

Spine: A leaf structure similar to the midrib that gives structure and shape to the leaf.

Spodosols: Light-colored acid soils formed from coarse silica parent material in cold, damp climates.

Spore: A cell from a fungus or similar organism that grows into a new organism when it is located in a site that is favorable to its growth; a male reproductive cell in a plant

Sprout method: A form of asexual reproduction of a forest in which all of the trees are cut and new forest growth is generated from the stumps of the harvested trees; also known as the coppice method .

Stamen: The male part of the flower consisting of the anther and the filament.

Stand A contiguous group of trees sufficiently uniform in species composition, arrangement of ageclasses, and condition to be a homogeneous and distinguishable unit.

Stand density management diagram: A graphical depiction of the relationships between stand density, top height, quadratic mean diameter and mean tree volume.

Stand improvement: Removal of competing vegetation from a forest that has matured beyond the sapling stage of development.

Stand management prescription (SMP):A site-specific plan describing the nature and extent of silviculture activities planned for a free growing stand of trees to facilitate the achievement of specified or identified social, economic or environmental objectives.

Stand table: A tabular display of the number of trees in a stand by diameter class.

Stand: A crop of trees in a particular area, such as a forest.

Stand: A population of trees in a given location.

Standard cord: A unit of measurement for pulpwood and firewood; the volume of wood in a pile 8 feet long, 4 feet height, and 4 feet wide; 128 cubic feet.

Standard parallel: A survey line that is established running east-west for each interval of 24 miles on either side of the baseline.

Starch: A molecule formed when large numbers of glucose molecules bond together in a long branching chain.

Stereoscope: A mapping instrument that is used to simultaneously view two aerial photographs at the same area from slightly different angles, creating an image in three dimensions.

Stigma: The female flower part that functions as a pollen receptor.

Stock table: The tabular counterpart to a stand table that displays stand volume by diameter class.

Stomata: The pores on the surface of leaves that control the movement of gases and out of plants.

Strata: Levels or layers of plant growth in a forest or other ecosystem.

Stratosphere: The portion of the atmosphere 10 to 25 miles above the earth's surface.

Style: A female flower part that connects the stigma to the ovary of the flower.

Succession: A sequential changes in vegetation often in response to environmental change.

Suppressed: A crown class that includes trees with crowns that are completely covered over by the crowns of one or more nearby trees.

Surface fire: A fire that burns the dry twigs, dead branches, grass, and leaves that lie on the forest floor.

Surface fuel: Fuels such as dry, un-decayed leaves and plant materials along with living plants.

Surface runoff: Water that flows directly to streams, rivers, and lakes across the surface of the soil. It is frequently accompanied by soil erosion.

Surface water: Water flowing in streams, rivers, and lakes as it moves across the land surface to the ocean.

Survival growth: The total difference in timber volume of survivor trees between the first and second measurements of a forest growth survey.

survivor trees: Trees that were measured in both surveys for the time interval during which forest growth is evaluated.

Sustain yield: The volume of timber products that can be harvested year after year without depleting the timber resource.

Sustained yield: The quantity or amount of wood that can be harvested per year over an indefinite period without completely harvesting a forest.

Symbiosis: A relationship between two organisms in which each organism receives benefits from its association with the other.

Systematic sampling: A method that is used to estimate the timber yield of a forest by measuring only those trees that have been selected at specific intervals along predetermined lines.

Salvage cut : Harvesting damaged or defective trees for their economic value.

Sanitation cut : Harvesting trees that are infected by, or highly susceptible to, insects or diseases, to protect the rest of the forest.

Sapling : A small tree, usually between 1 and 3 inches DBH, and 15 to 30 feet in height.

Sapwood : The light-colored, actively growing section of the tree, between the heart wood and the bark. Serves to conduct water and minerals to the crown.

Sawlog: A log large enough to produce a sawn product—usually at least 10 inches in diameter.

Scale stick: A flat stick, similar to a yardstick, calibrated so log volumes (usually in board feet) can be read directly when the stick is placed on the small end of a standard-length log.

Scaling: Estimating the usable wood volume in a log, following fundamental rules.

Scalping: Physically removing the sod or surface layer of debris, exposing mineral soil for tree planting.

Scarify: To disturb the forest floor and topsoil in preparation for natural regeneration, direct seeding, or planting.

Seedling: A tree, usually less than an inch in diameter, and no more than 3 feet in height, which has grown from a seed (in contrast to a sprout). Also refers to nursery grown trees that have not been moved in the nursery to obtain more growing space.

Seed-tree harvest: Removing nearly all trees from a harvest area at one time, but leaving a few scattered trees to provide seed for a new forest stand. Usually 5 to 10trees per acre are retained. These are removed later, after sufficient regeneration is established.

Seed year : A year in which a given species produces a large seed crop. Used in reference to trees that produce seed irregularly or infrequently.

Seed zone : Areas which have similar climate and elevation conditions. Used to specify where tree seed was collected and where trees from such seed will probably grow successfully.

Selection harvest : Harvesting individual trees or small groups (group selection) of trees at periodic intervals (usually 8 to 15 years)based primarily on their vigor and age. Selection harvesting perpetuates an uneven-aged stand.

Shade intolerant: A term applied to trees or shrubs that grow and reproduce better in direct sunlight than in the shade of other trees or shrubs. The opposite of shade tolerant. Tree examples include coastal Douglas-fir, western larch, lodge pole pine, and red alder.

Shade tolerance: The capacity to develop and grow in the shade of, and in competition with, other trees and shrubs. Examples of highly shade-tolerant tree species are western hemlock, western red cedar, and Pacific yew.

Shelter wood harvest: Harvesting trees in a series of two or more operations. New seedlings grow and become established in the partial shade protection of older trees. Harvests are usually 5 to 10 years apart, yielding an even-aged stand.

Shrub: A low-growing perennial plant with a woody stem and low branching habit.

Site: An area evaluated for its capacity to produce forest products. Evaluation is based on combined biological, climatic, and soil factors.

Site class: A grouping of similar sites that indicates relative productivity. The common system for the Douglas-fir region includes five site classes, in which Site I is the most productive and Site V is the least productive.

Site index: An expression of forest site quality based on the height of the dominant and co dominant trees in the stand at a specified age, usually 50 or 100 years.

Site preparation: Preparing an area of land for forest establishment. Methods used may include clearing, chemical vegetation control, or burning.

Skidding: Pulling logs with equipment or horses from the stump to the landing.

Skid trail: A road or trail over which horses or equipment drag logs from the stump to a landing.

Skyline logging: A logging method in which a traversing block or carriage rides on a sky line or cable. The line itself does not move; it is attached to a donkey drum or an anchorage at the landing, and to a tail tree or tail hold on the far end.

Slash : Non-merchantable residue left on the ground after logging, thinning, or other forest operations. Includes tree tops, broken branches, uprooted stumps, defective logs, and bark. Slash can have certain ecological benefits, such as adding nutrients to the soil or providing wildlife habitat.

Slope: The incline of the terrain usually expressed as the amount of incline in feet over a hundred feet of horizontal distance.

Snag: A standing dead tree.

Soil compaction: The process by which soil particles are squeezed or compressed, reducing air and water spaces.

Soil P^H: A measure of the acidity of the soil. Seven is neutral, less than seven is acidic, and greater than seven is alkaline.

Soil texture: Proportion of clay, silt, and sand in soil.

Spar: A portable tower used in cable logging to provide lift when yarding.

Species: A group of organisms (plants or animals) that are very similar in appearance and can interbreed freely with each other but not with other groups.

Species composition: In a forest, the mixture of tree species in a stand.

Sprout : A stem vegetative produced from a stump or the roots; sometimes a branch produced after the stem was formed.

Stand : A recognizable area of the forest that is relatively homogeneous and canbe managed as a single unit. Stands are the basic management units of the forest. Stand types include:

- All-aged—A stand that supports trees of all ages and usually all sizes. This stand type is rare. Contrast it with an even-aged stand.
- Even-aged—A stand in which trees are essentially the same age (within 10 to 20 years).
- Fully stocked—A stand where trees effectively occupy all growing space, yet ample room exists for developing crop trees.
- Mixed—A stand that has more than one species in the main tree canopy.
- Over mature—A stand of trees that is older than normal economic rotation age for the type.
- Overstocked—A stand that is overcrowded, thus reducing tree vigor.
- Pole—A stand in which the dominant trees are 5 to 9 inches in diameter.
- Pure—A stand in which at least 80% of the trees belong to a single species.
- Residual—The stand which remains after cutting.
- Sawtimber—Most trees in the stand are large enough in diameter (usually 10 to 12 inches DBH or larger) to be sawn into lumber.
- Understocked—A stand in which crop trees do not effectively occupy the growing space.
- Uneven-aged—A stand that supports trees of several age classes (technically, more than two age classes).

Stand density : A quantitative measure of stand stocking, or the number of trees for a given area.

Stand structure: A descriptive term to express the "fullness" of the ecosystem with living organisms, such as trees, shrubs, and forest floor species. Stand structure is defined in terms of both horizontal and vertical richness.

Stand table : A table by diameter classes of volume, basal area, or trees per acre existing in a stand or expected to exist at a certain time.

Stewardship : The responsibility to manage forest land with practical wisdom.

Surface runoff : Water that moves over the ground surface.

Sustainability: Forest development that incorporates the means to maintain biological diversity, resilience to stress, and ecosystem health and integrity, in the context of the ability to meet future as well as present human needs.

Sustained yield: Management of forest land to produce a relatively continuous flow of timber or revenue.

Stocking: A description of the number of trees, basal area, or volume per acre in a stand compared with a desirable level for best growth and management. Often stocking is used as a relative term, such as partially stocked, normally stocked, or overstocked.

Stomata: Small openings that allow moisture within the leaves and/or needles to escape, in a process called transpiration.

Strip cut: A clear-cut laid out as a long, narrow strip.

Stud log: A small log, usually 4 to 8 inches top DIB and 8 feet long usually cut into 2 x 4s.

Stumpage value: The monetary value of a tree or group of trees as they stand in the woods uncut (on the stump). Often this value is paid as a delivered log price less logging and transportation costs.

Succession: The replacement of one plant community by another, which may end in a climax community; Plant association.

Successional stage: A phase in the natural development of forest communities. Over time, favorable conditions are reached for the establishment of the next stage. Commonly used stages include stand initiation, stem exclusion, under stories initiation, and old-growth.

Scion: An aerial plant part, often a branch let that is grafted onto the root-bearing part (stock, rootstock) of another plant.

Seed: Morphologically the structure developed from the fertilized ovule in seed plants. It contains the embryo, which develops into the new plant, and usually endosperm, which forms the food supply. The seed is surrounded by a protective seed coat or testa. In a broad sense 'seed' often refers to the whole dispersal unit, e.g. in certain indehiscent fruits to the seed plus the whole or part of the fruit which continues to enclose the seed during collection and handling. Induces, for example, seed usually refers to the seed plus enclosing endocarp. Such 'seeds' may contain one or more morphological seeds which are then called kernels.

Seed collection area: An area with defined boundaries and altitude limits in which soil and climate are sufficient uniform to indicate high probability of reproducing a single ecotype.

Seed lot: A quantity of seeds of the same species, provenance, date of collection and handling history, and which is identified by a single number.

Seed orchard (SO): A plantation consisting of clones or seedlings from selected trees usually of proven genetic quality. The plantation is isolated to reduce pollination from outside sources, rogued of undesirable trees, and managed for early and abundant seed production.

- *Seedling seed orchard (SSO)*: Seed orchard raised from seedlings produced from selected parents through natural or controlled pollination.
- *Clonal seed orchard (CSO)*: Seed orchard raised from selected clones propagated by grafting, budding, air-layering or rooting of cuttings.
- *Extensive seedling seed orchard (ESSO)*: Stands established with special stock from a balanced mixture of seeds from at least 60 good parents (preferably of proven superior combining ability) and gradually culled. The family identity is not retained in the field.

Seed production area (SPA): Same as Seed Stand. A plus stand that is generally upgraded and opened by removal of undesirable trees and then managed for early and abundant seed production.

Seed source: The locality where a seed lot was collected; also the seed itself.

Seed stand: The stand of trees growing in a seed production area.

Seed year: In respect of any species, particular trees of irregular or infrequent seed production, a year in which it produces, either as an individual or a crop, an adequate amount of seed. Many periodic seeders produce heavy ('bumper') seed crops during their seed years.

Selection index (index selection): A form of multi trait selection that combines information on all traits of interest into a single index. This enables the breeder to assign a simple score to each individual based on heritability and economic importance. The selection index "I" is calculated as:I = a x X + b x Y + c x Z etc., where X,Y,Z etc. are the assessed or measured values of the individual traits (volume, straightness etc.), and a,b,c etc. are multiplication constants calculated on the basis of heritability and economic importance.

Selection intensity: The proportion of individuals in the population that are selected for breeding.

Select, superior or plus tree: A tree that has been graded and recommended for use in production or breeding orchards. It has a superior phenotype for growth, form, wood quality, or other desired characteristics and appears adaptable. It may not yet have been tested for its genetic value, but the chances of it having a good genotype are high for characteristics with a high heritability.

Self incompatibility (self sterility): The failure of gametes from the same plant to form a viable embryo.

Self-pollination: Pollination of a flower with pollen from the same tree or clone. The offspring maybe termed 'self'.

Self pruning ability: The ability of a tree to shed its lower limbs. The character is heritable and selected for in some species.

Sexual reproduction: The formation of new individuals of a species by the fusion of two normally haploid gametes to form a diploid zygote. Ant. asexual or vegetative reproduction.

Siblings (sibs): A group of individuals within a family. Full-sibs have both parents in common; half-sibs, only one.

Social cost of carbon: estimate of the economic damages associates with increase in carbon dioxide emissions.

Scrub: degraded forest lands having canopy density less than 10 percent.

Supporting services: ecosystem services that are necessary for the production of all other ecosystem services such as biomass production, soil formation and retention, nutrient cycling, etc.

Somatic cells (vegetative cells): Any other cell of the body than spores, gametes and their precursors. Ant. reproductive cells.

Somatic embryogenesis: Development of embryos from diploid somatic cells in suspension culture.

Species: A population of organisms that are able to interbreed freely under natural conditions. A species represents a group of organisms which has evolved distinct inheritable features and occupies unique geographical area. Species do not usually interbreed freely with other species.

Staminate: Having pollen-bearing organs (stamens) only. May apply to individual male plants of a dioecious species or to flowers, inflorescences, and strobili.

Stand: A population of trees possessing sufficient uniformity in composition, constitution, and arrangement to be distinguished from adjacent populations. See Population, Seed production area.

Sterility: (1) Absence or defectiveness of pollen, eggs, embryo, or endosperm, which prevents sexual reproduction. 2) The exclusion of contaminating microorganisms in laboratory work for example micro propagation. Sterility can be achieved by killing possible organisms by e.g. heating.

Sustainable use: The use of components of biological diversity in a way and at a rate that doesnot lead to long-term decline of biological diversity, thereby maintaining its potential to meet the needs and aspirations of present and future generations.

T

Tannin: A wood product that is used to cure animal hides into leather; also known as tannic acid.

Telophase: The last stage of cell division during which cell cytoplasm is divided as two new cells are formed.

Temperate forests: Temperate forests grow between the tropics and the polar region in both the northern and southern hemispheres. They include temperate deciduous forests, temperate broad leaf and mixed forests, temperate coniferous forests and temperate rainforests. Temperate forests have four distinct seasons, with a well defined winter.

Tensile strength: Strength across the long axis of a tree; the source of strength of wood used for timbers and structural beams.

Tensiometer: An instrument indicating the amount of moisture in the soil.

Terminal growth: Vertical growth in a tree.

Terraces: Level areas constructed along the contours of hills, often but not necessarily

Terracing: Dikes built along the contour of sloping farm land that hold runoff and sediment to reduce erosion.

Thermochemical liquefaction: A process by which wood chips are heated under high pressure in a hydrogen gas or syngas atmosphere to produce industrial oil products.

Thinning: Removing some of the trees form an overcrowded forest.

Thinning: The removal of some trees (usually of poorer quality) from a forest to reduce the competition for available resources.

Threshold/tipping point: A point or level at which ecosystems change, sometimes irreversibly, to a significantly different state, seriously affecting their capacity to deliver certain ecosystem services.

Total economic value (TEV):A framework for considering various constituents of value, including direct use value, indirect use value, option value, quasi-option value, and existence value.

Trees Outside Forests (TOF): Trees growing outside recorded forest areas.

Trade-offs: A choice that involves losing one quality or service (of an ecosystem) in return for gaining another quality or service. Many decisions affecting ecosystems involve trade-offs, sometimes mainly in the long term.

Timber cruiser: A person who estimates the volume of marketable timber by taking sample measurements from sites throughout a stand of trees.

Timber yield: The timber volume in a forest on a particular date.

Timber: A wood piece that measures greater than 5 by 5 inches and less than 8 by 8 inches.

Timber: Wood after it has been debarked, sawn and prepared for use in building or furniture manufacture.

Tissue culture: Producing new plants from very small pieces of plant tissue in a laboratory. Often used to grow species that are difficult to propagate by other means, or to produce desirable clones of a particular plant.

Tissue culture: The development of roots, stems, and leaves from callus tissue in a plant medium containing plant nutrient and hormones.

Tissue: A group of cells that contribute to a particular life function.

Topographic map: A map that show relative elevations.

Topography: The physical features of a surface area including relative elevations and the position of natural and man-made features.

Topography: The science of creating maps of the surface features of an area; the relative elevations of different features in landscape.

Total maximum yield load (TMDL): The total amount of load that will be tolerated each day for known pollutants in rivers, streams, lakes, and costal waters.

Tracheid: A long, tapered plant cell found in xylem tissue, having pits in the cell walls through which water is conducted.

Transgenic plant: Possess a gene or genes that have been transferred from a different species by genetic engineering.

Transpiration: Loss of water to the atmosphere from the leaf surface of plants.

Transpiration: The process by which water vapor is lost to the atmosphere from living plants. The term can also be applied to the quantity of water thus dissipated.

Transpiration: To lose water vapor through the stomata in the leaves of plants.

Trophic levels: A functional classification of species that is based on feeding relationships.

Tropical rainforest: A forest biome found in the tropics near the equator, in climates with continually warm to hot weather and very heavy rainfall. Tropical rainforests are green all year long, characterized by high species diversity, soils that are usually old and nutrient-poor, stratification into distinct layers of trees and shrubs and many epiphytes and vines. The largest tropical rainforest is in the Amazon region, which covers a large portion of South America.

Troposphere: The lowest layer of the earth's atmosphere, extending from the earth's surface up to about 7 to 17 km. Within the troposphere, temperature usually decreases at a regular rate with increasing altitude. Weather is confined to this turbulent layer of the atmosphere.

Tsunamis: A tsunami is an oceanic wave, also termed a seismic sea wave, generated most commonly by slippage of the Ocean floor along a fault in association with an earthquake of a rock slide.

Turbidity: Cloudiness caused by suspended solids or soil sediments in water.

Taper : The gradual diameter reduction of a tree or a log from the base to the top.

Tariff table : A tree-volume table based on DBH and total height.

Thinning : Tree removal in a forest stand that reduces tree density and tree-to-tree competition. Thinning encourages increased growth of fewer, higher quality trees. Thinning types include:

- Commercial thinning—Any thinning of merchantable trees.
- Crown thinning (also high thinning or thinning from above)—A thinning that removes the dominant and co dominant trees.
- Free thinning—To achieve specified stand structure.
- Low thinning (also thinning from below)—A thinning that removes the smallest, suppressed trees.
- Natural thinning—A natural process where trees die as a result of rooter crown competition.
- Precommercial thinning—Any thinning of no merchantable trees, often called PCT.
- Proportional thinning—A thinning that retains the same proportion of tree diameter classes.
- Row thinning—A plantation thinning that removes specific tree rows.
- Selection thinning—A thinning that selects and removes individual large trees.

Thinning shock : A condition where trees sometimes show signs of ill health or stress after a thinning operation that opens up the stand. See also Sun Scald.

Timber stand improvement (TSI) : Applying cultural practices to a young forest stand to improve the growth and form of trees and achieve a desired stocking and species composition.

Transplant: A very young tree or seedling lifted from a nursery seedbed and replanted at the nursery.

Tree: A woody plant having a well-defined stem, usually standing over 10 feet high at maturity.

Trim allowance : Excess log length to allow for losses in the manufacturing process, usually 6 inches on 16-foot logs and 8 to 12 inches on 32-foot logs. Trim allowance may vary among mills.

Tandem selection: Breeding and selection for one trait at a time until the desired level of improvements reached for that trait. Used in e.g. breeding for disease resistance and cold hardiness, but rarely used in forestry due to the slow succession of generations.

Taxon (taxa): Any formal taxonomic category, e.g. family, genus, variety.

Taxonomy: Classification of organisms, including identification of nomenclature, according to a natural (chiefly morphological) system that seems to depict evolution.

Tetraploid: Having 4n chromosomes.

Tissue culture: The growth of isolated plant cells or small pieces of tissue under controlled conditions in a sterile growth medium. The medium is designed to meet the requirement of the tissue involved,e.g. by application of hormones and nutrients.

See Micro propagation, Somatic embryogenesis, Tot potency, Vegetative propagation

Topcross test: Progeny test derived by crossing each parent with the same tester (topcross) pollen. The pollen may be a single lot or a mixture from a number of tester trees, but should not mask the contribution of the tested parents. The selections to be compared are sometimes placed in an isolation block where the tester is the only source of pollen.

Topophysis (Plagiotrophy): The phenomenon that occurs when scions, buddings, and rooted cuttings maintain for some time the branchlike growth habit they had as shoots on the ortet.

Totipotency: The capacity, exhibited by certain types of differentiated plant cells, to regenerate whole plants. To realize this potential the cell must be removed from the inhibiting influence of the rest of the plant body and given the appropriate stimuli, namely the correct balance of nutrients and growth substances Triploid. Containing 3n chromosomes.

U

Understory: Short trees in a forest that fill an intermediate stratum of vegetation beneath the canopy created by the branches and foliage of the tallest trees.

Understory: The small trees, shrubs, herbs and grasses that make up the lower layers of vegetation in a forest, woodland or other habitat.

Uneven-aged stand: The presence of trees of different ages and stages of maturity in a forest populations.

Urban forestry: An emerging branch of forestry practiced in or near cities.

Utisols: Soils formed in warm, humid climates, showing evidence of weather action in the alluvial deposits of clay and iron in the subsoil.

Understory : That portion of the trees or other vegetation below the canopy in a forest stand.

Uneven-aged management B Managing a forest by periodically harvesting trees of all ages to maintain a broad age (or size) class distribution in which there area greater number of trees in each smaller age-class than in the next older or larger class, up to some maximum age. This type of management is not common in the West. See also Selection harvest.

V

Vacuole: A cell structure that gathers excess water and wastes that are discharged through the cell wall.

Valuation, economic: the process of estimating a value for a particular good or service in a certain context in monetary terms.

Very Dense Forest (VDF): all lands with forest cover having a canopy density of 70 percent and above.

Vascular cambium: Meristem tissue that forms in a continuous ring between xylem and phloem tissues that gives rise to new growth of this tissue.

Vascular ray: A row of parenchyma cells that radiates to the center of the stem or trunk of a tree and that transport dissolved material across the woody section of the stem.

Vector: An insect or other organism that carries disease organisms in its body, from which it spread the disease to trees or other susceptible life forms.

Vegetative organ: A plant organ such as root, stem, or leaf.

Vegetative reproduction: Asexual reproduction of the plant from plant parts such as leaf, roots, or stems tissue.

Vegetative Strips: Long, narrow areas of any type of vegetation, usually planted along.

Vein: Leaf structure that connects to the vascular tissue in spines or midribs, allowing the flow of dissolved materials to the cells of the leaf.

Veneer: Thin sheets (1/4 inch thick or less) of high quality wood that are glued to the outer surfaces of lesser quality wood for the manufacture of plywood and construction of furniture.

Vessel element: A plant structure that become s hollow after it has died, facilitating the flow of dissolved nutrients in the stems, roots, and leaves of a plant.

Vessel: A plant structure that consists of vessel elements that have grown together, end to end, facilitating the flow of large volumes of dissolved nutrients within the plant.

Vulnerable species: Natural race, organisms or species that has a high risk of extinction in the wild in the medium term future. The probability of extinction in the wild within 200yrs may be 10%.

View shed : The landscape that can be seen from a particular viewpoint or along a transportation corridor.

Volume table : A table estimating the volume of wood in a standing tree based on its diameter and merchantable height.

Variance: A statistical concept denoting the numeric measure of variability. It is defined as the sum of the squared deviation of each observation from the arithmetical mean divided by one less than the total number of observations.

Variation: Difference in performance and characters of individuals due to internal and external factors. In a natural population the phenotypic variation is the product of developmental, environmental and genetic variation. In trials, sampling variation and experimental error make up the residual variation

Vegetative propagation: Propagation of a plant by asexual means, as in budding, dividing, grafting, rooting of cuttings, and air-layering. Hereditary characteristics of the resulting clone (ramets) are identical with those of the original plant (ortet).

Vigour: Healthy, profuse and rapid growth.

W

Warp: Distortion or twisting of lumber products as they lose moisture.

Water cycle: The movement of water in the form of vapor from the oceans to the clouds to the earth as precipitation, and back to the oceans through rivers and streams.

Water table: The level of the water-saturated underground areas below the surface of the soil.

Watercourse: A stream, river or canal along which water runoff flows towards the ocean.

Watershed: An area bounded by geographic features where precipitation is absorbed in the soil to form ground water that eventually emerges to become surface water, which ultimately drains to a particular stream, river, or other body of water.

Waterway: A river, canal or other navigable channel used for travel or transport.

Weed control: Reducing competition with forest trees for water and nutrients. This can be done by physical method or by using chemicals called herbicide.

Weed: A plant species growing where it is not wanted.

Wet land: A land area that is flooded by water during all or parts of the year.

Wet land: Any area of land that is regularly or occasionally inundated by water, including swamps, billabongs, mudflats, salt marshes and lakes.

Wetlands: Areas of permanent or intermittent inundation, whether natural or artificial, with water that is static or flowing, fresh, brackish or salt, including areas of marine water not exceeding 6 m at low tide. In urban areas wetlands are protected because they purify water, prevent floods and are called ‘the kidney of a city’.

White rot: A type of wood rot that occurs in trees as fungi break down the cellulose and lignin components of the cell walls.

Wildfire: Any fire that burns out of control or that has not been prescribed for a specific purpose.

Willingness-to-pay (WTP): estimate of the amount people are prepared to pay in exchange for a certain state or good for which there is normally no market price (e.g. WTP for protection of an endangered species).

Wilt: A fungal infections in the vessels of the xylem tissue where the flow of water and dissolved nutrients is blocked from the trunk and branches from the tree.

Wood chips: Chips produced from either soft or hardwood logs and turned into paper and other paper products.

Wood naval store: Wood products extracted from chipped or shredded wood obtained from pine stumps or logs by dissolving the wood components in organic solvents.

Woodland: A native plant community dominated by relatively openly spaced trees (covering up to 30 % of the land surface), often with a dense understory.

Woodlot: An area planted to trees for fuel, or timber.

Woodlots: Forest planting on farms for wood production.

World environment day: June 5 of each year; designated by the 1972 UN conference on the Human Environment as a day to focus on the environmental problems.

World wildlife fund (WWF) : It is an international conservation organization founded in 1961 and working globally to protect endangered wildlife and wetlands.

Watershed: An area of land that collects and discharges water into a single stream or other outlet. Also called a catchment or drainage basin.

Watershed analysis : The study of how a particular drainage network functions. An aspect of ecosystem management planning.

Wetlands : Marshes, swamps, and other water-saturated soils that offer important habitat for wildlife, significant support of nutrient cycling in ecosystems, and protection against the severity of storms and floods. Wetlands are among the most vulnerable lands to destruction and conversion toothier uses.

Wildlife habitat: The environmental conditions of a specific place occupied by wildlife species.

Wolf tree : A low-value tree occupying more space in the forest than its economic value justifies. Usually older, larger, or more branchy than other trees in the stand.

Woodlot : The wooded portion of a farm or ranch.

Wind pollination: Pollination by wind-borne pollen.

X

Xerophyte: Plants that grow in xeric or desert conditions.

Xylem: A woody tissue that conducts water with dissolved nutrients and plant materials from the roots to the stems and leaves.

Xenia: An observable effect of pollen on the endosperm or embryo, especially effects resulting from the dominance relationships and other types of gene action in the triploid endosperm. Because of Xenia, hybrid characteristics may be manifested in form, colour, etc. of the derived seed.

Y

Yard: A site to which logs are gathered in preparation for hauling them to a sawmill or processing plants; also known as a landing.

Yarding : Moving logs by a cable system from the stump to a landing.

Yield : The volume or number of stems that can be removed annually or periodically.

Yield table : A tabular statement which summarizes per unit area basis all the essential data relating to the development of a fully stocked and regularly thinned even aged crop at periodic intervals covering the greater part of its useful life.

Z

Zone map: A map that illustrates the severity of climatic conditions that can be expected in each region of the country.

Zooplankton: Animal species of plankton in freshwater and marine environments; that do not produce their own energy but feed on phytoplankton or smaller animal plankton. Their movements are largely dependent upon currents.

Zygote: The fertilized egg; sometimes also the individual developing from it. The zygotic chromosome number is normally diploid (2n).

1

Agroforestry

1. Taungya is a method of raising forest plantation in combination with_____________. (**Agricultural crops**)
2. Buffer strips left along streams to reduce erosion and provide shade for fish, which are known as _____________. (**Streamside Management Zone, SMZ**)
3. Agroforestry system journal is published in_____________. (**Netherland**)
4. Indian journal of agroforestry is a_____________journal. (**Quarterly**)
5. Taungya was developed by_____________. (**Brandis**)
6. The interaction in which both the member of the community in an ecosystem are benefitted is known as_____________. (**Mutualism**)
7. The tendency of neighbouring plants to utilize the same quantum of light, mineral nutrients and space is known as_____________. (**Competition**)
8. Allelopathy is the best example of_____________. (**Amensalism**)
9. D & D approach was developed by_____________. (**ICRAF**)
10. _____________design is followed in greenhouse experiments. (**RBD**)
11. ICAR-Central Agroforestry Research Institute is situated in_____________ (**Jhansi**)
12. Other name of shifting cultivation in Tamil Nadu is known as_____________. (**Poonam**)
13. _____________ and_____________ are the specific international and national institutes respectively for agroforestry research. (**ICRAF & NRCAF**)
14. In ICAR agroforestry is part of_____________ division. (**NRM**)
15. "Transforming lives and landscapes" is the motto of_____________ institute. (**ICRAF**)

16. Shifting cultivation is commonly called_____________ or_____________ system. (**Jhum or Slash & burn**)

17. In India homestead gardens are mainly found in_____________ parts of India. (**South**)

18. On economic basis agroforestry systems classified as_____________ ;_____________ and _____________. (**Commercial, intermediate & subsistence**)

19. Fallow areas left to grow selected trees-shrubs for regaining soil fertility in trees-crop rotation is_____________. (**Improved fallows**)

20. IIIrd World Agroforestry congress was held at_____________. (**New Delhi**)

21. Animals feed in fenced pastures with control access is termed as _____________. (**Paddock**)

22. Rotation of arable crops with two or more years of sown pasture is called as. (**Lay farming**)

23. The D & D procedure when applied in a region is known as_____________. (**Macro level**)

24. Vegetation density of _____________ per cent is considered best for windbreak design. (**60-75%**)

25. According to temporal arrangement of component, Taungya comes under the agroforestry system of_____________. (**Concomitant**)

26. *Acacia mearnsii* an important MPT, is native of _____________. (**Australia**)

27. Agroforestry practiced to fulfill the needs of farmer's family is known as _____________. (**subsistence system**)

28. According to agro-ecological zone, India comes under _____________. (**Humid or sub humid low land**)

29. In shifting cultivation, the regeneration or fallow phase is of _____________ years. (**5 to 20 years**)

30. The lopping method adopted for management of trees in agroforestry system consists of pollarding and _____________. (**Bushing/Trimming**)

31. The intermediate layer of homegarden is of_____________ meter height dominated by fruit trees. (**3-20**)

32. In North India, the commonly adopted agroforestry systems are _____________. (**Poplar**)

33. In protein banks, ____________ trees are to be considered as woody component. (**leguminous fodder**)

34. ____________ and ____________ plants are generally adopted in alley cropping system due to leguminous nature. (**Subabul and Gliricidia**)

35. Pioneering work on alley cropping technology was initiated at____________. (**IITA, Nigeria**)

36. Perkarangum is a ___________ type of agroforestry system. (**Home garden**)

37. Home garden is characterized by ____________ number of vertical strata. (**3-4**)

38. *Gliricidia sepium* an important MPT is native of____________. (**Central America or Mexico**)

39. Raising of grasses with trees is known as ____________ system. (**Silvopastoral system**)

40. In Taungya system the agricultural crops are generally grown upto ____________ years. (**2-3**)

41. The establishment and management of vegetation along denuded waterways to reduce agrochemical pollution are known as _________ system. (**Riparian Buffer System**)

42. The modified form of shifting cultivation is ___________. (**Taungya**)

43. The lopping method adopted for management of trees in agroforestry system consists of _________ and bushing. (**Pollarding**)

44. *Leucaena leucocephala* fixes nitrogen in the range of ____________ kg / ha/ year. (**100-500**)

45. Multipurpose woody hedgerows systems are generally adopted in the ____________ areas with hilly and sloping terrain. (**humid to subhumid**)

46. The role of MPTs in an agroforestry system can perhaps be clarified through the ____________ concept. (**ideotype**)

47. The dry matter production by any crop in an agroforestry system is____________ related to total amount of solar radiation intercepted. (**linearly**)

48. The intermediate layer of homegarden is of 3- 20 m height dominated by trees. (**fruit**)

49. Allelochemicals produced by trees are the ____________ plant metabolites. (**secondary**)

50. What is the full form of ICRAF. (**International Centre for Research in Agroforestry**)

51. Components of agroforestry are____________ ,____________ and/ or. (**Crop, tree, animal**)

52. The main criteria for good agroforestry design are____________, ____________ and ____________. (**Productivity, sustainability and adoptibility**)

53. Total geographical area of India is____________. (**329 m ha**)

54. A major cause of soil erosion is____________. (**Deforestation**)

55. Land use system involving trees combined with crops/or animals on the same unit of land is known as ________. (**Agroforestry**)

56. An agroforestry system always has____________ or more output. (**two**)

57. Agroforestry normally involves two or more species of plants at least one of which is always____________. (**woody perennial**)

58. Agroforestry is more____________ as compared to monocroping farms. (**complex**)

59. There are____________ agro climatic regions of India. (**15**)

60. According to Nair (1987) Agroforestry System can be classified on the basis of ____________ ,____________ ,____________ and____________. (**Structural, functional, socioeconomic and ecological**)

61. On the basis of structure Agroforestry systems grouped into____________ categories. (**2**)

62. The objective of improved fallow species in shifting cultivation is to depleted soil nutrients. (**Recover**)

63. Shifting cultivation is practiced extensively in the____________ hill region of India. (**North- eastern**)

64. Home garden is practised extensively in the states of____________ and____________ of India. (**Kerala and Tamil Nadu**)

65. Home garden are characterized by high species diversity and usually of canopy strata. (**Three**)

66. _____________ is a common example of the zonal pattern of agroforestry. **(Homegarden)**

67. The main feature of improved fallow system of Agroforestry is that trees and shrubs are not grown with_____________ on the same plot at the same time. **(crop)**

68. The Taungya is a Burmese word coined in Burma in the_____________. **(1850)**

69. The Taungya system was introduced into India by_____________ in 1980. **(Brandis)**

70. The first taungya plantation was raised in 1986 in_____________. (North Bengal)

71. Growing of trees with crops is called_____________ system. **(Agrisilviculture)**

72. System of production of crop under which plot of land are cleaned using axe and fire cultivated for a short period called_____________. **(shifting cultivation)**

73. Shifting cultivation is also called_____________ in north eastern hill region of India. **(Jhum)**

74. Shifting cultivation is also called_____________ in Andhra Pradesh. **(Podu)**

75. In south India, Taungya system of cultivation is called_____________. **(Kumari)**

76. Taungya system is practised in area with an assured annual rainfall of over mm. **(1200-1500 mm)**

77. In Taungya system, "Taungya" means and "ya" means. **(Hill & cultivation)**

78. In village taungya, each family has about 0.8 to 1.7 ha of land to raised trees and cultivate crops for to_____________ years. **(3-4)**

79. Taungya system are 3 types viz._____________, _____________ and_____________. **(Departmental, Leased and Village)**

80. _____________ is the system of Agroforestry in which various kind of tree species are grown mixed. **(Multipurpose tree gardens)**

81. Alley cropping is also known as_____________. **(hedge row intercropping)**

82. The primary purpose of alley cropping is to maintain or increase yield. **(Crop)**
83. Ideally hedge row intercropping should be positioned in_____________ direction. **(East-west)**
84. In alley cropping, the spacing used in field is usually tom between tree rows. **(4-8)**
85. In alley cropping, the spacing used in field is usually to m between trees within rows. **(0.25 to 2 m)**
86. On slopping land hedge row should always be placed on the_____________. **(Contour)**
87. In shelter belt, there are several_____________ established at right angle to the prevailing wind. **(rows of trees)**
88. In shelter belt, there are several rows of trees established at to prevailing wind. **(right angles)**
89. The purpose of shelter belt is to reduce the of the prevailing wind. **(velocity)**
90. Shelter belt have typical_____________ shape. **(triangular)**
91. Tall trees are raised in the _____________ of the shelter belt. **(centre)**
92. Ideally tree density in shelter belt approximately_____________. **(50-60 %)**
93. Shelter belt up to_____________ m width are consider ideal under Indian condition. **(50)**
94. In shelter belt, the ratio of height and width should be to_____________. **(1:10)**
95. The minimum length of shelter belt should be about to_____________ times of its height. **(25)**
96. Strips of trees and/or shrubs planted to protect fields, homes from wind is known as to_____________. **(wind breaks)**
97. On level ground a wind breaks will reduce the speed of wind for about_____________ times of the tree height. **(25)**
98. _____________ row wind break is generally adopted where land is valuable. **(single)**

99. A wind break 20 m tall will give some protection from 100 m on the upward side to ____________ m on the downwind side. (**500**)

100. Wind breaks of____________ to____________ rows are more effective for most farm situation. (**3 to 5**)

101. The shape of the shelter belt is generally____________. (**Triangular**)

102. In shelter belts the centre rows planted with____________ tree species and smaller bushy species in the____________rows. (**tall, outside**)

103. Normally tree with____________ vertical growth characteristics are ideal for wind breaks. (**Narrow**)

104. In wind breaks the wind velocity in the protected area may be reduced to____________ % of the open wind speed. (**25 to 75%**)

105. Primary function of most of the home garden is____________. (**Food production**)

106. In protein bank____________tree are planted on or around farm land and range land. (**Leguminous fodder trees**)

107. Home garden is a subsystem of____________. (**Agrosilvopasture**)

108. Planting nectar producing trees on the boundary of the agricultural field is called as ____________ ____________. (**Apiculture**)

109. Home garden is found extensively in____________ rainfall areas. (**high**)

110. Home gardens are characterized by high species diversity and usually vertical canopy strata. (**3-4**)

111. In____________ system various tree and shrubs preferred by fish are planted on the boundary and around fish ponds. (**Aquaforestry**)

112. ____________agroforestry systems are between commercial and subsistence scale of production and management. (**Intermediate**)

113. Freshly cut wood contains about____________% moisture. (**23-25**)

114. For cooking 1 kg of food____________ kg of wood is required. (**1.2 kg**)

115. Wood lots are____________ planting of trees. (**block**)

116. ____________ is the removal of branches in order to obtain trees with clear bole. (**Pruning**)

117. The removal of branches by physical and biotic agencies of the environment is known as____________ pruning. (**Natural**)

118. Removal of branches from the selected portions of the tree by mechanical mean is referred as_____________ pruning. (**artificial**)

119. Coppice shoots are arise from ____ buds near the base of a woody plant. (**dormant**)

120. Flexibility, speed and repetition are the key features of_____________. (**D & D**)

121. The D&D research exercise covering an entire ecological zone within a country is known as ________. (**Macro**)

122. The D&D team comprises_____________ number of specialist. (**5-10**)

123. A macro D&D is usually completed in about_____________ months. (**3**)

124. Tree species which given fuel, fodder, fibre, food and timber are called_____________. (**Multi Purpose Tree**)

125. In Andhra Pradesh and Odisha state, shifting cultivation is called_____________. (**Podu**)

126. Taungya system is followed in areas having rainfall of_____________. (**1200-1500 mm**)

127. In India average size of home garden is_____________. (**0.2-0.5 ha**)

128. Trees for fuel wood should have high calorific value and. (**Specific gravity**)

129. In agri-horticultural system, fruit trees should be pruned_____________ in a year. (**regularly**)

130. Roots of trees contain_____________ percentage of the total living biomass. (**20-25**)

131. _____________,_____________ and_____________ are the criteria of good agroforestry design. (**Productivity, sustainability, adoptability**)

132. In village taungya the land allotted to each family to raise the crop and tree has to about _____________. (**0.8-1.7 ha**)

133. Social forestry programme lunched in the_____________ year plan. (**Fifth five**)

134. _____________classification of agroforestry is more important and commonly adopted. (**Structural**)

135. Indian Grassland and Fodder Research Institute (IGFRI) is located at_____________. (**Jhansi**)

136. Forest products contribution to world gross domestic product (GDP) is_____________ %. **(1.7)**

137. The practice of forestry on lands out side the conventional forest areas for the benefit of the rural and urban communities is called as_____________. **(Social forestry)**

138. Social forestry was first coined by_____________. **(Westoby)**

139. The practice of forestry on farm lands is called_____________. **(Farm forestry)**

140. The practice of forestry away from the conventional forest area is_____________. **(Extension forestry)**

141. Social forestry was first adopted successfully in_____________ of India.**(Gujurat)**

142. Van-Mohotsav was started in the year _____________.**(1950)**

143. In India, plantation forestry was started in _____________.**(1842)**

144. _____________ type of branching should be preferred for roadside planting. **(Heliotropic)**

145. _____________ is the unit of management under JFM. **(Hamlet)**

146. The direct or indirect affect of one plant upon another through the production of chemical inhibitors is called _____________. **(Allelopathy)**

147. Allelochemicals produced by plants are _____________.**(Secondary plant metabolite)**

148. The broad and well known classification of agroforestry system was given by ______. **(Nair)**

149. _____________ is an important silvicultural operation practiced in agroforestry. **(pruning)**

150. National watershed development programme for rural areas had been lunched in the year_____________. **(1995)**

151. NRCAF is located in_____________. **(Jhansi)**

152. JFM concept initiated in India in the year_____________ .**(1970)**

153. Central Arid Zone Research Institute is located at_____________. **(Jodhpur)**

154. Social Forestry project phase-I lunched in_____________. **(1981)**

155. Social Forestry project phase-II lunched in_____________. **(1988)**

156. Van Mahostava was initiated by_____________. **(K.M. Munshi)**

157. JFM concept was initiated by_____________. **(A.K. Banerjee)**

158. Climate vegetation productivity index was given by_____________. **(Paterson)**

159. In JFM programme, the objectives are decided at_____________. **(Village level)**

160. In North India agroforestry the combination of_____________ _____________ is common. **(Wheat with poplar)**

161. When an immediate supply of a single necessary factor falls below the combined demand of the plant is called as_____________. **(Allelospoly)**

162. Energy plantation is established for_____________production. **(Fuel wood)**

163. _____________ practices is not included in the scope of social forestry. **(Protection forestry)**

164. _____________social forestry programmes is purely managed by private efforts. **(Farm forestry)**

165. The D&D survey in agroforestry was given by_____________. **(PKR Nair)**

166. Wasteland means land with_____________% productivity. **(<20%)**

167. Institute of tropical Forestry is situated at_____________. **(Puerto Rica)**

168. Head quarter of IUFRO is_____________. **(Vienna)**

169. "Indian Journal of Agroforestry" is published from_____________. **(Jhansi)**

170. "The Journal of Agroforestry Today" is published from_____________. **(Kenya)**

171. Nitrogen fixing ability of Casuarina is due to_____________. **(Frankia)**

172. Most common agroforestry system practiced in arid regions is_____________. **(Agrisilvicultural system)**

173. Social forestry was first recommended by _____________. **(National Commission on Agriculture)**

174. Head quarters of International Centre for Research in Agroforestry is located at _____________. (**Nairobi**)

175. The national commission on agriculture emphasized agroforestry education in the _____________ five year plan. (**Seventh**)

176. The last stage of water erosion is the formation of_____________. (**Ravines**)

177. A familiar procedures for the diagnosis of land management problems and potentials for designing an agroforestry solution is called as_____________. (**Agroforestry D & D**)

178. Agroforestry D&D covers entire ecological zone within a country is_____________. (**Macro D&D**)

179. Based on the nature of components, Nair classified the agroforestry system into _____________ groups. (**four**)

180. The first state in India, created a separate social forestry wing in the forest department was_____________. (**Gujurat**)

181. "Tree means water, water is bread and bread is life" was first time quoted by _____________. (**Westoby**)

182. NWDB stands for_____________. (**National Wasteland Development Board**)

183. NWDB classified wastelands into_____________ types. (**Two**)

184. Waste land is the land with_____________ per cent productivity. (**Less than 20**)

185. _____________ is a geo-hydrological unit that drains at a common point. (Watershed)

186. The drainage area of catchment is_____________. (**>1 lakh ha**)

187. The drainage area of sub-catchment is_____________. (**>40000 ha**)

188. The drainage area of sub-watershed is_____________. (**2000-4000 ha**)

189. The drainage area of mini watershed is_____________. (**400-2000 ha**)

190. The drainage area of micro watershed is_____________. (**<400 ha**)

191. The process of cycling of nutrients from soil to the plant and back to the soil is called as _____________. (**Nutrient cycling**)

192. The continuous circulation of water among the hydrosphere, atmosphere and lithosphere is called as_____________. (**Hydrological cycle**)

193. National Wasteland Development Board came into existence during the period of _____________. **(1985)**

194. Alley cropping was developed by_____________. **(B.T. Kang)**

195. The rate of litter decomposition in tropics is_____________ than temperate region. **(Faster)**

196. Crop land left without crops for period ranging from one season to several years is called as_____________. **(Improved fallow)**

197. Growing more than one crop on the same piece of land during one calendar year is termed as_____________. **(Multiple cropping)**

198. Radio isotope studies are helpful in agroforestry to known the_____________. (Root distribution pattern)

199. The law of self-thinning or -3/2 power law of self-thinning was given by. **(Yoda)**

200. Imposed mortality of plants due to competition is known as_____________. **(self-thinning)**

201. Concept of tolerance was given by_____________. **(Shelford)**

202. Growth of any organism is limited by the factor which is in least quantity was given by _____________. **(Leibig)**

203. Rotation of arable crops with two or more years of sown pasture is called as. (Lay farming)

204. _____________are the most common agroforestry system practiced in arid regions. **(Silvipastoral system)**

205. Repeated growing of the same crop on the same land is known as_____________. **(Monoculture)**

206. _____________is a type of social forestry programme which is purely managed by private efforts. **(Farm forestry)**

207. Protection and reforestation of degraded forests with the involvement of local people is called_____________. **(JFM)**

208. Benefit sharing concept is a core objective in_____________. **(JFM)**

209. The basic unit of D & D analysis is_____________. **(Land use system)**

210. _____________ ,_____________ and_____________ are the tool of agroforestry system evaluation. **(BCR, NPV and IRR)**

211. The national afforestation programme is being implemented through a_____________ tier structure. (**Two**)

212. International Institute of Tropical forestry (IITF) is situated at_____________. (**Río Piedras, Puerto Rica**)

213. _____________ is the international donor agencies for social forestry in India. (**SIDA**)

214. _____________ agroforestry systems is used sale of the production of output. (**Commercial**)

215. On the basis of nature of the components, AF system is classified into numbers. (**10**)

216. In Indian condition normal width of shelter belt is up to_____________. (**50 meters**)

217. According to temporal arrangement of components in an agroforestry system, coffee under shade trees comes under_____________. (**Coincident**)

218. *Leucaena leucocephala,* an important MPT in different AF models is native to. (**Central America/ Mexico**)

219. An interaction in AF system where the net effects are positive for both the species is known as_____________. (**Synergistic**)

220. The D&D procedure which is applied in the family farm is known as_____________. (**Micro level**)

221. The All India Coordinated Research Project (AICRP) on Agroforestry was launched in the year_____________. (**1983**)

222. The World Agroforestry Centre (WAC) is located at_____________. (**Nairobi, Kenya**)

223. The modified form of shifting cultivation is called as_____________. (**Taungya**)

224. Home gardens system of AF is practiced in_____________ state of India. (**Kerala/Tamil Nadu**)

225. According to temporal arrangement of components, annual crops under coconut come under the AF system_____________. (**Intermittent**)

226. The AF system in which trees and shrubs are planted in the boundary and around fish-ponds is known as_____________. (**Aquaforestry**)

227. *Acacia auriculiformis,* an important MPT is native to____________. (**Australia**)

228. An interaction in AF system where the net where the net effects are positive for one species and zero for the other species is known as ____________. (**Commensalistic**)

229. The D&D procedure applied in the village is known as ____________. (**Meso level**)

230. The minimum length of a shelter belt should be about ____________ times of its height. (**25**)

231. The survey type research done for agroforestry is known as____________. (**Exploratory Research**)

232. The intensification of cultivation in shifting cultivation can be evaluated by ____________ factor. (**land use**)

233. The role of MPTs in AF system can perhaps be clarified through the ____________ concept. (**ideotype**)

234. The fallow periods of shifting cultivations in recent times are of ____________ years. (**2-5**)

235. Alley cropping practiced appears to be a rational alternative land use for improving ____________ sustainability. (**Agricultural**)

236. Home gardens are the agroforestry systems practiced in the ____________. (**humid and sub-humid lowlands**)

237. In India the organized agroforestry research began in the year ____________. (**1979**)

238. The suitable tree species for fallow system should be ____________ in nature. (**nitrogen fixing**)

239. ____________ agroforestry system is the use of land towards satisfying basic needs and is managed by the owner. (**Subsistance**)

240. The name of ICRAF was changed into WAC in the year ____________. (**2002**)

241. WAC stands for __________. (**World Agroforestry Center**)

242. First agroforestry seminar in India was organized at ____________ in 1985. (**Imphal**)

243. First of all post graduation education in agroforestry was started in ____________. **(Dr. Y.S. parmar university of Horticulture & Forestry, Solan)**

244. __________ & __________ are two fundamental attributes of all agroforestry systems. (**Production & Protection**)

245. The process of taking nutrients from deeper soil profile and depositing them on the surface layer is referred to ______________. (**Nutrient pumping**)

246. Social forestry was first adopted successfully in the state ____________. (**Gujarat**)

274. Taungya system is also called as ____________ in East Africa. (**Shamba**)

248. ____________ & ____________ are forms of heading back carried out on the main trunk (stem). (**Coppicing & Pollarding**)

249. Wind breaks planted north-south gives better __________ while planted in east-west gives better ____________ to crops. (**Protection & Shading**)

250. The four key criteria characterize agroforestry practices are ________, ________, ________, ____________. (**Intentional, Intensive, Interactive, Integrated**)

251. Heliotropic type of branching should be preferred for ________ plnting. (**Roadside**)

252. The process by which the complex organic substances are broken down to simple inorganic elements is called as _________. (**Nitrification**)

253. ___________ is the international donor agency for social forestry in India. (**SIDA**)

254. The cross sectional shape of any shelter belt is ______________. (**Neloid**)

255. ____________ is the first state in India, created a separate social forestry wing in the forest department. (**Gujarat**)

256. Central Arid research Institute is located at ____________. (**Jabalpur**)

257. ___________is called as green gold. (**Bamboo**)

258. ____________ is the most commonly used tree species for shade in tea estates. (***Grevillea robusta***)

259. The queen of timber is _____________. **(Teak)**

260. The practice of forestry with the object of obtaining maximum gain per unit area is called _____________. **(Intensive Forestry)**

261. Based on arrangement of components, agroforestry systems are classified into two categories namely _____________ & _____________. **(Spatial & Temporal arrangements)**

262. Erosion permitting and erosion resisting crops grown alternatively at right angle to the slope of the land is called as _____________. **(Strip Planting)**

263. Total number of biomes in the world is _____________. **(Twelve)**

264. _________ country in the world which has the maximum per capita forest area. **(Canada)**

265. The watershed is considered to be synonymous with _____________ and __________. **(Catchment & Drainage basin)**

266. The pattern of water flow in a given region or basin is called as _____________. (Hydrology)

267. The main host plant of silk worm is _____________. **(Mulbery)**

268. The main aim of raising crops in the departmental taungya is to ___________. **(Reduce the weed growth)**

269. NAP stands for _____________. **(National Afforestation Programme)**

270. The value of a wood as a fuel depends mainly on its ___________. **(Combustability)**

2

Silviculture

1. The "silviculture of Indian trees" in three volume was published in 1921, by ______________. (***R.S. Troup***)
2. ______________ silvicultural system aims at opening of the canopy for regeneration and even-aged condition of the crop. (**Uniform shelter-wood system**)
3. Irregular involutions and convolutions on the bole just above the base of a tree are called as ______________. (**Fluting**)
4. Stool shoots refers to the production of coppice shoots from______________ of stumps. (**Near the top**)
5. Geographical area covered by forest area in India is______________ per cent. (**21-22%**)
6. Indian forests were classified into _______ number of groups by Champion and Seth in 1968. (**16**)
7. The most favourable soil for the growth of *Shorea robusta* (Sal) is______________. (**Well drained deep moist sandy loam**)
8. The clear felling system was introduced for the first time by______________. (**Heinrich Vancotta**)
9. The book "Genera Plantarum" was written by______________. (**Englaer and Prantl**)
10. "Pricking" is a term used in raising plants in______________ area. (**Main plantation**)
11. Generally size of the forest nursery bed is______________. (**10 × 1.2 m**)
12. Sunken beds are preferred for forest nursery in______________area. (**Dry**)
13. ______________ meter wide nursery beds are prepared for Deodar in the hills. (**1.2**)
14. Acclimatization of plantlets in a place to face the harsh condition after transplanting in the fields is called as______________. (**Hardening off**)

15. *Tectona grandis* is frost_____________ species. (**Tender**)
16. *Casuarina equisetifolia* is a_____________ nitrogen fixing tree. (**Non-leguminous**)
17. Bamboo belong to family_____________. (**Poaceae**)
18. _____________plant produced seed only once in its lifetime. (**Monocarpic**)
19. _____________ test of a seed determines seed is capable of germinating into a seedling possessing all essential structures needed to develop into a normal plant. (**Viability test**)
20. Normally the moisture content of seeds to be stored is_____________ per cent. (**10-12**)
21. _____________ _____________ means seed capable of germination under favourable condition. (**Seed viability**)
22. Mycorrhiza is a symbiotic association between_____________ and_____________ of higher plants. (**Fungi and Roots**)
23. Particle size of silt is_____________. (**0.002 to 0.02mm**)
24. Plants grow on salty soil are_____________. (**Halophytes**)
25. A species which has been grown out of its geographical range is called_____________. (**Exotic**)
26. _____________ started Van Mahotsava to promote farm forestry. (**K.M. Munshi, 1950**)
27. The culmination stage in plant succession for a given ecosystem is called_____________. (**Climax**)
28. _____________ is a highest elevation in mountains at which tree can grow. (**Timber line**)
29. In degraded hills, trees are planted_____________ contour bunds. (**Along**)
30. Dispersal of seeds by wind is called_____________. (**Anemophily**)
31. The science of silviculture pertains to_____________. (**Trees**)
32. Thinning of plantation pertains to_____________of selected trees. (**Removal**)
33. Forest Survey of India brings out the ‘State of forest report’ in every_____________ year. (**Two**)

34. In scrub forest canopy density is less than____________. **(10%)**
35. Canopy density for very dense forests is____________. **(>70%)**
36. Aerial roots of *Ficus bengalensis* is known as____________. **(Prop roots)**
37. The series of changes occurred in the vegetation of pond over long period is referred as ____________. **(Hydrosere)**
38. Decomposition of litter in tropical forest is____________ than temperate forest. **(Faster)**
39. When one organism is benefited without affecting the other is called____________. **(Commensalism)**
40. The mature crop removed in one operation is called____________. **(Clear felling system)**
41. Clear felling system is generally regenerated by____________. **(Artificially)**
42. The young crop is protected against adverse climatic factor such as frost, cold, wind, insolation etc. under____________. **(Shelter wood system)**
43. The clear strip system results in production of____________ crop. **(Even-aged)**
44. Seeding felling carried out before____________felling. **(Secondary)**
45. Foresters recognize____________as the basic unit of forest. **(Stand)**
46. Site index is a term which is the relationship between____________ of trees. **(Height to age)**
47. Within the forest, trees that have tall, well formed crowns or tops that receive light from above and the sides are known as____________ trees. **(Dominant)**
48. The ability of tree to withstand shade is called shade____________. **(Bearer)**
49. The art of growing and reproducing a forest is called____________. **(Silviculture)**
50. The most limiting element of all the environmental factors on the growth of trees is ____________. **(Moisture)**
51. A forest which has been reproduced through sprouting is called a____________ forest. **(Coppice)**

52. A stand of trees produced after a clear cut is most likely to be________ . **(Even-aged)**

53. _____________ indicates the productive capacity of a specific area of forest land for a particular tree species. (**Site quality or index**)

54. Most pine tree cones require_____________ years to mature, while most hardwood tree fruits mature in_____________ years. (**2, 1**)

55. The periodic removal of individual trees in order to produce better yields is referred as _____________. (**Selective cutting**)

56. New terminal growth of pines is called_____________. (**Candles**)

57. The leaves and branches of a tree present above the bole is referred as_____________ _________. (**Crown**)

58. Overcrowding reduces the_____________of the tree. (**Diameter**)

59. The term_____________is defined as the reproduction and management of the forest to obtain continuous high yields of forest products. (**Silviculture**)

60. Under ideal conditions it requires 8 hours per day to plant _______ number of seedling by using a hand planting bar. (**500-800**)

61. The best time of the year to plant seedling is_____________. (**Early monsoon**)

62. The proper depth to plant seedling is the same depth as these were in the_____________. (**Nursery**)

63. Temporary planting of several trees in long narrow rows is called_____________. (**Heeling-in**)

64. Another name for the hand planting bar is the_____________. (**Dibbler**)

65. An excellent tool for pruning branches less than 1 inch in diameter is the. (Lopping shears)

66. The plant hardiness zone map developed by the USDA, dividing the country into 10 zones based on_____________. (**Average winter temperature**)

67. The closeness of tree tops is called _____________. (**Crown density**)

68. A cutting which removes trees of poor quality, undesirable species or those in poor condition is called_____________. (**Improvement cutting**)

69. Cutting burnt or damaged trees before they become completely worthless is called _____________. **(Salvage cutting)**

70. Cutting large trees which are shading small trees is called_____________. **(Release cutting)**

71. Removing a part of the stand to promote faster growth of the remaining stand is called _____________. **(Thinning)**

72. Removal of highest quality and most valuable matured trees is known as _____________. **(Selection cutting)**

73. Removing trees damaged by fire, insects or disease is called_____________. **(Salvage cut)**

74. Removal of mature timber in a series of cutting so that the remaining trees might reseed the area is called_____________. **(Shelterwood cutting)**

75. Removal of lower branches of a tree is called_____________. **(Pruning)**

76. A mixed stand is one in which less than_____________ per cent of the trees are of one species. **(80)**

77. Tree stand developed almost totally from sprouts is called_____________ forest. (Coppice)

78. Pruning is done to produce_____________. **(Knot free lumber)**

79. Trees characterized by having unisexual flowers of both sexes produced on the same tree are_____________. **(Monoecious)**

80. Cutting made past the sapling stage for the purpose of improving the composition or the quality of the growing timber stand is called_____________. **(Improvement cutting)**

81. Primary purpose of flower is_____________. **(Reproduction)**

82. A totally occupied forest stand that allows plenty of room for development of crop trees is called_____________stand. **(Fully stocked)**

83. The removal of live or dead branches from standing trees is called_____________. **(Pruning)**

84. The first plants to occupy a field after a fire or when it is no longer cultivated, are called _____________. **(Pioneer species)**

85. Forest biomass is the weight of_____________. **(Vegetative matter)**

86. A_____________is a woody plant with a DBH greater than 1 inch but less than 4 inches. **(Sapling)**

87. The number of trees in a fully stocked stand____________ as the trees increases in diameter. (**decreases**)

88. Silvics is the science of tree____________ and species distribution in the forest. (**Growth**)

89. To be certified as a tree farm, one must manage____________ or more of forest land to produce timber. (**10 acres**)

90. A timber harvesting method used to allow faster growth of remaining trees, to remove diseased trees or to remove undesirable trees before the final harvest is called ____________. (**Intermediate cuttings**)

91. Thinning is designed to reduce the time period of the____________. (**Final harvest**)

92. In the____________ method, the entire stand of trees are cut except for a few which are selected for their quality and seed producing ability. (**Seed tree cutting**)

93. The purpose of____________ harvest method is to produce a large immediate income, provide for the natural reseeding of the area and provide partial shelter for the seedlings as they begin to germinate. (**Shelterwood cutting**)

94. A tree which is under the level of the forest crown and receives no direct sunlight is considered a____________ tree. (**Suppressed**)

95. Theremovaloftreesfromtheforest,includingurbanforests,____________ global warming and the amount of carbon dioxide in the atmosphere. (**Increases**)

96. A deciduous tree survives winter by____________. (**Dropping its leaves**)

97. A very complex community of associated trees, shrubs, other plants and animals is ____________. (**Forest**)

98. Tree that bears seeds in cones and that have long, narrow leaves is a ____________. (**Conifer**)

99. Applying tree seeds to the harvested area or prepared beds is called____________. (**Direct seeding**)

100. Neither natural seeding or____________ will result in trees being planted in rows unless a grain drill is used. (**Direct seeding**)

101. A silvicultural prescription usually contains____________ relating to objectives. (**Cultural operation**)

102. A process which removes damaging, dominant trees in an all-aged forest stand is ____________. (**Liberation cut**)

103. Coppice shoots and root suckers are example of____________ reproduction. (**Vegetative**)

104. Thinning overstocked stands in order to manage or control pests is a____________ pest management strategy. (**Silvicultural**)

105. Trees in the 4 inch to 10 inch dbh range are called____________. (**Pole timber**)

106. Rain, snow, frost and hail are forms of____________. (**Precipitation**)

107. Home owners can reduce cooling cost by planting tree on the____________and ____________ sides of the house. (**South and West**)

108. The practice most commonly used to produce high grade stands is____________. (**Improvement**)

109. The process to remove damaging dominant trees in an all-aged stand is called____________. (**Liberations cut**)

110. An achene is____________ seed. (**Unwinged**)

111. ____________ is the best measurement of site quality. (**Tree height**)

112. Leaf litter and____________ on the forest floor act as moderators of soil temperature. (**Humus**)

113. Using the natural shape and texture of a tree to complement the architecture of a building or to enhance the vertical elements of a structure is an example of ____________ benefits of urban trees. (**Aesthetic**)

114. Tree species that have male and female flowers on separate trees are called ____________. (**Dioecious**)

115. Using containerized seedlings rather than bare root seedlings usually results in improved rates of____________ and____________. (**Survival and Growth**)

116. The term social forestry is introduced by____________. (**Westoby**)

117. Nursery which is not usually irrigated is called____________. (**Dry nursery**)

118. Viability of Neem seed is_____________ days. (**One month**)

119. The concept of "Van Mohostava" was initiated by_____________. (**K.M. Munshi, 1950**)

120. Autumn tints are_____________ colour change before fall. (**Leaf**)

121. _____________state has largest percentage of forest area. (**Mizoram**)

122. Eucalyptus introduced into India for first time at_____________ _____________ in 1790. (**Nandi Hills**)

123. Sandal is partial_____________ parasite. (**Root**)

124. The altitudinal range of sub-alpine region is_____________. (**2900-3300m**)

125. Trees planted in degraded lands come under_____________ forestry. (**Protection**)

126. Community forestry deals with planting trees on_____________ lands. (**Panchayat**)

127. International day for conservation of biological diversity celebrated on_____________. (**22 May**)

128. Flowers and fruits developed from stem of trees is known as_____________. (**Cauliflory**)

129. Biggest CO_2 sink is_____________ forest. (**Tropical rain**)

130. _____________ type of forest are described in Vishnu Puran. (**13**)

131. *Acacia catechu* is a frost_____________ species. (**Resistant**)

132. Frost frequently and more intensely occurring in an area is called_____________. (**Frost pockets/ frost hole**)

133. Azatobactor and Clostridium are_____________ bacteria. (**Nitrogen fixing symbiotic**)

134. Optimum temperature necessary for seed germination is_____________. (**20-35 °C**)

135. Aerial seeding is done in_____________ area. (**Inaccessible hilly terrain**)

136. In degraded hilly lands trees are planted_____________ the contour bunds. (**Along**)

137. Raised beds are made in _______ area. (**High rainfall**)

138. Genetic improvement in poplar can be made from_____________. (**Clonal propagation**)

139. The most important use of poplar is ______. (**Matches box**)

140. Eucalyptus is cut for coppicing and shoots are arranged from 10 cm below the. (**Stump**)

141. _____________ type of wood readily catches fire and burns with low flame. (**wood containing resin**)

142. Teak is managed in dry teak forests under_____________ system. (**Selection**)

143. The regeneration of fir is inhibited due to_____________. (**Over humus**)

144. The practice of forestry with the object of producing timber and other forest product as business enterprise is_____________. (**Commercial forestry**)

145. Restocking of felled or otherwise cleared woodland by artificial means is. (**Reforestation**)

146. A subsidiary crop of low plantations in plantation of soil cover between and below soil is called_____________. (**Cover crop**)

147. _____________ number of plants per ha in a 2 × 2m spacing. (**2500**)

148. Taungya system in ancient times was practised as_____________. (**Shifting cultivation**)

149. Silviculture means cultivation of_____________. (**Trees**)

150. Epicornic branches originates from_____________. (**Adventitious buds**)

151. A tree which has more than one leader stem is called_____________. (**Forked**)

152. Pneumatophores are commonly seen in_____________trees. (**Mangrove**)

153. *Alnus nepalensis* is a non-leguminous_____________ tree. (**Nitrogen fixing**)

154. Generally teak flowers during_____________months. (**June-August/ September**)

155. Teak is naturally found in_____________forests. (**Tropical deciduous**)

156. Sal is normally found in_____________ forests. (**Tropical moist deciduous**)

157. Sal flowers during_____________ months. (**February – March**)

158. High forests are regenerated through_____________. (**Seed**)

159. Champion and Seth (1968) classified the Indian forests into_____________ major groups. (**Five**)

160. Champion and Seth's classification of Indian forests is based on_____________. (**Ecological**)

161. Endemics mean the plant species which has distribution is limited to a_____________. (**Specific area**)

162. Sholas are_____________ forests. (**High altitudinal broad leaved**)

163. *Pinus roxburghii* is commonly found in_____________forests. (**Sub-tropical pine**)

164. In clonal forestry trees are propagated through_____________ means. (**Vegetative**)

165. Tropical wet evergreen forests in Western Ghats are managed under_____________ system. (**Selection**)

166. Azadirachtin is obtained from_____________. (***Azadirachta indica***)

167. Silviculture systems aiming at diffused regeneration are_____________ _____________ and _____________ systems. (**Selection, Group selection**)

168. An operation carried out for the benefit of forest crop at any stage of its life cycle between seeding and mature stage is_____________. (**Tending**)

169. Restocking of blanks in an artificially regenerated area with fresh sawing or planting is called_____________. (**Beating up**)

170. Trees or shrubs grown to foster the growth of another and more important tree crop in early stages are known as_____________. (**Nurse crop**)

171. Succession which is initiated under extremely dry conditions is known as_____________. (**Xerarch**)

172. Culminative stage of plant succession for a given environment is called. (**Climax**)

173. Generally Bamboo flowers once in_____________ years. (**25-40**)

174. In silvicultural systems canopy is opened to get_____________. (**Regeneration**)

175. *Casuarina equisetifolia* fixes atmospheric nitrogen in association with_____________. (**Frankia**)

176. A plant which avoids calcareous soil is called_____________. (**Calcifuge**)

177. A plant which tends to be restricted to calcareous soil is_____________. (**Calciphile**)

178. A woody plant that climbs up on tropical tree is called_____________. (**Liana**)

179. The species that invade new territory to form a colony is_____________. (**Colonist**)

180. Sandal derive its nutrient requirement through its parasitic nature with the help of _____________. (**Haustoria**)

181. If the canopy density is more than 70% the forest is_____________. (**dense forest**)

182. Neem day is celebrated on_____________ _____________.(**8th August**)

183. Presence of_____________ is an indication of good site for Sal. (***Flemengia* spp.**)

184. Presence of *Glyeasmis pentaphylla* and *Helectris isora* is an indication of good site for _____________. (**Teak**)

185. Presence of *Xylia xylocarpa* is bad indicator for_____________. (**Teak**)

186. *Viola canescens* is an indicator species for_____________ and_____________. (Deodar and Kail)

187. The study of life history and general characters of tree with reference to environment is known as_____________. (**Silvics**)

188. Subabool contains_____________ toxic compound. (**Mimosene**)

189. Toxin compound in fodder is_____________. (**Tannin**)

190. Wrenching is carried out in nursery for to cut_____________ and roots. (**Lateral & Bottom**)

191. Forest is derived from Latin word_____________. (**foris**)

192. Thumb rule thinning is called_____________ thinning. (**Mechanical**)

193. Ordinary thinning is carried out mainly in_____________ species. (**Light demander**)

194. Bacteria involved in conversion of ammonical to nitrate form of N is __________________________. (**Nitrobacter**)

195. ____________ amount of solar energy used in primary production. **(2%)**

196. As a result of competition, the green of the individuals in a population is reduced and even the plant continues to thrive which is known as____________. **(Plastic response)**

197. Imposed mortality of plants by plants is known as____________. **(Self thinning)**

198. The law of self thinning or the -3/2 law of self thinning was given by____________. **(Yoda)**

199. Concept of tolerance is given by____________. **(Sheford)**

200. Growth of any organism is limited by the factor which is in least quantity was told by ____________. **(Leibig)**

201. Stump planting in teak was initiated by____________. **(Bourdilloni)**

202. In India plantation forestry was started in____________. **(1842)**

203. ____________silvicultural system is followed for Bamboo. **(Culm selection)**

204. Natural Teak was managed under____________ system. **(Selection)**

205. Natural regeneration in sal is best achieved under____________ system. **(Shelter wood)**

206. A patch usually 20m x 20m in size is burnt and artificially regenerated in a felled coupe to supplement natural regeneration is known as____________. **(Rab)**

207. A plant adopted to sunny sites is known as____________. **(Psamnophyte)**

208. Study of timely appearance of characteristic periodic event such as leaf shedding, appearance of flower etc. in life of a plant is known as____________. **(Phenology)**

209. Sites inside the forest set apart for the use of cattle during season of fodder and water scarcity in villages and outlaying forests is known as____________site. **(Penning)**

210. The healing over or closing of wounds in plants is____________. **(Occlusion)**

211. *Casuarina equisetifolia* is native to____________. **(Indonesia)**

212. *Eucalyptus* is native to____________. **(Australia)**

213. Casuarina was introduced first to_____________. (**Karwar coast**)

214. *Acacia auriculiformis* is native to_____________. (**Australia**)

215. In India *Acacia auriculiformis* was introduced at_____________. (**West Bengal**)

216. The solar energy flow per unit area is called as_____________. (**Light intensity**)

217. *Dalbergia sissoo* is frost_____________ species. (**Hardy**)

218. Effect of light on the form of trees is_____________. (**Photoperiodism**)

219. The classification of Indian forests according to Champion and Seth is based on _____________ _____________ approach. (**Ecosystem**)

220. TZ test is a method for determining the_____________ of a seed. (**Viability**)

221. C:N ratio is measure of_____________status in soil. (**Nitrates**)

222. Generally temperature under trees and in forests is_____________ °C less than the open. (**4-6**)

223. Standard nursery bed size is_____________. (**12m × 1.2m**)

224. Neem has_____________ type of crown with ever green tree is suitable for first row planting. (**Umbrella**)

225. The forestry which deals with growing of trees in non-traditional area is called _____________ _____________. (**Extension forestry**)

226. The average size of tree patta could be_____________ ha. (**0.4**)

227. "Social forestry is a forestry which aims at producing flow of protection and recreation benefits for the community" given by_____________. (**Westoby**)

228. The person given more emphasis on village forests and the village forests are used as fuel and fodder resource are to manage by people themselves was_____________. (**Brandis**)

229. Fast growing species is one which yields an increment of_____________. (**10 m^3/ha/yr**)

230. _____________ major groups of forests are recognised in India. (**Five**)

231. Plants observed in 4000 m^2 plot is only 500 after a year of planting with a spacing of 2m × 2m then_____________ _____________ per cent was survived. (**50**)

232. Lands having 33-100 % slope is well suited for____________ crop. **(Forest)**

233. In Burma taungya system was developed for _____ plantations by Brandis. **(Teak and)**

234. JFM concept was first implemented at____________. **(Arabari)**

235. ____________ law is used to estimate the proportionate distribution of trees in selection forests. **(DE Liocourts law)**

236. Rate of photosynthesis increase with increase in temperature up to____________. **(30 °C)**

237. Intensity of light on the southern aspects compare to northern aspects will be ___________. **(Higher)**

238. Maximum thinning was developed by____________. **(Gerhardt)**

239. The term forest succession was first used by____________. **(Thoreau)**

240. The C : N ratio considered suitable for measuring nitrate status of the soil is. **(10 : 1)**

241. For preparation of yield tables, the optimal number of sample plots for each thinning regime is____________. **(600)**

242. Site index is the relation between____________ and____________. **(Age and Height)**

243. The Forest Survey of India (FSI) conducts forest cover survey once in every ____________ years. **(Two)**

244. The poly climax theory in plant succession was developed by____________. **(Tansley)**

245. Bamboo seed retains its viability for a period of____________ weeks. **(2-3)**

246. In *Azadiracta indica*____________ number of seeds per kilogram is present. **(3000 – 3300)**

247. Teak forests in India are generally found up to elevation of _______ meter. **(900 meter)**

248. *Acacia auriculiformis,* an exotic species is native to ____________. **(Australia)**

249. *Bambusa vulgaris* is popularly known as ____________. **(Golden/ Striped Bamboo)**

250. In uniform crops of sal, first mechanical thinning is advised at the age of ______ years. **(10 years)**

251. The most common soil type on which sandal grows well is ____________. **(Red ferruginous Loam type)**

252. According to leafless character, neem is __________ in nature. **(Evergreen)**

253. *Prosopis juliflora* may be grown in soil having pH upto _______. **(11.5)**

254. *Populus deltoides* belongs to family __________. **(Salicaceae)**

255. As per coppicing ability, *Dalbergia latifolia* is a _____________. **(Strong coppicer)**

256. *Pinus wallichiana* otherwise known as ____________. **(Blue Pine/ Kail)**

257. Naturally sandal is profusely regenerated through seeds and _____. **(Root- suckers)**

258. *Swietenia mahogany* is ______ to India which is deciduous or semi-deciduous in nature. **(exotic)**

259. Optimum growth of *Picea smithiana* is found on soils with ______ condition. **(moist and rich humus)**

260. Teak practically absent when pH of soli is _______. **(below 6.0 and beyond 8.5)**

261. Siris is a tree of mixed deciduous forests types ________. **(dry and moist)**

262. The general rotation period of sissoo in forest working plan is fixed between __________. **(50-60 years)** 270 Number of seeds present/ gm of spruce is _____________. **(60-65)**

263. The dying- back of sal seedlings occurs in areas due to Low rainfall and __________. **(Underlying kankar)**

264. Eucalyptus seedlings raised in nursery are ready to plant after the age of __________. **(6-9 months)**

265. *Bambusa bambos* flowers gregariously after a cycle of _____ years. **(30-45 years)**

266. Deodar belt is common in the altitudinal range of _______ meter. **(1800-2600)**

267. Good seed year for *Santalum album* is ____________. **(Every year)**

268. Variety of *Acacia nilotica* is ______. ***(ramkanta/teli/kaora)***

269. *Populus deltoides* is popularly known as _______. **(Eastern Cottonwood)**

270. *Samanea saman* otherwise known as ________. **(Rain Tree)**

271. *Acacia mangium* is popularly known as _______. **(Hickory Wattle)**

272. *Dalbergia latifolia* is otherwise known as _______. **(Indian Rosewood)**

273. Scientific name of Whistling Pine is _______. ***(Casuarina equisetifolia)***

274. *Eucalyptus globulus* is popularly known as ______. **(Blue Gum)**

275. *Pterocarpus marsupium* is otherwise known as _____. **(Indian Kino Tree)**

276. Scientific name of Himalayan Longleaf Pine is ______. ***(Pinus roxburghii)***

277. Scientific name of Woman's Tongue tree is ________. ***(Albizia lebbeck)***

278. *Prosopis juliflora* is popularly known as _______. **(Mesquite)**

279. Sal forests in India are generally found up to elevation of_______ meter. **(1200 to 1500 meter)**

280. *Swietenia mahogany,* an exotic species is native of ______________. **(West Indies)**

281. *Dendrocalamus strictus* is popularly known as ____________. **(Male/ Solid Bamboo)**

282. ETPs method is generally adopted for tree species ____________. ***(Populus deltoides)***

283. *Acacia auriculiformis* is harvested after ______ years. **(10/12 years)**

284. According to leafless character, arjuna is __________ in nature. **(Nearly Evergreen)**

285. The synonymous name of *Terminalia tomentosa* is ______________. ***(Terminalia alata)***

286. *Prosopis juliflora* belongs to family __________. **(Mimosaceae)**

287. As per coppicing ability, *Madhuca indica* is a ____________. **(Bad Coppicer)**

288. *Pinus roxburghii* other wise known as ____________. **(Chir Pine)**

289. Sandalwood is an _______ root parasitic plant. (**Obligatory**)

290. Best growth of *Acacia nilotica* is found in the alluvial and ______ soil (**Black-cotton**)

291. Number of seeds present/ gm of sissoo is ______. (**50-55**)

292. The tree species, in which dying back of seedlings are generally found ______. (***Shorea robusta***)

293. Neem seedlings raised in nursery are ready to plant after the age of ______. (**12 months**)

294. *Bambusa bambos* flowers gregariously after a cycle of _____ years. (**30-45**)

295. Good seed year for *Abies pindrow* is _________ years. (**10**)

296. Variety of *Populus deltoides* is _________. (***occidentalalis***)

297. *Pterocarpus santalinus* is popularly known as _______. (**Red Sanders**)

298. *Acacia mangium* is popularly known as ________. (**Hickory Wattle**)

299. *Eucalyptus camaldulensis* is otherwise known as _____. (**River Red Gum**)

300. Scientific name of red gum tree is _______. (***Eucalyptus grandis***)

301. *Juniperus macropoda* is commonly known as ________. (**Himalayan Pencil-Cedar**)

302. ________ tree species is known as frywood tree. (***Albizia lebbeck***)

303. The part of the earth where life exists is called the_____________. (**Biosphere**)

304. A shrub is usually not more than_____________ meters in height. (**Six**)

305. The branches originating in clusters from dormant or adventitious buds on the trunk of a tree is_____________. (**Epicormic branches**)

306. The total surface area of the earth is approximately_____________ million m³. (**510**)

307. Tropical rainforest occurs in area where annual rainfall is usually more than _____________ mm and is evenly distributed almost throughout the year. (**2500**)

308. The total biomass per hectare in the rain forest is the maximum and is about _____________ tonnes per ha. (**1000**)

309. A large area in India is occupied by tropical_____________ forest. (**Deciduous**)

310. The condition of the tree with dead upper branches projecting above the green portion of the crown is_____________. (**Stagheaded**)

311. The roots produced from parts of the plants other than the radicle is called as _____________. (**Adventitious roots**)

312. The special roots produced from its branches and adventitious roots remain suspended in the air till they reached the ground is called as_____________. (**Prop roots**)

313. The adventitious roots which emerge from the butt of a tree above ground level is termed as_____________. (**Stilt root**)

314. A spike like projection of the roots of swamp tree is known as_____________. (**Pneumatophore**)

315. The average rainfall in tropical deciduous forest varies from to mm. (**750 to 2500**)

316. _____________ forest consists of evergreen with salt tolerant trees. (**Mangrove**)

317. Lignotuber is found in_____________. (**Eucalyptus**)

318. Mimosin is present in_____________. (***Leucaena leucocephala***)

319. Most suitable silvicultural system of deodar in hilly terrain for control of soil erosion is _____________. (**Selection system**)

320. "Silviculture of Indian Trees" was written by_____________. (**Banerjee**)

321. Forests of *Acacia arabica* are frequently termed as_____________. (**Babul meadow**)

322. Highest calorific value of wood is found in_____________ species. (***Casuarina equisetifolia***)

323. The growing point consists of_____________ cells which have the capacity to divide and give rise to new cells. (**Meristematic cells**)

324. The word "forest" is derived from Latin word_____________ which meaning outside the village boundary. (**Foris**)

325. _____________is referred to an area occupied by different kind of trees, shrubs, herbs, and grasses and maintained as such. (**Forest**)

326. The practice of forestry with the object of producing timber and other forest products as a business enterprise is known as____________. **(Commercial forestry)**

327. Even aged forest is also called____________ forest. **(Regular)**

328. True even aged forest can only be____________forest. **(Man-made)**

329. A forest is called____________ aged when trees vary widely in age. **(Uneven)**

330. When regeneration of forest obtained from seeds called____________. **(High forest)**

331. When the regeneration is obtained naturally, the forest is called ____________ forest. **(Natural)**

332. When the regeneration is obtained artificially, the forest is called ____________ forest. **(Man made)**

333. ____________ are composed almost entirely one species, usually to the extent of not less than 80 per cent. **(Pure forest)**

334. A normal forest is an ideal forest with regard to growing stock, age classes distribution and____________. **(Increment)**

335. Social forestry is the practice of forestry which aims at meeting the requirement of ____________ ____________ and____________ population. **(Rural and Urban)**

336. The practice of forestry on land outside the conventional forest area for the benefit of local population has been called____________. **(Social forest)**

337. ____________is defined as the practice of forestry in all its aspect on farms or village land generally integrated with other farm operations. **(Farm forest)**

338. ____________ forestry is defined as the practised of forestry with the object of developing and maintaining forest of high scenic value. **(Recreational)**

339. ____________ ____________ forestry is referred to the practice of forestry whose objective is to maximize production per unit area by applying outside silvicultural and managerial techniques. **(Intensive)**

340. The most primitive member of plant kingdom is perhaps____________ which appeared on the earth about 1500 million year ago. **(Algae)**

341. In_____________ Puran, there is a description of 13 types of forest occurring in different parts of the country. **(Vishnu)**

342. _____________ Appointed as first inspector general of forests of India. **(Dr. Dietrich Brandis)**

343. About_____________ per cent of the forest of the country are owned by the government. **(95%)**

344. The coppice shoot arising from the base of the seedling that have been cut or burnt back is_____________. **(Seedling coppice)**

345. A large number of non-wood products are commonly_____________ products. _______**(Non-forest)**

346. Katha and cutch are obtained from_____________ trees. (***Acacua catechu*/Khair**)

347. Indian copal tree is_____________. (***Vateria indica***)

348. Stilt roots present in_____________ forests. **(Mangroves)**

349. In teak pollination occurs by_____________. **(Insects)**

350. _____________ proposed Mosaic theory of succession. **(Aubreville)**

351. _____________ and_____________ are exuded by trees a result of wound or injury to the bark of wood. **(Gum, resin)**

352. The term_____________ refers to certain aspect of theory a practice of raising forest crops. **(Silviculture)**

353. Rotationiscountedperiodinyearsfrom_____________ to_____________. **(Regeneration, harvesting)**

354. The word_____________ which represent the environment of forest. **(Site)**

355. _____________for a given period is perhaps the best indicator for site quality. **(MAI)**

356. In India, site quality classes are determined by consiering_____________ of various tree species. **(Top height)**

357. A large part of solar radiation is reflected back is called as_____________. **(Albedo)**

358. In plants light is essential for_____________ activity. **(Photosynthesis)**

359. An average green leaf absents about____________per cent of light in photosynthesis and hardly____________ per cent of light is utilized in chemical reaction. **(80-85, 1)**

360. Plants like____________ may develop chlorophyll in absence or under reduced light conditions. (**Algae**)

361. Optimum range of temperature for growth lies between____________ to____________ °C. (**25 to 30**)

362. Apical meristem is responsible for____________. (**Height growth**)

363. Lateral meristem is responsible for____________. (**Diameter growth**)

364. At higher temperature above 50 °C____________ of protoplasm takes resulting in death of plants. (**Coagulation**)

365. *Acacia catechu* is a frost____________species. (**Hardy**)

366. *Tectona grandis* is a frost____________ species. (**Tender**)

367. A rainy day is a day in which____________ mm or more rainfall occurs. (**2.5**)

368. The number of rainy days in India varies from less than____________ days in Rajasthan to over____________ days in Western Ghat part of Assam and Andaman &Nicobar Island. (**10, 150**)

369. FSI is located at____________. (**Dehradun**)

370. The type of frost common in Dehradun valley is____________. (**Pool frost**)

371. *Acacia nilotica* is a drought____________ species. (**Hardy**)

372. The first stage of regeneration felling under a shelter wood system is ____________. (**Seeding felling**)

373. The removal of last seed or shelter trees after regeneration has been effected under a shelterwood system is known as ____________. (**Final felling**)

374. A regeneration felling carried out between the seeding felling and final felling under a shelter wood system is called as ____________. (**Secondary felling**)

375. *Dalbergia sisoo* is a ____________ drought hardy species. (**Moderately**)

376. *Tectona grandis* is a drought ____________ species. (**Sensitive**)

377. Indian butter tree is_____________. (***Madhuca latifolia***)

378. Chilgoza is obtained from_____________. (***Pinus gerardiana***)

379. Prop root is common phenomenon in_____________. (***Ficus spp.***)

380. The term 'Forest succession' was first used by_____________. (**Thoreau**)

381. Father of Botany is _________. (**Theophrastus**)

382. A forest fire spreading slowly over the ground with low flame is termed as. (**Creeping fire**)

383. The climax stage of estuarine succession is_____________. (**Evergreen forest**)

384. Texture of soil refers to particles ________. (**Size**).

385. The availability of water in soil for plant growth decrease with decrease in_____________ .(**Moisture content**)

386. Root development ceases of water content falls below_____________ percent. (5)

387. The decomposed organic matter is called_____________. (**Humus**)

388. Humus is regarded as the_____________of soil. (Life blood)

389. Plant materials like leaf litter and other plant residue contain about_____________ and _____________ per cent carbon respectively. (**0.05 to 3.5, 45-50**)

390. _____________ refers to the height of a place from mean sea level. (**Altitude**)

391. Rainfall is more on_____________ side of the hill than on_____________ side. (**Windward, Leeward**)

392. _____________ represent a symbiotic relationship in which some fungi are associated with roots of trees. (**Mycorrhiza**)

393. _____________ plants which grow on the trunk and branches of other plants for the purpose of support. (**Epiphytes**)

394. _____________ _____________ refers to steady increase in the temperature of Earth's surface and the lower atmosphere. (**Global warming**)

395. _____________ is defined as a belt of tree and/or shrubs maintained for the purpose of shelter from the wind, sun, snow drift etc. (**Shelter belt**)

396. The tree crown and other woody vegetation intercept as much as_____________ per cent of the total precipitation. (**15-20**)

397. _____________ refers to that part of the bole which is free from branches. (**Clear bole**)

398. Standard timber bole is the length of bole from ground level up to the point where the average diameter over bark is_____________ cm. (**20**)

399. If the canopy density is between 0.75 to 1.0, then the forest is referred as_____________. (**Dense**)

400. *Denrocalamus strictus* is a good associate of_____________. (**Teak**)

401. The clear felling or clear strip system is suitable for_____________ _____________ species. (**Light demander**)

402. The mono climax theory of plant succession was given by_____________. (**Clements**)

403. The up and down movement of wind in valley is known as_____________. (**Dadu**)

404. _____________ is known as first silviculturist of India. (**Troup**)

405. As per Benthem and Hooker's classification, there are_____________ non-standard exceptions to the family name ending 'aceae'. (8)

406. Indian biodiversity contribution to world biodiversity is_____________ per cent. (**7%**)

407. _____________forest type are there in the world. (**5**)

408. The natural system of classification is based on_____________. (**form**)

409. The largest family among the Indian flora is_____________. (**Asteraceae**)

410. _____________ per cent of regeneration of a sample plot could be taken for a successful regeneration of that area. (**60-70%**)

411. *Eucalyptus globulus* plantation in the Nilgiri hills was commenced in the year. (**1843**)

412. A change from one silvicultural system to another is necessitated and this is usually referred as_____________. (**Conversion**)

413. First forest policy of Independent India was introduced in_____________. (**1952**)

414. The national afforestation programme is being implemented through a tier system. (**Two**)

415. The period during which the change from one silvicultural system to another is effected as_____________. (Conversion period)

416. RIL stands for_____________. **(Reduced impact logging)**

417. _____________ is a silvicultural system in which equal or equi-productive areas of mature crop are successively clear felled in one operation. **(Clear felling system)**

418. The canopy is uniformly opened up over the whole area of a component to obtain uniform regeneration_____________. **(Uniform system)**

419. The group system is developed by_____________. **(Karl Gayer)**

420. The group system is also known as_____________. **(Gayers/ Bavarian femelschlag)**

421. In group system canopy has wavy profile usually known as_____________. **(Gayer's line)**

422. The wedge system is developed by_____________. **(Eberhard)**

423. A silvicultural system in which the crop to be regenerated is opened up irregularly and the crop is uneven aged is called as_____________. **(Indian irregular shelter wood system)**

424. The Indian irregular shelter wood system is otherwise called as_____________. **(Punjab shelter wood system)**

425. The improvement felling is applied to_____________. **(Mixed uneven aged forests)**

426. In coppice system the number of shoots is reduced only to one per stool the special operation known as_____________. **(Singling out operation)**

427. A plantation map showing areas suitable for different species is_____________. **(Suitability map)**

428. The regeneration is obtained under shelter of the over wood such system are called as _____________. **(Shelter wood system)**

429. The arrangement of ovules within the ovary is termed as_____________. **(Placentation)**

430. A pair of appendages that present at the point of attachment of a leaf to the stem is _____________. **(Stipulates)**

431. Trees, with clean bole in later part of there lives, due to some adverse facts, develop small branches which are called_____________ branches. **(Epicormic)**

432. _____________ is an outgrowth from the base of the tree connecting with the roots. (**Buttress**)

433. Plants grows in marshy places develop special kinds of roots called _____ roots. (Respiratory)

434. Inner wood of the trees are called_____________. (**Heart wood**)

435. _____________ part of wood is darken in colour. (**Heart wood**)

436. Growth means a_____________ and_____________ increase in size and form attended by increase in weight. (**Permanent, Irreversible**)

437. On the basis of position meristems may be of_____________ ,_____________ , and_____________. (**Apical, Intercalary and Lateral**)

438. In tree the period from germination to the stage that the young trees have a few leaves called_____________. (**Recruit**)

439. The period from recruit stage to a height of 1 m of the young plant is called. (**Seedling**)

440. Growth in volume is the product of growth in diameter and_____________ of trees in a crop. (**Height**)

441. Wider growth rings indicate rate of growth_____________and it is associated with poor timber properties. (**Faster**)

442. Annual growth rings are generally used for find up out_____________ of the tree. (**Age**)

443. _____________stage starts with germination and extends up to a period till the plant reaches a height of one meter. (**Seedlings**)

444. _____________ stages of trees characterized by very little or almost no growth and progressive decline in the quality of wood. (**Over mature**)

445. A crop is said to be_____________ when it consists of mostly sapling and poles. (**Young**)

446. When the stand is formed by a_____________species it is called pure stand. (**Single**)

447. Density refers to relative_____________ of trees in a stand. (**Closeness**)

448. A crown free to grow at the tip and is not hedged in by its neighbours is called _____________. (**Free crown**)

449. A crown which is over topped and/or shaded by other trees is called_____________. (**Confined**)

450. The science which deals with interrelationships with the organisms with environment is called_____________. (**Ecology**)

451. The word ecology is derived from the Greek word_____________. (**Oikas**)

452. The study of an individual organism in relation to its environment is called _____________. (**Autoecology**)

453. The study of community and the interactions of the organisms which consists it is known as_____________. (**Synecology**)

454. _____________ succession refers to the replacement of one set of biotic community or biota by another set of different nature. (**Primary**)

455. Succession, which is subsequent to a disturbance, which destroys existing biotic community either partially or totally is called_____________ succession which is more_____________ than primary succession. (**Secondary, Rapid**)

456. _____________ succession involves the predominance of green plants in the initial several stages. (**Autotrophic**)

457. _____________ succession involves the predominance of heterotrophic organism like fungi, bacteria and animals at the initial stages. (**Heterotrophic**)

458. _____________succession involves the replacement of one community by another due to modification of the environment by community them several. (**Autogenic**)

459. _____________ involves the replacement of one community by another due to force other than the effects of communities on the environment. (**Allogenic**)

460. Several activities such as overgrazing, scrapping, shifting cultivation, industrial pollution etc. may cause_____________ succession. (**Induced**)

461. The series of stages in the succession are called_____________. (**Seres**)

462. Four stages of succession are_____________ ,_____________ , and. (**Pioner, Consolidation, Subclimax, Climax stage**)

463. Pioneer stage refers to the primary stage when the bare site in occupied by the _____________ _____________. (**Vegetation**)

464. The species which grow in the area for the first time is called_____________ species. (**Pioneer**)

465. The species which occupy the new sites for the first time are calledspecies. (**Colonizers**)

466. The stage where vegetation is said to be in equilibrium with environment and remain stable indefinitely by reproduction itself efficiency, known as. (**Climax**)

467. Several stages refers to_____________ stage in vegetation. (**Development**)

468. _____________ is a stage in succession of forest just preceding the climax community. (**Sub-climax**)

469. _____________ is an art of judging character from features of face or form of body. (**Physiognomy**)

470. _____________ structure means stratification in forest dimensions especially height and spacing. (**Stand**)

471. Dynamics refers to the consideration of mutual interaction in forest community and the_____________. (**Habitat**)

472. _____________ Is the the combine influence of climatic, edaphic, physiographic and biotic factors towards vegetation development, growth and distribution. (**Site factor**)

473. Wet evergreen forest include forests occurring in area having mean annual rain fall over _______ mm. (**2500**)

474. Moist and semi-moist forests occurs in areas where mean annual rainfall between_____________ mm to_____________ mm. (**1250 to 2500**)

475. The operations are carried out in the forest crop at different stages of growth in order to provide a healthy environment for there development called_____________. (**Tending operation**)

476. _____________ are the unwanted plants. (**Weeds**)

477. When weeds are controlled by some mechanical means called _____________________. (**Mechanical method**)

478. Under biological method of weed control *Opuncia spp.* has been successfully controlled with the help of a_____________ insect in forest of south India. (**Cockineal**)

479. Cleaning are done mainly to reduce_____________ competition and for minimizing the root competition. (**Light**)

480. In cleaning_____________species are usually removed. (**Inferior**)

481. The most common method of cleaning is_____________ in which the inferior growth removed with some sharp implements. (**Cutting**)

482. Intensive cleaning is not recommended in_____________ area. (**Dry**)

483. Light demander plant species require_____________ cleaning. (**Intensive**)

484. When a felling made in an immature stand for the purpose of improving the growth and forms of the trees called_____________. (**Thinning**)

485. Thinning is a tending operation carried out in a crop beyond the_____________stage up to the beginning of_____________ period. (**Sapling, Regeneration**)

486. Thinning principles have been developed on the basis of natural development of the _____________. (**Stand**)

487. _____________ is the fittest, the most important biological law of nature. (**Survival**)

488. Improvement in the_____________ of the crop can be obtained by removing dead dying and diseased trees. (**Hygiene**)

489. According to yield table about_____________ per cent of the yield is obtained from thinning. (**25-50**)

490. Thinning helps in_____________ the rotation. (**Shortening**)

491. Reduction in number of trees in thinning, increase the_____________ of the remaining trees. (**Diameter**)

492. Density has a direct influence upon the_____________. (**Growth**)

493. Diameter increment can be accelerated upon the_____________. (**Thinning**)

494. Thinning may also improve the_____________ of wood. (**Quality**)

495. The basic principle involved in thinning is to_____________the number of stem per hectare. (**Reduced**)

496. _____________ thinning is usually applied in young crops or young plantations before the crown differentiation has taken place. (**Mechanical**)

497. In method where trees are cut in line or row throughout the area is called as ___________ thinning. (**Row**)

498. _____________ thinning also called as low thinning or German thinning or thinning from the below. (**Ordinary**)

499. In ordinary thinning, the trees are removed from_____________ crown class. (**Lower**)

500. _____________ thinning is the most commonly used thinning practices in forestry. (**Ordinary**)

501. Ordinary thinning is most suites for_____________ demander species. (**Light**)

502. _____________thinning is simple in execution and only ordinary skill is required for selection of trees. (**Ordinary**)

503. _____________ thinning is useful in forests infested with climbers and where there is risk of crown fire. (**Ordinary**)

504. _____________ thinning is a method of thinning in which, the less promising ones being removed in the interest of the best individuals. (**Crown**)

505. Crown thinning is known as_____________ thinning because of its origin. (**French**)

506. _____________ thinning trees from upper crown classes are removed in order to fever the growth and development of most promising trees of the same class. (**Crown**)

507. In_____________ thinning , the object is to promote the growth of trees which show sign of success in the competition and promise a good value as a future crop. (**Crown**)

508. Crown thinning is best suited for_____________ shade tolerant species like sal. (**Moderately**)

509. _____________ thinning is suited for areas where there is no demand for small size timber. (**Crown**)

510. _____________ thinning is especially suited for areas where there is denser of site deterioration due to removal of trees of lower crown classes. (Crown)

511. Crown thinning provides better_____________for growth and development of retained dominant trees. (**Environment**)

512. _____________ thinning requires greater skill in execution. (**Crown**)

513. Free thinning is a modification of crown thinning and is also called stem silviculture. (**Single**)

514. Free thinning is also called_____________ thinning. (**Elite**)

515. In_____________thinning method, elites are first selected in number appropriate to the size or age of the crop with special regard to their stem form and to their even spacing. (**Free**)

516. In_____________ thinning, thinning is carried out before the competition among individual trees has set in the competition is never allowed to take place. (**Advance**)

517. _____________is the number of trees corresponding to a given standard average diameter. (**Stand density index**)

518. For a given site quality the optimum number of trees per unit area is found to be inversely proportional is the sequence of the height of the stand is called. (**N/H^2 relationship**)

519. Thinning is an irregular crop is usually called_____________. (**Selection thinning**)

520. Factors affecting thinning practices are_____________ ,_____________, _____________. (**Nature of species, Age, Site quality**)

521. Light demander tree species are less tolerant to_____________ than shade bearers. (**Crowding**)

522. Light demander species are usually responding better to_____________ thinning. (**Ordinary**)

523. Shade bearing species are usually responds better to_____________ thinning. (**Crown**)

524. If thinning is done early and heavily the stem tend to become_____________ and poor shape. (**Branchy**)

525. On_____________site heavy thinning is not recommended. (**Poor**)

526. On steep slope, drier area and poor site_____________ thinning should not be done. (**Heavy**)

527. Good sites may support_____________ number of trees than poor sites. (**Large**)

528. _____________ is defined as the planned interval in years which elapsed between successive thinning in the same crop. (**Thinning cycle**)

529. Thinning cycle is shorter during _____________period of growth. (**Early**)

530. Thinning cycle is_____________during middle aged to mature. (**Longer**)

531. Thinning intensities is described as light or heavy according to_____________ of trees that are removed. (**Number**)

532. _____________law is used to estimate the proportionate distribution of trees in selection forests. (**De liocourts law**)

533. The removal of less valuable trees in a crop with the interest of better growth of more valuable individual is called_____________. (**Improved felling**)

534. The practice of forestry with the object of obtaining the maximum in volume and quality of products per unit area through the application of best techniques of silviculture and management is_____________. (**Intensive forestry**)

535. The theory and practice of all that constitutes, the creation, conservation and scientific management of forest and utilization of their resources is known as_____________. (**Forestry**)

536. Improved felling is usually carried out in mixed forest beyond the_____________ stage. (**Sapling**)

537. Improvement felling is usually prescribed in forests which are in Condition. (**Poor**)

538. Intensity of light on the southern aspects compare to northern aspects will be. (**Higher**)

539. The exotic tree which has natural hybrid in India is_____________. (**Eucalyptus**)

540. _____________ is defined as the elimination of branches in order to obtain trees with clean bole. (**Pruning**)

541. The elimination of branches by physical and biotic agencies of the environment is called _____________. (**Natural pruning**)

542. The simplest method of obtaining natural pruning is to develop and maintain _____________ stocking in the main crop. (Dense)

543. Clear bole trees on shorter rotation can be obtained by_____________ pruning. (**Artificial**)

544. In artificial pruning, the use of ______ has been found useful than bladed instruments. (**Hand saw**)

545. In young plantation, artificial pruning should be carried out up____________ of the total height of the tree. (**1/2 to 3/5**)

546. The object of pruning is to obtain____________ free timber. (**Knot**)

547. Renewal a forest crop by natural or artificial means is known as____________. (**Regeneration**)

548. A species which is native to a specified region is called____________. (**Indigenous**)

549. Climbers are particularly harmful when the tree are____________. (**Young**)

550. The natural forest influenced by human activity it may or may not be a climax forest is known as____________. (**Virgin forest**)

551. The mature crop is removed in a series of operation in which the first stage is seeding felling and the last stage is the final felling is referred as. (**Shelter wood system**)

552. A survey for the assessment of established and unestablished regeneration generally by sample enumeration is____________. (**Regeneration survey**)

553. .Any unrooted portion of a plant used for grafting or budding on a rooted stock is called ____________. (**Scion**)

554. A rooted plant on which a scion is grafted is called____________. (**Stock**)

555. The removal of less valuable trees in a crop in the interest of better growth of the more valuable individuals is____________. (**Improvement felling**)

556. To mark with stakes the position of plants in a plantation is called as____________. (**Staking out**)

557. Counter trenches are usually made on slopes up to____________ per cent. (**20%**)

558. A thinning in which the trees to be cut are selected by some rule of thumb is called ____________ or otherwise called as. (**Mechanical thinning or Stick thinning**)

559. The method of thinning in common use that consists in the removal of inferior individuals of a crop is called as____________ or otherwise called as____________. (**Ordinary thinning or German thinning**)

560. A method in which thinning is primarily directed to the dominant trees in a regular crop is known as____________ or otherwise known as. (**Crown thinning or French thinning**)

561. Free thinning is developed by____________. (**Heck**)

562. Advanced thinning is developed by____________. (**Craib**)

563. The stand density index was developed by____________. (**Reineke**)

564. The extent to which the crop is thinned is known as____________. (**Thinning intensity**)

565. The whole set of thinning carried out in a crop from the earliest stage to maturity is called as____________. (**Thinning regime**)

566. CVP index refers to____________,____________and____________. (**Climate, Vegetation and Productivity**)

567. The silvicultural system which is adopted for bamboo forest is. (**Culms selection system**)

568. Pure forest are those which cover per cent of the crop consisting of a single species. (**>80 %**)

569. ____________and____________ classified the Indian forests into major sixteen groups in 1968. (**Champion and Seth**)

570. A plant which tends to be restricted to calcareous soil is called as____________. (**Calciphile**)

571. A technique of propagation plants under a controlled environment using aseptic plant cells, tissues in an artificial growing medium is called____________. (**Micropropagation**)

572. The micropropagation is synonymous of____________. (***In vitro* propagation**)

573. The cotyledons are pushed upwards by the rapid elongation of the hypocotyls is called ____________ type of germination. (**Epigeous**)

574. Cotyledons are remaining in the soil or just in its surface of earth is called type of germination. (**Hypogeous**)

575. Sal one of the important timber yielding tree species found in India up to altitude of ____________ ____________ Feet. (**4000-5000**)

576. The tree species popularly known as rose wood is____________. (***Dalbergia latifolia***)

577. The scientific name of teak skelitonizer is____________. (***Eutectona machaeralis***)

578. *Anogeisus latifolia* belongs to family____________. (**Combretaceae**)

579. *Leucaena leucocephala*, an important foreign exotic tree species is native of. (**Central America/ Mexico**)

580. The gregarious flowering of *Dendrocalamus strictus* occurs after a period of years. (**30-45**)

581. The *Acacia auriculiformis* is harvested after____________ years. (**10-12**)

582. The ETP technique of planting is generally adopted for____________. (**Poplar**)

583. The silvicultural system adopted generally for light demander species is____________. (**Clear felling/Clear Strip**)

584. The system used for forest situated in inaccessible areas is____________. (**Selection system**)

585. The minimum height of bamboo culms cutting for silvicultural system is cm. (**15**)

586. On the basis of systems of concentrated regeneration silvicultural system is divided into clear felling system and____________. (**Shelter wood system**)

587. The group selection system belongs to the system of____________. (**Diffused regeneration**)

588. The clear felling system, areas having____________ are clear felled in one operation. (**Equi-productive**)

589. The first stage of regeneration felling under a shelter wood system is____________. (**Seedling felling**)

590. The areas allotted to a PB is equal to the ratio of production of felling series and years in the period to____________. (**Rotation**)

591. Felling cycles under Indian conditions should be of____________ years in selection system. (**10 to 20**)

592. The shelter wood coppice system was developed by____________. (**Trevor**)

593. In CWS system the crop produced in the under wood is____________ in nature. (**Even-aged**)

594. The silvicultural system generally practiced for *Syzygium cumini* is_____________. (**CWS system**)

595. The forest developed from seedling crop is known as_____________. (**High forest**)

596. Species which are the hosts of lac insects are generally adopted_____________ system. (**Coppice with reserve system**)

597. *Avicennia* in mangrove forest is practiced under silvicultural system of _____________. (**Clear felling**)

598. Seeding felling is the first stage of regeneration felling under_____________ system. (**Shelter wood**)

599. Indian conditions felling cycles in selection system should be of_____________ years. (**10 to 20**)

600. The forest developed from root suckers is known as_____________. (**Coppice forest**)

601. Gayer's line character is generally seen in the system of_____________. (**Group**)

602. In CWR system the crop produced is_____________in nature. (**Uneven-aged**)

603. Improvement felling comes under the silvicultural system of_____________. (**Accessory**)

604. The wedge system was developed by_____________. (**Eberhard**)

605. The time which elapses between main felling on the same area is known as_____________. (**Felling cycle**)

606. The time taken for conservation of one silvicultural system to another is years. (**100**)

607. The casuality replacement during the first year planting is also known as_____________ or _____________. (**Beating up or infilling**)

608. Singling operation is carried out in a plantation where the seedlings with stem and attained height of more than_____________ meter. (**Forked/ multiple & 2m**)

609. The meaning of "Taungya" is_____________ and was originated in the country of _____________. (**Hill cultivation & Myanmar**)

610. Nurse crop are generally adopted to protect the____________ and____________ tree species against adverse climatic conditions. **(Shade bearer & Frost tender)**

611. Weeding and pruning operations in a plantation are normally operated during the season of____________ and____________ respectively. **(Rainy & Winter)**

612. In ordinary and crown thinning____________and____________ type of trees are removed respectively in the first thinning operation. **(Diseased & Dominant)**

613. Generally double trench is made in____________ and____________ areas with small amount of soil. **(Stony & Slopy)**

614. Teak stumps are prepared with keeping shoot and root length ofcm andcm respectively. **(2-3 & 20-22)**

615. Conifer seedlings are generally planted at the age of____________ to____________ years. **(2.5 to 4.5)**

616. Thinning is an felling in an immature stand for the purpose of improving theand ____________ of the trees that remain without permanently breaking the canopy. **(Growth & Form)**

617. Thick climbers are controlled by cutting, first at____________ and____________ m above the first cut. **(Base & 1m)**

618. Suitable thinning method adopted for Deodar and Chir Pine is____________ and ____________ respectively. **(Ordinary & Crown)**

619. Two tree genera mainly preferred for afforestation in cold desert areas of India are ____________ and____________. **(Salix & Poplar)**

620. *Pinus roxburghii* and *Tectona grandis* are generally planted at spacing of____________ m^2 and____________m^2 respectively. **(3×3 & 2×2)**

621. Dibbling method is usually adopted for the tree species of____________ and. **(*Juglans regia & Quercus incana*)**

622. Pre-monsoon planting will not be successful if it is dry in the month of. **(May-June)**

623. Bare root planting should be done at the growing season period of____________. **(Beginning)**

624. In a 4 acres land Eucalyptus is planted by quincunx method of 2m × 2m spacing, then total number of plants established will be___________. **(5000)**

625. The total length of stumps of different species varies between_____________. **(20-25 cm)**

626. Selective thinning is adopted for the forest of_____________. (**Uneven-aged**)

627. The removal of less valuable trees in a crop in the interest of better growth of the more valuable individual is known as_____________. (**Improvement felling**)

628. Codominant and dominant trees usually attains the height of predominant trees of_____________and_____________respectively. (**5/6th & 3/4th**)

629. The most promising raw material tree species, for pulp and paper industries are _____________ and_____________ _____________. (**Poplar & Eucalyptus**)

630. Energy plantation are raised with a spacing of less than one meter and rotation period of _____________ years. (**2-4**)

631. The tree species recommended for afforestation on salt affected soil in India is. ***(Butea monosperma)***

632. During Autumn and Winter best way to identify trees will be to compare _____________ and_____________ characteristics. (**Twig & Bark**)

633. On deciduous trees_____________ is the most important identification feature in the dormant winter period. (**Bark**)

634. The Indian rubber plant is_____________. (***Ficus elastica***)

635. _____________ is the name of Indian laburnum tree. (***Cassia fistula***)

636. The living fossil tree is_____________. (***Ginko biloba***)

637. pine is called living fossil. (**Wollemi pine**)

638. Sal is belongs to family_____________. (**Dipterocarpaceae**)

639. The term commercial forest means that the land is capable of producing _____________ _____________ trees. (**Economically useful**)

640. Clear strip system is otherwise known as ________ system. (**Progressive strip**)

641. In coppice system the new crop originates mainly from ______ coppice and rotation of coppice in short. (**Stool**)

642. The main objective of coppice of two rotations system is to produce _____ and poles. (**Timber**)

643. Area with common forest occurrence, _______ system is practiced. (**Shelter wood coppice**)

644. CWR system is not applicable when the valuable species are ________ in nature. (**shade bearers**)

645. In _____ system, fellings are carried out on the principle of selection system. (**coppice selection**)

646. Generally the crops in clear felling systems are raised by the _____ method. (**Taungya**)

647. In Punjab shelter wood system, trees up to ______ cm diameter are retained as part of future crop. (**40**)

648. In modified shelter wood system, regeneration fellings are done in ______ direction in strip wise. (**East-west**)

649. *Prosopis juliflora* is excellent, for _______ plantations in semi-arid and marginal lands. (**Energy**)

650. Weeding for two year old plantation is carried out _____ times in a season in general. (**two**)

651. The trench size of sunken-mound technique is of ______ and part of the dug up soil is put back in the trench. (**1.8 m × 30 cm × 30 cm**)

652. *Azadirachta indica* and *Santalum album* may also be propagated through _____. (**root sucker**)

653. Deodar is found generally in association with _____ in rich, moist soil with high elevation areas. (**Spruce**)

654. *Acacia nilotica* is a _____ as well as drought hardy tree species.(**frost tender**)

655. The first plantation of _______ was raised in the Nilgiri hills of peninsular India. (***Eucalyptus globulus***)

656. *Gmelina arborea* managed by ______ system. (**Coppice with standard**)

657. *Schleichera oleosa* is generally used for ____ culture. (**Lac**)

658. Unit of silviculture is ________. (**Stand**)

659. Seed dispersal in oak is by agent _______. **(Gravity)**

660. Good seed year for sal is _______ years. **(2-3)**

661. Sal seeds remains viable for ______ days. **(7)**

662. An ectomycorrhizal fungus forming symbiotic relationship with most of the temperate trees is_____________. **(Pisolithus)**

663. olume of soil under the influence of roots of growing plants is referred to as. **(Rhizosphere)**

664. The site of nitrogen fixation in Frankia associated plants is_____________. **(Vesicles)**

665. Instead of root hairs, pines have_____________ attached to the rootlets which absorb water and nutrients. **(Mycorrhizae)**

666. A species of less value than the principal species is termed as_____________. **(Accessories species)**

667. Without light, plant become pale yellow and have long thin internodes, a condition known as_____________. **(Etiolation)**_____________

668. When regeneration is obtained from seed forms then the crop is called as. **(Seedling crop)**

669. The shoot portion of seedling keeps on dying year after year while the root portion remains alive this phenomenon is termed as_____________. **(Dying back of seedling)**

670. Oak species are permanently found in_____________ forests. **(Temperate broad leaf)**

671. ___________ is one of the most important single factor which determines the distribution of vegetation. **(Temperature)**

672. A species which are capable of persisting and developing under shade is called __________. **(Shade bearer)**

673. The species which require abundant light for their optimum growth and development are called __________. **(Light demander)**

674. The species which require shade atleast in early stages for their optimum growth and development are called __________. **(Shade demander)**

675. A reserved forest is an area with complete protection, constituted according to chapter _________ of Indian Forest Act-1927. **(II)**

676. A protected in an area with limited degree of protection, constituted according to chapter __________ of Indian forest act-1927. **(IV)**

677. __________ forest is a stable forest assigned to a village community under the provision of Chapter III of Indian forest act-1927. **(Village)**

678. CVP index (climate, vegetation & productivity index was given by __________ in 1956. **(Paterson)**

679. The direction towards which a slope faces is defined as __________. **(Aspect)**

680. When a stem shows irregular innovation and swelling it is called __________. **(Fluting)**

681. The symbiotic association between certain fungi and roots of higher plants which enhances the uptake of water and nutrients is called __________. **(Mychorrhiza)**

682. From the time the young tree reaches 1m height till the lower branches begin to fall and having dead bark on the stem is called __________. **(Sapling)**

683. The ratio of Sand: Soil : FYM should be __________ for soil filling polybags in Nursery. **(1 : 3 : 1)**

684. __________ refers to area where the mother trees of the seeds are growing. **(Provenance)**

685. __________ species is one of that grows naturally in the country or in a region. **(Indigenous)**

686. __________ species is one which is grown outside the limits of its natural range. **(Exotic)**

687. The plant whose stem is always green and tender and height is usually not more than one meter is __________. **(Herb)**

688. The plants which have naked seeds are belong to the subdivision of __________. **(Gymnosperms)**

689. The plant in which the seed are enclosed in the fruits are belong to a sub-division of __________. **(Angiosperms)**

690. The average height of a regular crop is determined by __________. **(Loreys formula)**

691. The estimated light used in photosynthesis is less than __________. **(2%)**

692. The relative length of day and night to which the plant is exposed is called as __________. **(Photoperiod)**

693. Frost occurring on nights with clear sky produced by loss of heat by radiation is called as __________. **(Radiation frost)**

694. A frost produced by cold air brought from elsewhere is called as __________. **(Advective frost)**

695. A definitely relatively localized necrotic lesion primarily of the bark and cambium is known as __________. **(Canker)**

696. A plant that can subsist in dry situation is termed as __________. **(Xerophyte)**

697. A plant whose normal habitat is neither very wet nor very dry is called as __________. **(Mesophyte)**

698. A plant whose grows in moist sites is __________. **(Hygrophyte)**

699. A plant which grows wholly or partially immersed in water is __________. **(Hydrophyte)**

700. An organism that draws a part or whole of its nourishment from another living organism is known as __________. **(Parasites)**

701. The progressive dying usually back towards from the tip of any portion of the plant is known as __________. **(Die back)**

702. The annual or periodical removal of exploitable trees in an uneven aged forest is known as __________. **(Selection felling)**

703. All cuttings and grafts derived from one tree are collectively called as __________. **(Clone)**

704. The killing of young seedling by certain fungi that cause decay of the stem or roots is called as __________. **(Damping off)**

705. Turning a crop or plant parts into manure under while green or soon after maturity for the purpose of soil improvement is called as __________. **(Green manuring)**

706. The planned interval which elapses between successive thinning in the same area is called as __________. **(Thinning cycle)**

707. A cutting through back and outer living layers of wood in a continuous incision all round the bole of a tree is known as __________. **(Girdling)**

708. In coppice system generally the rotation of the coppice is __________. **(Short)**

709. Woody climber is known as __________. **(Liana)**

710. Symbiotic bacteria is __________. (**Rhizobium**)

711. __________ percentage of regeneration of a sample plot could be taken for a successful regeneration of that area. (**60-70**)

712. The beef wood is also known as ________. (Casuarina)

713. Sucking roots of parasite is known as __________. (**Haustoria**)

714. The cotyledons are pushed upwards by the rapid elongation of the hypocotyls is called __________ type of germination. (**Epigeous**)

715. New terminal growth on pines is called _________. (**Candles**)

716. Basal area of woodlot represents __________. (**Stand Density**)

717. Litter from leaves, especially from needles, at various stages of decomposition is called ____________. (**Duff**)

718. The function of root hairs is to __________. (**Absorb water & nutrients**)

719. Instead of root hairs, pines have _________ attached to the rootlets which absorb water and nutrients. (**Mycorrhizae (Fungi)**)

720. _______ and _______ are the main constituents of wood. (**Cellulose and Lignin**)

721. Seed formation depends on the amount of reserve ___________ and __________ in a tree. (**Carbohydrates and Nitrogenous materials**)

722. Quantity of seed required for a nursery bed can be worked out by __________. (**W = (A × D × 100)/ (P × N)**)

723. The term 'forest succession' was first used by __________. (**Cawles**)

724. The father of Indian Forestry is ___________. (**Dietrich Brandis**)

725. ___________ is the upper limit of the tree growth beyond which the growth of the trees is not seen. (**Tree Line**)

726. Rain forest occurring above the cloud level is known as ________. (**Cloud Forest**)

727. In India the tree line is found around __________ from MSL. (**3500M**)

728. The decrease in diameter of the stem of a tree from the base upwards is known as __________. (**Taper**)

729. Flame of the forest is ___________. (**Butea monosperma**)

730. The seed leaves of flowering plants are known as __________. (**Radicals**)

731. Driving force of ecosystem is ___________. (**Solar Energy**)

732. __________ group of plants possesses both animal and plant characteristics. (**Myxomycetes**)

733. ______ colour of light is least effective in driving photosynthesis. (**Green**)

734. Koppen classified world climate into __________ group. (**Five**)

735. The relation of a site in relation to sun and wind is called __________. (**Exposure**)

736. The structure of the surface layer of the soil is called _______. (**Soil tilth**)

737. A sequence of soil derived from the same age and same parent material is called _________. (**Catena**)

738. Kanker pan is commonly found in ___________ soil. (**Alluvial**)

739. Succcession takes place where ther is no previous vegetation is called ___________. (**Primary succession**)

740. A succession which takes place after the destruction of the whole or part of the original vegetation is called __________. (**Secondary**)

741. The whole process where a plant establishes itself in a new area from germination to reproduction is called __________. (**Ecesis**)

742. Grouping followed by establishment of scattered colonizing invaders as a result of propagation is called __________. (**Aggregation**)

743. The effect of vegetation on the site is called __________. (**Reaction**)

744. A major unit of vegetation comprising the climax communities of an area uniform in its major physiognomic features is termed as _________. (**Formation**)

745. Mosic theory was proposed by __________. (**Aubreville**)

746. Theory of vegetation gradient and climax patterns was proposed by _________. (**Whittiaker**)

747. The climax which owes distinctive characters due to climatic factors is referred as __________. (**Climatic climax**)

748. A plant which differs from the climatic clima due to the influence of special soil factors is defined ___________. (**Edaphic climax**)

749. The plant community immediately preceding the climatic climax is called _________. (**Pre climax**)

750. A plant community more exacting than the climatic climax is defined __________. (**Post climax**)

751. A climax which differs from the climatic climax of the area due to biotic factors is referred as __________. (**Biotic climax**)

752. Vegetation first classified according to temperate zones by __________. (**Schimper**)

753. Classification of vegetation based on thermodynamic principles was done by ________. (**Shanbhag**)

754. A year in which a given species bear seed abundantly called as _________. (**Seed year**)

755. In a year which a given species produces abundant first year seedling is called __________. (**Seedling year**)

756. The height of stumps in simple coppice system for Eucalyptus hybrid is ______ cm. (**15-25**)

757. Wagner's Blender Saumschlag is otherwise called as __________ systems. (**Strip selection system**)

758. The form or character of the new cropproduced in any silvicultural systems depends on __________. (***Spacing***)

759. In culm selection system, the felling cycle will be ________ years. (**Four**)

760. Chipko movement is related to __________ conservation. (**Forest**)

761. Conservation of species in its natural habitat is ________. (**In-situ**)

762. Development of botanical garden is ____ type of conservation. (**Ex-situ**)

763. ____________ system is called as Andaman's canopy lifting shelterwood system. (**Indian Irregular shelterwood systems**)

764. Deodar forests of Utter Pradesh and Himachal Pradesh are generally worked under _______ system. (**Indian Irregular shelterwood systems**)

765. CIFOR stands for ______. (**Center for International Forestry Research**)

766. CIFOR is located at ______________. (**Bagor, Indonesia**)

767. ___________ silvicultural system is followed for black wattle in the nilgiris. (**Clear felling followed by natural regeneration**)

768_____ colour in the regeneration map represents good regeneration. (**Blue**)

769. Sandal wood forest is worked under ____________ systems. (Natural Selective)

770. ______ colour in the regeneration map represents fair regeneration. (**Red**)

3

Forest Mensuration

1. Callipers are used to measure_____________. (**Diameter**)
2. Biltmore stick is used to measure_____________. (**Diameter**)
3. Wedge prism with 5 BAF and tallied 50 numbers of trees/ acre then basal area per acre is_____________ __________. (**250**)
4. A ring count is most accurate in aging tree by using increment borer on _____________ of the trunk. (**Lower region**)
5. False ring is a/an_____________ ring for a given year. (**Second**)
6. A formula to determine diameter quotient for each form class is called_____________. (**Hojer's formula**)
7. The instrument used to determine the age of a tree is_____________. (**Pressler's increment borer**)
8. _____________ ,_____________ and_____________ are the variables required for general volume table preparation. (**Diameter, height and form**)
9. The sampling method used in non-random sampling are_____________ ,_____________ and _____________. (**Selective, Systematic and Sequential**)
10. _____________ formula is used for determining increment per cent. (**Schneider's**)
11. Diameter or girth measurements are taken in India is_____________ m from ground level. (**1.37**)
12. _____________ sampling consist of choosing samples according to the subjective judgement of the observer. (**Selective**)
13. _____________ stage sampling is commonly used in forest survey. (**Two**)
14. Absolute form quotient was developed by_____________. (**Tor Jonson**)

15. The metric system is introduced in India during____________. **(1962)**
16. Sampling intensity of tropical moist deciduous forest is____________ per cent. **(2.5%)**
17. One inch is equal to____________ centimetre. **(2.54)**
18. One foot is equal to____________ inches. **(12)**
19. ____________instrument is used to measure height or diameter at any height of a tree. **(Dendrometer)**
20. Limit of standard timber is upto__________diameter over bark. **(20 cm)**
21. Stump analysis is done to estimate____________of felled tree. **(Age of tree)**
22. Stem analysis is done to estimate for____________. **(Past rate of growth)**
23. ____________ is a measure of density and is claimed to be independent of site quality and stand age. **(Crown competition factor)**
24. Commonly used reducing factor in India is____________. **(Site quality)**
25. The growth of a stand is a function of____________ and____________ ____________. **(Stand density and site quality)**
26. ____________ is the number of trees per hectare represented by each tree tallied. **(Tree factor)**
27. A series of sampling points are selected randomly or systematically distributed over the entire area to be inventoried is called as____________. **(Horizontal point sampling)**
28. Basal area of standing trees is calculated by____________. **(Wedge prism/ Spiegel Relaskop)**
29. On sloppy land tree height is measured in____________ side. (Up-hill)
30. The volume of logs in India are calculated using quarter girth formula, which is ____________. **$\{(g/4)^2 \times L\}$**
31. Hypsometer, Clinometer, Haga Altimeter, Spiegel Relaskop are the instruments used for measuring the____________ of trees. **(Height)**
32. The average of all CAI is known as____________. **(MAI)**
33. When the CAI is equal to the MAI, the MAI is____________. **(Stationary)**
34. Schneider's formula is____________ for finding the increment per cent of standing tree. **(400/nD)**

35. Where the density is not uniformly normal, but is variable, a correction factor is called_____________ _____________. (**Reducing factor**)

36. The yield is regulated by periods and prescribed, as nearly as possible by_____________ only. (**Volume**)

37. _____________ of tree implies, cross sectional area taken at breast height. (**Basal area**)

38. One cubic meter of timber is equal to_____________ cft. (**35.3147**)

39. Girth of a tree is always recorded in_____________ and to the nearest of _____. (**Meters, Centimetres**)

40. Standard length of log is meter except the last section. (**3**)

41. The formula for basal area is_____________. (**$\pi d^2/4$**)

42. _____________ Form factor, the basal area is measured at breast height. (**Artificial form factor**)

43. Girth/diameter and diameter/girth ratio for a circular bole is_____________ _____________ and _____________ respectively. (**3.142 & 0.3182**)

44. Considering a leaning tree towards observer as straight gives_____________ estimate of height. (**Over**)

45. "Beyond the canopies" is motto of_____________ institute. (**CIFOR**)

46. The vertical length of a tree from the tip to the lowest green branch on the bole is known as_____________ _____________. (**Crown length**)

47. Artificial form factor is also known as ________form factor. (**Brest height**)

48. One square meter is equal to_____________ square feet or_____________ sq. yards. (**10.76 & 1.195**)

49. Distribution and representation of age and/or size classes of trees in a stand is called _____________. (**Stand structure**)

50. In India, multiple yield table was prepared for_____________ _____________ species. (***Cedrus deodara***)

51. The theory of limiting density or stand density index was given by_____________. (**Reineke**)

52. Quarter girth formula gives_____________ per cent of tree volume. (**78.5**)

53. ____________showing the distribution of stems by diameter class for each series of crop diameters. (**Stand table**)

54. In India commonly used diameter and height class are____________ and____________ ____________ respectively. (**2,5,10 cm & 1,3,5 m**)

55. The ratio between the mid diameter and dbh of a tree is known as____________ ____________. (**Form quotient**)

56. Quarter girth formula is otherwise known as____________ ____________. (**Hoppus rule**)

57. Bark thickness can be estimated by____________ ____________ instrument. (**Swedish bark gauge**)

58. ____________ ____________ instrument is particularly used for vertical point sampling. (**Conimeter**)

59. It is a measure of the relative productive capacity of a site for a particular species is called as____________. (**Site quality**)

60. Vertical point sampling was developed by____________. (**Hirata**)

61. Densitometer is used to measure____________. (**Canopy area**)

62. The method of fractional quality was developed by ____________. (**Lauri**)

63. Crown volume of a pine tree can be calculated by formula ____________. (**$\pi D^2 H/12$**)

64. Timber calculators generally give idea about ____________. (**Volume of logs**)

65. The future growth of stands can be predicted by ____________. (**Yield Table**)

66. In India, minimum number of trees used for general volume table preparation is ______. (**1000**)

67. Form class of trees is expressed by ____________. (**Form Quotient**)

68. The sampling intensity adopted for tropical moist deciduous forest is ______%. (**2.5**)

69. Plot radius factor in horizontal point sampling method is used for checking of________. (**Borderline trees**)

70. Table of basic average is calculated during the preparation of ____________. (**General Volume Table**)

71. If BAF value is 4 then value of constant 'k' is __________. **(0.04)**

72. The formula which gives maximum error for volume estimation of a log is __________. **(Smalian's Formula)**

73. The most convenient method for age determination of standing trees is ________. **(Increment Boring)**

74. Form class of trees is expressed by ____________. **(Form Quotient)**

75. The most reliable method for past rate of growth determination of tree is __________. **(Stem analysis)**

76. The value of fractional quality lies between _________. **(0 to 2)**

77. The instrument used for upper stem diameters measurement by using laser technology is __________. **(Criterion Dendrometer)**

78. The unit of form height is _________________. **(meter)**

79. If the value of D/ G is 0.2024, then the shape of the tree at measuring site is _____. **(Irregular)**

80. In leaning condition, the DBH of trees is measured in the side of _____________. **(Along the tree stem)**

81. The theory which justifies the tree form is _____________. **(Metzger/ Girder/ Mechanistic theory)**

82. The height class generally used when trees are high of 15m to 25 m is _________. **(3 m)**

83. The volume of hemispherical shape crown can be calculated by using formula _______. **($\pi D^2L/12$)**

84. Mensuration word derived from _______ word. **(Latin)**

85. For conoid shape tree/ stem, the form class generally used is _________. **(0.50)**

86. The preferred value of breast height form factor for cubic paraboloid shape tree is _________. **(0.6)**

87. Basal area of a tree is generally measured at __________. **(1.37 m)**

88. Diameter of a tree should be measured when buttress is extended upto 1.4m is. **(Slightly above 1.4 m)**

89. The average of crown length and crown width is ____________. **(Crown diameter)**

90. If diameter is 20 cm then girth of that tree will be ________ cm. (**62.8**)
91. The length of pole used for single pole method is of _______ m. (**1.5**)
92. Sum of crown length and crown height is the __________. (**Tree height**)
93. Horizontal Point Sampling is a type of sampling where the probability of selecting a tree is proportional to its ______. (**Basal area**)
94. The artificial form factor of a conoid shape tree ranges between _____. (**0.3 to 0.45**)
95. A log is said to be cull when it is less than ____ per cent sound. (**50**)
96. Calibration distance factor is _____ times of plot radius factor. (**1000**)
97. The height of dominant and _____ trees is considered for site index calculation. (**Codominant**)
98. A log is said to be cull when it is less than ___ ft long and ___ inch in diameter respectively. (**8 & 6**)
99. In vertical line sampling trees selected with probability proportional to ____ only. (**height**)
100. Form height is the product of _____ and height of the trees. (**form factor**)
101. The solid volume of firewood can be calculated by using _____ method. (**xylometric or specific gravity**)
102. Past growth of trees can be determined by_________. (method of control)
103. _____ sampling method in forestry is used for test of hypothesis in research. (**Sequential**)
104. Money yield table is generally prepared from________. (**Stand table**)
105. The height class more frequently used are of_______. (**1m, 3m and 5m**)
106. The value MAI is always ______during the growth period of trees. (**Positive**)
107. The height of trees also be determined by non- instrumental method is ______. (**Single pole/ Shadow method**)
108. Non-instrumental methods for height measurement of trees are based on principles of __________. (**Geometric principle of similar triangles**)
109. Stand table is used for determination of _______ crops. (**Financial value**)
110. Tariff table is prepared by taking ______ as a variable. (**GBH**)

111. A tree is cylindrical bole when the form factor value is nearer to ____. (**one**)

112. A tree should be harvested at _______ stage, where there is maximum volume of wood to be produced. (**Culmination**)

113. Fuel woods are generally sold on the basis of _______. (**weight**)

114. Fiber saturation point the moisture content lies between _____ per cent. (**20 to 30**)

115. The volume loss in sawn timber due to quarter girth formula is_____________per cent. (**36.4**)

116. In absolute form factor basal area is measured at_____________ height of the trees. (**Convenient**)

117. The length of logs for transportation is of_____________ meter. (**4.5**)

118. The parameter determined by vertical point sampling of trees is_____________. (**Height**)

119. The age of standing trees can be determined by the method of_____________. (**Increment boring**)

120. The future growth of stands can be predicted by_____________. (**Yield table**)

121. The full form of 3P is_____________. (**Probability Proportional to Prediction**)

122. The DBH of a leaning tree in sloping ground is measured in_____________ side. (**Uphill**)

123. Total height of tree is equal to sum of crown length and_____________. (**Crown height**)

124. The tree species for which multiple yield table is prepared in India is_____________. (***Cedrus deodara***)

125. The sampling intensity used for tropical wet evergreen forest is_____________ per cent. (**10**)

126. The shape of sampling plot generally used in India is_____________. (**Rectangular**)

127. The width of strips in strip enumeration is of___________ meter. (**20/40**)

128. The variation in volume of aggregate check shouldn't exceed_____________. (**≤ 1%**)

129. The size of circular sample plots should be of____________ hectare. **(0.05 to 0.1)**

130. When the crop density is 1.5 then it is known as____________. **(Over stocked)**

131. According to 'Girder theory' the growth materials of tree are deposited at____________. **(Base)**

132. The sampling size decreasing in each phase of forest inventory is known as ______.____________ **(Multistage sampling)**

133. If g = GUB, g'= GOB and t = bark thickness then GOB is equal to____________ .**(g' = g + 2πt)**

134. The volume of standard stem timber is calculated up to____________. **(20 cm DUB)**

135. The standard length of last section of log is not more than____________ meter. **(4.5)**

136. The value of fractional quality lies between____________. **(0 to 2.0)**

137. Stem analysis in trees is done to determine____________. **(Past rate of growth)**

138. In artificial form factor basal area is measured at____________of the tree. **(1.37m)**

139. The vertical point sampling method was developed by____________. **(Hirata)**

140. The age of felled trees can be determined by the method of____________. **(Stump analysis)**

141. The future growth of stands can be predicted by____________. **(Yield table)**

142. The full form of PPS is____________ ____________. **(Probability Proportional to Size)**

143. One cubic feet (ft^3) is equal to____________ cubic meter (m^3). **(0.0283)**

144. In volume table preparation relative check difference shouldn't exceed____________. **(3.0)**

145. The type of thinning grade adopted for single yield table preparation is____________ .**(C-grade)**

146. By taking girth as variable the volume table prepared is known as____________. **(Tariff table)**

147. The form of trees can be studied by_____________. **(Form Factor/ Form Quotient)**

148. Flury calipers are used to measure_____________. **(Diameter only)**

149. The maximum spread of the crown along its widest diameter is_____________. **(Crown width)**

150. Timber calculators generally give volume of round timber for length of_____________. **(1 to 25 m)**

151. The number of bamboo culms in the forest can be estimated by_____________ ________ sampling method. **(Multiphase sampling)**

152. Crown ratio is the ratio between_____________ and height of the tree. **(Crown length)**

153. Dendrometer is generally used for measurement of_____________ of trees. **(Upper stem diameters)**

154. The measurement point of a tree is circular when d/g is equal to_____________. **(0.3182)**

155. The height class used for matured trees of heights more than 25m is_____________. **(5 meter)**

156. The method of fractional quality was developed by_____________. **(Lauri)**

157. The volume loss in round timber due to quarter girth formula is_____________ per cent. **(21.5)**

158. The parameter measured in vertical point sampling is_____________. **(Height)**

159. Smythie's hypsometer measure height correctly up to_____________ meter. **(30)**

160. 5 cft of wood is equal to_____________ cum. **(0.141)**

161. The most reliable method for past rate of growth determination of tree is_____________. **(Stem analysis)**

162. The conversion factor multiplied to get stacked volume to solid volume in Oak is _________. **(0.50)**

163. The number of sample trees selected for biomass estimation of a stand is_____________. **(30)**

164. The variation in volume of relative check shouldn't exceed_____________ per cent. **(3)**

165. In Indian condition, the girth class used for medium sized tree is______________. **(15 cm)**

166. The maximum value of form factor should be less than______________. **(One)**

167. Usual diameter classes used in India are______________. **(2,5,10 cm)**

168. Usual girth class for small, medium and large trees are used such as______________. **(5, 15 and 30 cm)**

169. Forest mensuration aims at relative accuracy with respect to and______________. **(Time and money)**

170. Standard volume tables in forestry are based on______________variables. **(Two)**

171. Distance between ground and crown point is known as______________. **(Bole height)**

172. In India breast height is measured at______________ meter, where as in Europe, UK, and other commonwealth countries measured at______________. **(1.37 m and 1.3 m)**

173. Proportion of error involved in use of calliper and tape is __________. **(21:1)**

174. Instruments used for height measurements based on geometric principle are ______________ ______________ ,______________ and______________. **(Altimeter, Clinometers and Hypsometer)**

175. Measurement of the lean of tree was given by______________. **(Krishnaswamy)**

176. Height from ground level upto point of average diameter over bark is 20 cm is called as______________. **(Standard timber bole height)**

177. Instruments working based on similar triangles methods are______________ ______________, ______________ ______________ and______________ ______________. **(Christens hypsometer, improvised callipers and smithies hypsometer)**

178. The instruments which is used to measurement of diameter at any convenient height is ______________ ______________ ______________. **(Wheeler pentaprism)**

179. One acre is equal to ___________ square feet. **(43,560)**

180. DBH is measured______________ ______________ feet from the ground. (**4.6**)

181. ________________number of cubic feet in a cord. (**128**)

182. ________________ cubic inches are there in a broad foot. (**144**)

183. A chain is a unit of land measurement________________ feet in length. (**66**)

184. The basal area or cross-sectional area of trees at breast height is commonly used to determine the________________of a stand of trees. (**Degree of stocking**)

185. Estimating the amount of board feet, cords, or pieces in a timber stand is called ________________ ________________. (**Cruising**)

186. ________________ tool is used for timber cruising. (**Wedge prism**)

187. The haga altimeter will measure________________. (**Tree height**)

188. Rate of taper of a log is called________________. (**Form**)

189. Form factor is used for estimate the________________ of a standing tree. (**Actual volume**)

190. Form factor gives idea about laws of________________. (**Growth**)

191. Basal are is measured at constant proportion 1/10th , 1/20th etc. For ________________________factor. (**Normal/true form**)

192. Intervals in which the range of form quotients of trees is divided for classification and used is called________________. (**Form class**)

193. Diameter taper table otherwise called as________________. (**Ordinary taper table**)

194. All logs are________________ meter length except the top end which is generally not more than________________ meter. (**3 and 4.5 m**)

195. Smalian's formula which over estimates the volume of tree is________________. (**S1+S2/2 *L**)

196. Huber's formula for volume estimation of log is________________. (**Sm * L**)

197. Quarter girth formula exploits the relationship between________________ and________________ ________________ of wood. (**Girth and Basal area**)

198. Quarter girth formula is________________ ________________. **{(g/4)^2 * l}**

199. A log of marketable size which is made unmarketable because of rot, defect or poor form is called a________________. (**Cull**)

200. Sawn timber volume are given upto a thin end diameter to which conversion is done is called_____________ _____________. (**Sawn outturn table**)

201. Volume table gives volume in round to various thin end diameter is called. (**Assortment table**)

202. In India height of a tree is used for classification of_____________. (**Site**)

203. _____________ number of sample plot is required in each thinning regime for one species. (**600**)

204. Minimum size of sample plot is_____________ hectare. (**0.1 to 0.2 ha**)

205. Selection of sample trees should not be less than_____________ trees. (**Twelve**)

206. Horizontal point sampling was developed by_____________. (**Bitterlich**)

207. Vertical point sampling was developed by_____________. (**Hirata**)

208. Site quality concept was given by_____________. (**Paterson**)

209. Site quality expression with a decimal subdivision is called as_____________. (**Fractional quality**)

210. Standard volume table gives volume of standard timber which include volume of _____________ _____________. (**Stump**)

211. A hand held instrument used for measuring tree heights, slopes and vertical angles is _____________. (**Clinometer**)

212. A wedge prism is used to measure the_____________ of trees in a stand. (**Basal area**)

213. _____________is the weight of complete trees including roots, stump, stem, limbs, branches and leaves. (**Biomass**)

214. The most accurate method of selling pulpwood is by_____________. (**Weight**)

215. Chains are there in a_____________ mile. (**80**)

216. Square chains are there in an_____________ acre. (**10**)

217. A broad foot is_____________ cubic inches. (**144**)

218. Yield table is not applicable to_____________ forest. (**Unevenaged**)

219. _____________ is a tabular statement which summarizes per unit area basis all the essential data relating to the development of a fully stocked

and regularly thinned even aged crop at periodic intervals covering the greater part of its useful life. (**Yield table**)

220. Single yield table is applicable only to__________. (**C grade thinning**)

221. Portion of the tree stem which is unmerchantable is called____________. (**Cull**)

222. Merchantable height can be estimated in the field with a____________. (**Hypsometer**)

223. An area 10 square chain in size equals to____________ acre. (**1**)

224. A bark gauge measures____________ of the tree. (**Bark thickness**)

225. The tools that will be most useful in locating your position in the forest would be a/an ____________. (**GPS receiver**)

226. When purchasing a wedge prism you find each has a number, such as 5, 10, 15, etc. these numbers refer to____________ ____________. (**Basal area factors**)

227. When performing a forest inventory with a wedge prism you are selecting trees by ________. (**Diameter**)

228. A wedge prism is used in a type of inventory called____________. (**Point sampling**)

229. A table showing the distribution of stems by diameter classes is called as____________. (**Stand table**)

230. Christens hypsometer is used to measure the____________. (**Height**)

231. ____________ is a method of sampling in which the number of observation in the sample is not determined in advance but sampling units are taken successively from a population. (**Sequential sampling**)

232. ____________sampling units are selected according to a predetermined pattern without resource to random selection. (**Systematic**)

233. Simple angle gauge is used during ____________ sampling process. (**Horizontal Point Sampling**)

234. Normal form quotient was proposed by ________. (**A. Schiffel**)

235. The most accurate formula for determination of volume of log is ______. (**Newton's formula**)

236. Tor Jonson formula used for determination of ______ of trees. (**Form**)

237. Hojer's formula used for determination of ______ of trees. (**Form**)

238. Behre's formula used for determination of ________. (**Tree form**)

239. Timber calculator otherwise known as ________. (**Timber Measurement Table**)

240. Timber calculator gives idea about ______ of round as well as sawn timber. (**Volume**)

241. Timber calculators are prepared by using ______ formula for volume estimation. (**Quarter girth formula**)

242. Tree quality volume table was developed by __________. (**Chaturvedi**)

243. Tree quality volume tables are applicable for ______ forests, (**Mixed**)

244. Vartex Laser Hypsometer is generally used for measurement of _______ trees, (**Height**)

245. Improvised calliper is used for measurement of ______ trees. (**Height**)

246. Electronic clinometer is used for measurement of ________ trees. (**Height**)

247. HEC-R clinometers are generally used for measurement of ______ trees. (**Basal Area**)

248. Flurry callipers are used for measurement of ________. (**Diameter**)

249. Ruler is used for measurement of _________ of trees. (**Upper stem diameter**)

250. Wheeler pentaprisms normally used for the measurement of ______. (**Upper stem diameters**)

251. Christen's hypsometer measures correctly height of trees up to _____ m. (**30**)

252. For preparation of local volume table _____ variable is to be measured. (**DBH**)

253. Volume table prepared by using three variables in the country of ______. (**Sweden or England**)

254. Volume table based on three variables otherwise called as ________. (**Form class volume table**)

255. Form in the form of ________ is considered as third variable for preparation of volume table. (**Form quotient**)

256. The most important variable used for preparation of volume table is ______. (**Diameter**)

257. Tariff tables are generally used in _________. (**Europe**)

258. Indian condition volume tables are prepared by taking ______ and ______ as variables. (**Diameter, Height**)

259. True form factor also known as __________. (**Pressler's normal form factor**)

260. Basal area for absolute form factor is measured at ________. (**Convenient height**)

261. _____________ is the most important measurement from the point of view of forest management. (**Volume**)

262. The standards of weights and Measure Act was formed in the year ___________. (**1956**)

263. The circumference or perimeter of a circle is ___________. (**2πr**)

264. ___________ is the universal adopted standard height for measuring girth, diameter and basal area of standing trees. (**Breast Height**)

265. Diameter measurements should be recorded in ___________ and to the nearest multiple of ____________. (**Centimeters and Two millimetres**)

266. D.O.B can be converted into D.U.B by deducting __________ bark thickness. (**Twice**)

267. G.O.B can be converted into G.U.B by the formula= ____________. (**g = g'-2πt**)

268. The formula of form factor is ____________. (**F=V/ (S × h)**)

269. The formula for Form quotient is ___________. (**mid diameter/d.b.h**)

270. The ratio of the diameter of a stem at any given height to its breast height diameter is called ____________. (**diameter quotient**)

271. The horizontal projection on the ground of the tree crown is referred as ___________. (**Crown cover**)

272. ___________ formula is used to calculate the error in the volume calculated by Smalian or Huber formula. (**Newton's**)

273. The difference between volumes obtained by Smalain's and Newton's formula is ____________. (**$\Delta = (1/3) \times (s_1 + s_2 - 2Sm)$**)

274. The difference between volumes obtained by Smalain's and Newton's formula is ____________. (**$\Delta = (1/3) \times (s_1 + s_2 - 2Sm)$**)

275. A table showing a species's average contents of trees, logs or sawn timber for one or more given dimensions is called ___________. (**Volume table**)

276. Diameter corresponding to the mean basal area of a uniform, generally pure crop is called ___________. (**Crop diameter**)

277. Diameter corresponding to the mean basal area of a group of trees or a stand is called ___________. (**Mean diameter**)

278. Boring is done at _________height above ground in case of smaller tree and at ___________ for bigger tree. (**30 cm and d.b.h**)

279. Diameter corresponding to the mean basal area of the _________ number of biggest trees per hectare in a uniform, generally pure crop is called top diameter. (**250**)

280. Height corresponding to the men diameter of a group of trees or the crop diameter of a stand is called ___________. (**Mean height**)

281. In India, sample plots are generally thinned to ________ grade ordinary thinning. (**C or D grade**)

282. Volume of a sample plot can be calculated by using a computer programme named as ___________, which is used in FRI for *Eucalyptus hybrid* sample plot volume calculation. (**FORTRAN**)

283. To determine optimum intensity in various forests the sampling error should be less than ___________ %. (**10**)

284. National Inventory Design cocept was given by ______________. (**FSI**)

285. The difference between the population mean and sample mean is referred as ____________. (**Sampling error**)

286. Basal area is expressed in __________ while land area is expressed in __________. (**m² & hectare**)

287. The formula of basal area factor of wedge prism is ___________. (**BAF= 10,000/ {1 +4(L/D)²}**)

288. A slope angle of 25° cause an error of abbot ________ % in counting of tally trees by wedge prism. (**10**)

289. The wedge prism is held in a ___________ position in level terrain and at ________ to the line of sight on sloping ground. (**Vertical, right angle**)

290. ___________ instrument is widely used in point sampling. (**Spiegel Relaskop**)

291. Tele Relaskop is used to calculate accurate _________ & __________ of single standing tree. (**Height & Diameter**)

292. 1 diopter is equal to an angle of ____________ or _________. (**0.57° or 34.36 minutes**)

293. The branch of forestry which deals with the determination of dimension, form and age and increment of single trees, stands or whole woods either standing or after felling is called ____________. (**Dendrometry**)

294. Volume of stem wood in round between 20cm and 5 cm diameter over bark is called ____________. (**Standard stem small wood**)

295. Spectral signature of green vegetation in a satellite captured image is _________ colour. (**Red**)

296. Indian Institute of Remote Sensing is located at __________. (**Dehra Dun**)

297. The ideal time for taking Aerial photograph is ___________. (**9 to 11 AM**)

298. One acre is equal to ___________ square feet.(**43560**)

299. General volume tables which are applicable to large geographical area are based on ___________ variables. (**Three**)

300. A log rule is a table or formula used to determine tree___________. (**Volume**)

301. Land cover classification using three levels of the scale of 1:100000 was developed by ____________ agency. (**NRSA**)

302. Relationship of height of the form point above ground level to the total height of the tree is _______________. (**Form point ratio**)

303. A tree which has more than one leader stem is called __________. (**Forked**)

304. Site index is the relation between _________ and __________. (**Age & Height**)

305. The situation of objects with reference to site factors such as climate, topography and soil is called as ___________. (**Location**)

4

Forest Management, Policy and Law

1. The Indian Institute of Forest management is located at_____________. **(Bhopal)**
2. First forest policy in independent India was enacted in_____________. **(1952)**
3. A plan of operations for the management of the forest is known as_____________. **(Working plan)**
4. _____________ are the yields harvested as a matter thinning operation. **(Intermediate yields)**
5. Indian Wildlife Protection Act was enacted in the year_____________. **(1972)**
6. The present National Forest Policy envisages having_____________per cent of forest and tree cover. **(33%)**
7. Forest conservation act was enacted in the year_____________. **(1980)**
8. Indian forest act was enacted in the year_____________. **(1927)**
9. Density of tree in a forest stand is normally expressed as_____________ _____________. **(Basal area of the stand in square feet.)**
10. A forest balance between timber growth and harvest is referred to as_____________ timber growth. **(sustain yield)**
11. A log rule is a table or formula used to determine tree_____________. **(Volume)**
12. The most expensive timber management activity is_____________. **(Site preparation)**
13. Determine site index the two measurements needed are_____________ and_____________. **(Age & height)**
14. Highest authority of forest in state is_____________. **(PCCF)**
15. Joint forest planning and management scheme has been passed in NFP of_____________. **(1988)**

16. 1988 policy stress on_____________. **(Environmental stability)**
17. According to NFP, 1952 the minimum forest area for environmental stability should be_____________ _____________ per cent. **(33)**
18. _____________ type of forest shows legal ownership. (Reserve forests)
19. MAI of Indian forests is_____________. **(0.7-0.9 m^3/ha)**
20. World average MAI is ___________. **(2.1 m^3/ha)**
21. Trespass in reserved forest would be punishable under_____________ of IFA, 1927. **[26(1)]**
22. The forest act which is in force today was amended in the year_____________. **(1927)**
23. Basic prerequisites to achieve the objective of sustainable yield is_____________, _____________ and_____________. **(Normal increment, age gradation, growing stock)**
24. Thinning which stimulates CAI in light demander is_____________. **(Crown thinning)**
25. Smallest functional territorial unit is_____________ which is the foundation of Indian forest administration. **(Beat)**
26. Smallest permanent working plan unit in India is_____________. **(Compartment)**
27. _____________ _____________ law is used to estimate the proportionate distribution of trees in selection forests. **(DeLiocourt's Law)**.
28. In a normal forests with normal series of age gradations and normal increment at the end of rotation is equal to_____________. **(r × i)**
29. Working plan is generally prepared for the period of_____________ years. **(10)**
30. Von Mantel's formula for the determination of annual yield is_____________. **(Y_a=2NGS/r)**
31. Stock map are prepared by working plan ranges with of_____________. **(1:15,000)**
32. Average annual increment for any short period is called as_____________. **(Periodic Annual Increment)**
33. The CAI and MAI cross each other at the_____________. **(Rotation age)**

34. Broad principle on the forest of India are managed is based on_____________. (**Sustain yield**)

35. Swiss method of yield regulation is mainly used for_____________ only. (**Increment**)

36. Public forests in which state governments have a proprietary interest are divided in three categories namely_____________ ,_____________ _____________ and_____________. (**Reserved forest, Village forest and Protected forest**)

37. Land rights to forests declared to be reserved forests or protected forests are owned by_____________. (**Govt. of India**)

38. The national park and wildlife sanctuaries of India are declared by_____________. (**Central Govt.**)

39. The procedure for settlement of right for reserved, village and protected forest is provide in_____________of the Indian forest Act. (**Section 3-34**)

40. The recognition of Forest Rights Act states that non tribal forest dwellers are only those who have been living in the forest for_____________ years. (**75**)

41. The Elephant Preservation Act was enacted in_____________ year but is not in force. (**1879**)

42. The provision related to cattle trespass can be found in_____________ of Indian Forest Act 1927. (**Section 70 and 71**)

43. Cattle trespass Act was enacted during_____________ year. (**1871**)

44. The National Afforestation and Ecodevelopment Board (NAEB) was set up in the year _____________. (**1992**)

45. The compensatory Afforestation fund bill was introduced in the year_____________. (**2008**)

46. CAMPA stands for_____________. (**Compensatory Afforestation Fund Management and Planning Authority**)

47. The 1952 forest policy suggested that the minimum one third of the country's total land area under forest with_____________ in hills and_____________ in plains (**60% and 20%**)

48. _____________ type of rotation are recognised in forestry. (**Six**)

49. The bulk of Indian forests are_____________ per cent irregular. (**About 70**)

50. The first national forest policy was formulated based on the recommendation given by_____________. (**Dr. Voelker**)

51. _____________ number of national forest policy so far implements in our country. (**Three**)

52. The first Indian forest act was drafted in the year_____________. (**1865**)

53. Forest conservation act was enacted in the year. (**1980**)

54. Conservation of ecological fragile ecosystem and preservation of biological diversity in terms of flora and fauna was emphasized by_____________forest policy. (**1988**)

55. National Forest policy, 1988 envisages that_____________ per cent of the total geographical area of India should be under forest. (**33%**)

56. The forest are brought on concurrent list by 42nd amendment to the constitution in the year_____________. (**1976**)

57. Theft in forest can be charged under the section (IPC)_____________. (**378**)

58. Declaration of National parks is covered in_____________ section of wildlife protection act,1972. (**Section 35**)

59. The amount of timber or other produce which may be removed from the forest, annually or periodically is_____________. (**Yield regulation**)

60. If the annual felling exceed the annual growth, the growing stock or the working capital is_____________. (**Reduced**)

61. _____________is a very important unit in the management and administration of forests. (**Range**)

62. In shelter wood system of management annual coupe is called_____________. (**Periodic Block**)

63. Unit of working plan is_____________. (**Division**)

64. The constitution of a working plan division is generally decided by the_____________. (**Conservator of forest**)

65. Preparation of stock maps in the scale of_____________. (**1:15000**)

66. Preparation of management maps in the scale of_____________. (**1:50000**)

67. Preparation of regeneration survey map in the scale of____________. **(1:5000)**

68. Preparation of working plan map in the scale of____________. **(1:60000)**

69. Preparation of enumeration map in the scale of____________. **(1: 50000)**

70. Preparation of forest type map in the scale of____________. **(1:50000)**

71. In regular shelter wood system age class are recognised by____________ ____________. **(Area methods)**

72. The Von Mantel's formula is otherwise known as____________ ____________. **(Formula of glorious simplicity)**

73. ____________ ____________ is the working plan officer. **(Deputy Conservator of Forests)**

74. The Environment (Protection) Act is attributed to the year____________. **(1986)**

75. First working plan in India was prepared for____________ forest. **(Nilambur)**

76. First Indian forest act was drafted in the year____________. **(1865)**

77. Joint forest management concept was first implemented at____________. **(Arabari)**

78. Joint forest management was first introduced in____________. **(West Bengal)**

79. ____________ play a vital role in the implementation of JFM. **(People participation)**

80. Wildlife Protection Act of 1972 was amended during the year____________. **(1991)**

81. The measure of immunization against communicable diseases of livestock is taken within____________ kilometre of sanctuary. **(5)**

82. Trade of commerce in wild animals, animal articles are tropics dealt in____________ section of wildlife protection Act, 1972. **(Section 39)**

83. The officer in-charge of administrative unit section is____________. **(Forester)**

84. Legally the forest in the country are classified as____________ and_____________. (**Reserved forest and Protected forest**)

85. Forest ranger under_____________ section has got the power for evicting any person unauthorised occupying any land in reserve forest. (**68A**)

86. Age at which MAI is maximum and equals CAI is termed as_____________. (**Rotation of maximum volume production**)

87. Concept of progressive yield was started in Germany by_____________. (**Hartig**)

88. Scientific forest management in India was started by_____________. (**Brandis**)

89. The concept of "normal forest" was first introduced by_____________ foresters. (**German**).

90. In Howards modification crop is measured down to diameter corresponding to the rotation of_____________. (**Half**)

91. The choice of the type rotation will depend on the_____________. (**Object of management**)

92. The principle of normal forest is_____________. (**Sustain yield**)

93. Forest and wildlife were brought under concurrent list by the parliament in 1976 in seven schedule of_____________ amendment. (**42nd**)

94. Smallest functional territorial unit_____________ _____________. (**Beat**)

95. In India working plan period generally followed by_____________ year. (**Ten**)

96. Working plan aimed for_____________ principle. (**Sustained yield**)

97. The practical application of the scientific, technical and economical principles of forestry is called_____________ _____________. (**Forest management**)

98. Under Section 20 of the Indian forest Act, 1927 deals with the_____________ _____________. (**Reserved forest**)

99. Functional classification of forest is mentioned in the_____________ forest policy. (**1952**)

100. It shall be the duty of every citizen of India to protect and improve the natural environment including forests, lakes, rivers and wildlife and to

have compassion for living creatures, so article____________ of the constitution of India does it appear. (**Article 51 A**)

101. ____________ is the smallest permanent working plan unit of management. (**Compartment**)

102. The main territorial division of the forest is called____________. (**Block**)

103. The unit of forest management is called____________. (**Working circle**)

104. Working circle may be divided into____________. (**Felling series**)

105. Annual felling areas is otherwise called as____________. (**Coupe**)

106. ____________ ____________ is the backbone of forest management. (**Sustained yield**)

107. Felling series in uniform system is called____________ ____________. (**Periodic block**)

108. In forest, the standard colour given for the regeneration status Excellent (81-100 %) is ____________. (**Green**)

109. In forest, the standard colour given for the regeneration status Fair (21 - 40 %) is____________. (**Red**)

110. In forest, the standard colour given for the regeneration status Deficient (0 - 20 %) is____________. (**Black**)

111. The material that a forest can yield annually in perpetuity is known as____________. (**Sustain yield**)

112. Sustain yield principle is applicable to____________ ____________ forestry. (**Production**)

113. A written scheme of management aiming at continuity of policy, controlling the treatment of a forest is called____________. (**Working plan**)

114. An age class with one year as the interval is called____________. (**Age gradation**)

115. Common practice in forestry is to adopt a combination of____________ ____________ and ____________. (**Maximum volume production and financial rotation**)

116. The period which a forest crop takes between its formation and final felling is known as ____________. (**Rotation**)

117. National forest policy, 1988 envisages that____________ per cent of the total geographical area of India should be under forest. (**33**)

118. Social forestry programme was launched in the year plan. (**Fifth five year plan**)

119. ____________type of rotation is useful for aesthetic and recreational purpose. (**Silvicultural rotation**)

120. ____________ percentage of forest owned by central government. (**95**)

121. GATT stands for____________ ____________. (**General Agreement on Trade and Tariff**)

122. GATT was set up in the year____________. (**1984**)

123. Masson's formula is also known as____________ or____________ ____________. (**Exploitation per cent or Masson's ratio**)

124. Masson's formula for yield is____________. (**200/r**)

125. Masson's formula is based on____________. (**Volume**)

126. Smythie's safe guarding formula is based on____________. (**Number of trees**)

127. Smythie's Modification formula for yield is____________ ____________. (**2v/r-x**)

128. Normally felling cycle in selection forests is____________ years. (**5-10**)

129. Rotation is applicable only to____________. (**Regular crops**)

130. The rotation which coincides with the natural lease of life of a species on a given site is called____________. (**Physical rotation**)

131. ____________ is the rotation through which a species retains satisfactory vigour of growth and reproduction on a given site. (**Silvicultural rotation**)

132. ____________is the rotation in which a species yields the maximum material of a specified size for economic conversion. (**Technical rotation**)

133. Howard's formula for yield regulation is____________. (**8v/3r**)

134. Brandis diameter class method is otherwise called as____________. (**Indian method**)

135. Pressler's formula (Quality increment percent)to find out increment per cent is. (**V-v/V+v * 200/n**)

136. Third world forestry congress was held in____________. **(Helsinki)**

137. ____________ is called the trinity of Norms for normal forest. **(Osmaston)**

138. Technical rotation is adopted particularly for____________. **(Industrial firms)**

139. National parks and Wildlife sanctuary comes under____________ section of wildlife protection act, 1972. **(u /s 35)**

140. Man and Biosphere programme was started in the year____________. **(1971)**

141. The challans of forest offences are prepared under section____________. **(26)**

142. The "Chapter of Indian Forestry" was based on the report given by____________. **(McClelland)**

143. Forest ranger under____________ section has got the power for evicting any person unauthorized occupying any land in reserve forest. **(68A)**

144. In working plan, tropical wet evergreen are coloured with____________. **(Violet)**

145. The forests are brought on concurrent list by 42nd Amendment to the constitution in the year____________. **(1976)**

146. The yield from a normal forest is called____________. **(Normal yield)**

147. Government of India brought forests and wildlife into the concurrent list in the year ____________. **(1976)**

148. The unit of yield regulation in regular forest is ____________. **(Annual Coupe)**

149. The rotation which yields the highest net return on the invested capital is ____________. **(Price Increment)**

150. Declaration of 'Reserved Forest' is possible by IFA, 1927 under sections ____________. **(3 to 27)**

151. Functional classification of forest was given in the National Forest Policy of __________. **(1952)**

152. The increase in price, independent of quality increment is known as ______________. **(Economic/ Financial increment)**

153. First forest policy in India after independence was enacted in the year____________. **(1952)**

154. The period required to regenerate the whole of a P.B. is called _______. **(Regeneration period)**

155. The sum of all the trees growing in the forest, or a specified part of it is called as _____. **(Growing Stock)**

156. Need for the sustained supply of timber and other forest produce was stated in NFP _____________. **(1988)**

157. Working plan will be generally prepared for the period of _______ years. **(10)**

158. Point of intersection of CAI and MAI curves indicate _______. **(Maximum volume production)**

159. National Commission on Agriculture submitted report in the year ____. **(1976)**

160. In irregular shelter wood system of management yield regulation is carried out by ________ basis. **(Volume)**

161. Indian Forest Act contains, _________ number of sections. **(86)**

162. Increment per cent of a stand can be determined by using _____ formula. **(Pressler's formula)**

163. Conservation of natural heritage is one of the basic objectives of NFP ______. **(1988)**

164. Single periodic block is also known as___________________. **(Floating Periodic Block)**

165. Biolley's Check method is generally used for determination ______________ of forest. **(Increment/ Yield)**

166. An age class with one year age interval is called as _________________. **(Age- gradation)**

167. The total increment up to a given age divided by that age is called ______________. **(Price Increment)**

168. Rotation of a forest crop is otherwise known as __________________. **(Production period)**

169. The total increment up to a given age divided by that age is called ______________. **(MAI)**

170. Indian Forest Act of 1927 has total ________ number of chapters. (**13**)

171. ______________is also considered as sustained yield, provided that the period is short. (**Periodic Yield**)

172. Smallest functional territorial unit of administration is _________. (**Beat**)

173. The amendment by which forest and wildlife were brought into the concurrent list is _____________. (**42**)

174. The term rotation strictly applicable to ________ crop. (**Regular**)

175. Indian Penal Code was enacted in the year ________. (**1862**)

176. In Indian condition, working plan generally prepared for the period of ____________ years. (**10**)

177. The head of the forest department of a state is ______. (**PCCF**)

178. The compartments are preferably designated by numerals of__________. (**Arabic**)

179. Reserved forest declared under forest act of_____________. (**1927**)

180. Protected forest declared under forest act of_____________. (**1927**)

181. Village forest declared under forest act of_____________. (**1927**)

182. A forest is to be declared as protected forest under sections_____________ of IFA, 1927. (**29-34**)

183. A forest is to be declared as reserved forest under sections_____________ of IFA, 1927. (**3-27**)

184. A forest is to be declared as village forest under section _____________ of IFA, 1927. (**28**)

185. A sanctuary is declared under_____ act. (**Wildlife Protection Act, 1972**)

186. __________ % of Indian forests are under the state ownership. (**95.8%**)

187. Head quarter of Inspector General of Forest is situated in ___________. (**New Delhi**)

188. __________ rotation applicable in case of protection and amenity forest, park lands, roadside avenues. (**Physical**)

189. The formula of Soil Expectation Value is _____________. (**S_e= X/0.0p**)

190. The period during which a change from one silvicultural system to another silvicultural system is effected is called __________. (**Conversion period**)

191. Conversion period is usually ________ than the rotation period. (**Less**)

192. A forest which has normal increment, normal distribution of age gradation and normal growing forest is called ________. (**Normal Forest**)

193. Mean annual increment of fast growing species is ___________. (**10 m^3/ ha**)

194. Determination of increment in irregular crop is done by using __________ formula. (**Andre's**)

195. Growing obtained from the __________ is more accurate than that obtained from the MAI formula. (**Yield table**)

196. For determination of Normal Growing Stock in selection system __________ evolved a formula which was based on CAI. (**Munger**)

197. The diameter or girth decided upon as the normal size for felling in order to fulfill the objects of management is called ___________. (**Exploitable size**)

198. The FSI conducts forest cover survey once in every _________ years. (**Two**)

199. In India __________ method was widely adopted for fuel wood plantation.(**Annual coupe by gross area**)

200. Floating Periodic Block method is not suitable for ___________ forests. (**Irregular**)

201. Mason's formula is also known as ______________. (**Exploitation percent or Masson's ratio**)

202. In French method of yield regulation, rotation 'r' is divided into three equal parts viz. young, medium and old in the ratio___________. (**1:3:5**)

203. ICFRE was established during the year ___________. (**1986**)

204. In fast growing species if younger plantation, the height increment must not be less the __________ per annum. (**60 cm**)

205. Cumulative weightage value of 0-20 makes the regeneration status __________ in regeneration survey. (**Deficient**)

206. ___________ country has recorded maximum growing stock per ha. (**Brazil**)

207. The scheduled tribes and other traditional forest dwellers Act was enacted in _______ year. (**2006**)

208. _________ are forest lands owned by Government but not constituted into reserved, protected or village forest. (**Unclassed forests**)

209. ______________ is the head of the forest administration in India. (**Inspector General of Forest**)

210. Felling series is divided into number of _____________. (**Annual coupes**)

211. The main objective of yield regulation is to achieve ___________ from the forests. (**Sustained yield**)

212. Presslers formula to find out increment percent is ____________. [**{(V-v)/(V+v)} × {200/n}**]

213. Howard's modification is based on __________ only. (**Volume**)

214. It is a measure of the relative productive capacity of a site for a particular species is called as ___________. (**Site quality**)

215. In working plan, Himalayan dry temperate forest are colored with _________. (**Yellow**)

216. In working plan, tropical thorn forest are colored with _________. (**Red**)

217. Burma formula for yield is ______________. (**9v/4r**)

218. Floating periodic block is also known as _____________. (**Quartier Blue Method**)

219. The discount rate at which NPV is zero is called ___________. (**IRR**)

220. Joint forest management evolved in early _________. (**1970**)

221. The project tiger was launched in the year ___________. (**1973**)

222. World forest day is celebrated on ___________. (**21st March**)

223. Bombay National History Society was started in the year _________. (**1983**)

224. National park and wildlife sanctuary comes under ___________ section of wildlife protection act, 1972. (**u/s 35**)

225. The measures of immunization against communicable disease of livestock are taken within __________ km of sanctuary. (**5 km**)

226. The Indian Board of Wildlife (IBWL) was set up by the government of India in the year __________. (**1952**)

227. GATT (General Agreement on Trade and Tariff) was set up in the year _________. (**1984**)

5

Forest Product Utilization

1. The process of harvesting, conversion and disposal of standing forest produce is called _____________. (**Forest utilization**)
2. Palmarosa grass oil is known as_____________. (**Geranium Oil**)
3. *Terminalia chebula* fruit tans is commonly called as_____________. (**Myrobalans**)
4. The gum resin_____________ is obtained from *Garcinia morella*. (**Gamboges**)
5. Black dammer is_____________. (***Canarium strictum***)
6. Rock dammer is_____________. (***Hopea odorata***)
7. White dammer is_____________. (***Vetaria indica***)
8. True dammer is_____________. (***Agathis loranthifolius***)
9. The Indian copal tree is_____________. (***Vetaria indica***)
10. The Indian gum arabica or babul gum is obtained from_____________ species. (***Acacia nilotica***)
11. Gurjan is obtained from_____________ species. (***Dipterocarpus turbinatus***)
12. Agar wood oil extracted from_____________. (***Aquilaria agallocha***)
13. Mint oil obtained from_____________. (***Mentha arvensis/Mentha piperita/Mentha citrata***)
14. The valerian oil is obtained from_____________ parts of plants. (**Root**)
15. The kuth oil is obtained from_____________. (***Saussuria lappa***)
16. *Dendrocalamus strictus* is referred as_____________. (**Male bamboo**)
17. _____________ is the biggest of the Indian bamboo. (***Dendrocalamus giganteus***)
18. _____________ is the Indian butter tree. (***Madhuca latifolia***)

19. _____________ part of the chinchona plant which yields quinine. (**Bark**)
20. The NTFP are abundant in_____________forest. (**Dry deciduous forest**)
21. NTFP contributes_____________ per cent of the total revenue from forests. (**60%**)
22. The NTFP are also called_____________. (**NWFP/MFP**)
23. Timber which are capable of withstanding rapid seasoning are termed as_____________. (**Non-refractory**)
24. _____________ species is the world's best aircraft timber. (***Ochrama pyramidalis***)
25. _____________ and_____________ that attacks timber in sea water are collectively called shipworms. (**Crustaceans and Mollusca**)
26. _____________number of bamboo species are found in India. (**136**)
27. Henna dye is obtained from_____________ species. (***Lawsonia innermis***)
28. _____________ gum is obtained from *Lannea coromandelica*. (**Jhingan gum**)
29. _____________ system is profitable for maximum and sustained in lac production. (**Coupe**)
30. The wastages in felling and conversion will be_____________ in hilly region. (**More**)
31. Logging training centre was established at and established by_____________. (**Batote and H.G. Winklemann**)
32. The transport constitutes_____________per cent of the total cost of delivery. (**60 –70**)
33. The characteristics of red colour resulting from chewing of a pan is due to the combination of_____________ and_____________. (**Katha and lime**)
34. The commercial resins are collected from_____________. (**Artificial wounds**)
35. The leaves of several species of Indigofera contain a soluble colourless glucoside is known as_____________. (**Indican**)
36. To impart smoothness to paper the dry web is passed through a tack of caste iron rolls is known as_____________. (**Calenders**)

37. Kokam butter is obtained from____________ species. (***Garcinia indica***)
38. Addition of non-fibrous mineral matters to stock prior to sheet formation is known ____________. (**Loading**)
39. The resin exuded by *Shorea robusta* is known as____________. (**Ral**)
40. *Caesalpinia sappan* yields a red dye is known as____________. (**Brazilin**)
41. ____________ myrobalan is obtained from *Emblica officinalis*. (**Emblic**)
42. The beleric myrobalan is obtained from____________ species. (***Terminalia bellirica***)
43. ____________ myrobalan is obtained from the *Terminalia chebula*. (**Chebulic**)
44. Bark of *Terminalia arjuna* content____________ per cent of tannin. (**20-24**)
45. Konnai is the bark of____________species. (***Cassia fistula***)
46. Waxes are usually found on the____________ of the leaves and fruits. (**Epidermis**)
47. ____________ oil is obtained from the seeds of *Hydnocarpns kurzii* species. (**Chaulmugra**)
48. ____________ is known as bullet wood. (***Mimusops elengi***)
49. After desiccation the cane is bleached by fumigation with____________. (**Burning sulphur**)
50. The chief constituent of oil is obtained from *Citrus limonia* is____________. (**Citral**)
51. oil is obtained from the heart wood and berries of *Bursera delpechiana* species. (**Linoloe**)
52. The cinnamon oil is distilled from leaf and bark of____________ species. (***Cinnamomum zeylanicum***)
53. Camphor oil is obtained from the wood and leaves of____________ species. (***Cinnamomum camphora***)
54. *Eucalyptus globules* is commonly called as____________. (**Blue gum**)

55. Agar oil is distilled from the resinous portion of the wood of_____________species. (***Aquilaria agallocha***)
56. The chief constituent of sandal wood oil is_____________. (**Santalol**)
57. The citronella oil is obtained from_____________ species. (***Cymbopogon nardus***)
58. Lemon grass oil is obtained from_____________species. (***Cymbopogon flexosus***)
59. The common cane of north India is_____________ _____________. (***Calamus tenuis***)
60. The regularly tapping resin tree is_____________ species. (***Pinus roxburghii***)
61. The blue pine tree is known as_____________. (***Pinus wallichiana***)
62. *Anogeissu latifolia* is known as_____________. (**Axel wood**)
63. _____________ is known as ebony tree. (**Diospyros melanoxylon**)
64. Khair tree is_____________. (***Acacia catechu***)
65. _____________ is known as chilgoza pine. (***Pinus geardiana***)
66. Gum kino is obtained from_____________. (***Pterocarpus marsupium***)
67. _____________gum is derived from *Anogeissus latifoloa.* (**Dhaura**)
68. Salai gum is obtained from_____________. (***Boswellia serrate***)
69. Bengal kino is obtained from_____________species. (***Butea monosperma***)
70. _____________ and_____________ are the two important varieties of *Cymbopogon martini.* (**Motia and sofia**)
71. _____________ species is known as ben teak. (***Lagerstroemia lanceolata***)
72. _____________ is known as yellow silk cotton. (***Cochlospermum religiosum***)
73. *Bombax ceiba* is known as_____________ tree. (**Indian kapok**)
74. ___________ method is usually used for the extraction of stem fibre. (**Retting method**)
75. Separation of fibres along the grain is called_____________. (**Check**)

76. Patches of bark either partially or wholly embedded inside the wood is called_____________. (**Inbark**)

77. A partial or complete separation between adjoining layers of fibres is called_____________. (**Shakes**)

78. Tree that shows departure from its cylindrical shape is called_____________. (**Flute**)

79. A type of growth in which the fibres take a spiral course about the bole is called _____________. (**Spiral grain**)

80. Separation of fibres along the longitudinal directions is called_____________. (**Crack**)

81. Palmrosa oil is the product of_____________ species. (***Cymbopogon martini***)

82. The sisal fibre is produced from_____________ species. (***Agave** sisalana*)

83. Manilla hemp is a commercial name of *Musa textilis* and it is indigenous to. (**Philippines**)

84. Sun hemp is extracted from_____________species. (***Crotolaria juncea***)

85. The commercial name of cane is_____________. (**Rattans**)

86. The new method of resin tapping is_____________. (**Bore Hole method**)

87. Flowering pattern of *Melocanna bambusoides* is_____________. (**Gregarious**)

88. Shellack is obtained from_____________ insect. (***Laccifera lacca***)

89. Kino gum is obtained from_____________ species. (***Pterocarpus santalinus***)

90. _____________ bamboo is called as golden or yellow bamboo. (***Bamboosa vulgaris***)

91. The superior lac is produced from_____________ strain. (**Kusumi**)

92. Felling cycle of bamboo is_____________ year. (**3-4**)

93. Turpentine is obtained from_____________. (**Resin**)

94. The product of destructive distillation of wood and consists in burning wood without contact with the air is_____________. (**Charcoal**)

95. Lac is a resinous_____________of insects. (**Secretion**)

96. The mature lac is otherwise called as_____________. (**Brood lac**)

97. Fibres generally occur as_____________. (**Sclerenchyma**)

98. _____________ bamboo gives annual flowering. (***Ochlandra strictus***)

99. Resin from broad leaved species is known as_____________. (**Dammer**)

100. The method of resin tapping commonly adopted is the French is called _____________ method. (**Cup and lip**)

101. The champ oil is obtained from the flower of_____________ species. (***Michelia champaca***)

102. The_____________ dye is obtained from the heartwood of *Pterocarpus santalinus* species. (**Santaline**)

103. The annatto dye of commerce is extracted from_____________ species. (***Bixa orellana***)

104. The portion of axe head in front of the eye is known as_____________. (**Blade**)

105. The portion of axe head behind the eye is known as_____________. (**Back**)

106. Resin is obtained from pine tree by_____________ method. (**Tapping**)

107. Lac is a valuable resin, secreted by the lac insect_____________. (***Laccifer lacca***)

108. The best quality lac is produced from_____________ species. (***Schleichera oleosa***)

109. _____________ _____________is the improved method of katha and cutch extraction. (**Factory method**)

110. The alternative species for manufacturing of cutch and katha is_____________ species. (***Acacia chundra***)

111. The katha and cutch are obtained by boiling_____________ part of *Acacia catechu* in water. (**Heartwood**)

112. _____________ species yields good quality resin but is limited in distribution. (***Pinus kesiya***)

113. The lac larva come out from the lac encrustations in large numbers and the phenomena is called as_____________. (**Swarming**)

114. _____________ and_____________ are the two distinct strains of lac insects in India. (**Kusmi and rangeeni**)

115. The Indian lac research institute is situated at_____________. (**Namkun, Bihar**)

116. _____________ is the only species that yield oleo-resins which is regularly tapped in India. (***Pinus roxburghii***)

117. The principal colouring matter of the seeds of *Bixa orellana* is_____________. (**Bixin**)

118. Dyes of yellow and red shade are obtained from the roots of_____________ species. (***Punica granatum***)

119. National Newspaper and paper Mills established at_____________. (**Nepanagar**)

120. Total cost of felling, conversion and transport is approximately per cent of the total cost of timber in the market. (**50**)

121. The edge of the tooth which faces the cutting direction is called_____________. (**Face**)

122. The distance between the two adjacent teeth is known as_____________. (**Space**)

123. The entire opening between two adjacent teeth is called_____________. (**Gullet**)

124. The width of the saw cut is known as_____________. (**Kerf**)

125. The thickness of the blade is called_____________. (**Gauge**)

126. Bow man saw is_____________man saw. (**One**)

127. _____________ saw will have tooth pattern of reverse teeth cutting. (**Pruning**)

128. Wedges are used while felling to prevent_____________. (**Jamming**)

129. The best felling season is_____________. (**Winter**)

130. Felling should begin with_____________ of the slope. (**Top**)

131. _____________ is made to decide the direction of fall. (**Notch/Undercut**)

132. The undercut should be_____________ diameter of the tree. (**2/3**)

133. The rounded logs split with axe are called_____________. (**Hakries**)

134. A piece of short lengths with square surface is called_____________. (**Blocks**)

135. In India_____________ _____________ organisation is the responsible for grading of timbers. (**Indian Standards Institution**)

136. Intelligent choice of timber for specific purpose is called_____________. (**Grading**)

137. Most satisfactory system of felling and extraction is worked by_____________. (**Purchaser alone**)

138. _____________ is the system of sale where the amount fixed with the actual produce extracted or sold. (**Payment on return**)

139. Sale of forest produce based on individual consultation is called_____________. (**Sale by private bargain**)

140. _____________ is the system of sale where the exact quantity of produce is not known. (**Lump sum sale**)

141. _____________ is the method of sale of forest produce where keen competition exists. (**Sale to higher bidder**)

142. In quarter sawn timber_____________ will be exposed. (**Medullary rays**)

143. To extract timber from nallas where quantity of water is less,_____________ _____________ used. (**Wet slides**)

144. Major hazard in the river floating is_____________. (**Jamming**)

145. _____________ dye is obtained from *Lawsonia inermis*. (**Henna**)

146. _____________ species is most suitable for cricket bat. (***Salix alba***)

147. Silk worms are fed on_____________leaves. (**Mulberry**)

148. Preserved fodder in dry form is called_____________. (**Straw**)

149. In logging_____________ imply preparation of timber to commercial sizes. (**Conversion**)

150. _____________ products has the largest share of forest based industries. (**Paper and pulp**)

151. The objective of wood seasoning is_____________. (**Drying timber**)

152. Allelochemical produced by walnut is_____________. (**Juglone**)

153. Honey bee species which cannot be domesticated is ________. (***Apis dorsata***)

154. _____________ oil seed of plant species is being projected for use as a substitute for diesel. (***Jatropha curcus***)

155. Apiculture is the practice of_____________. (**Bee-keeping**)

156. In Uttarakhand, extensively distributed conifer tree i.e.____________ which yields turpentine. (***Pinus roxburghii***)

157. Non-timber forest produce are also referred as_____________. (Minor forest produce)

158. _____________ wood comprise of glued and compressed wood pieces of different species. (**Composite**)

159. Logging is referred as_____________ of trees. (**Felling**)

160. _____________ is extracted from heartwood of Khair. (**Katha**)

161. Bidi is made from leaf of_____________. (***Diospyros melanoxylon***)

162. Country liquor is distilled from_____________ flower. (**Mahua**)

163. The length of the chain is measured from of one handle to of other handle. (**Outside, inside**)

164. The tribes are generally dependent on_____________ for livelihood. (**Forests**)

165. Bamboo is an important long fibred cellulosic raw material for industries. (**Paper**)

166. The word 'buck' is often used in the discussion of forest harvesting activities which means _________. (**To cut trees into logs**)

167. The scaling diameter for saw logs is normally measured at_____________ _____________. (**The big end inside the bark**)

168. The term stumpage refers to the value of trees standing in a_____________. (**Woodlot**)

169. The practice of measuring the volume of logs or bolts after they are harvested is called _____________. (**Scaling**)

170. A tool with a pointed tip used to push logs is the_____________. (**Cant hook**)

171. Most cruising sticks use_____________ inch distance from the user's eye to the tool to get accurate reading. (**25**)

172. _____________ wood is used in fine furniture because of its 'bird's eye' grain. (**Sugar maple**)

173. _____________ is the most durable part of a wooden fence post. (**Heartwood**)

174. A standard cord of unsplit fuel wood contains_____________ cubic foot. **(128)**

175. The first step in forest harvesting is_____________. **(Felling and bucking)**

176. _____________ is cutting down tree into log length. **(Bucking)**

177. Selling standing timber is called a_____________ sale. **(Stumpage)**

178. A bolt is a log which is_____________than a standard log. **(Shorter)**

179. Logging costs typically represent_____________ per cent of the total cost of sawn timber delivered to a mill. **(33)**

180. The light coloured part of a cross section of a tree which is responsible for conducting water and minerals to the crown of the tree is the_____________. **(Sapwood)**

181. When timber buyers openly bid against one another at a timber sale is called as _____________. **(Auction sale)**

182. Diameter measurements for saw logs are taken at_____________ of the log. **(Thick end)**

183. The debris left after logging, such as limbs, tops, culls and knockdown trees is called _____________. **(Slash)**

184. _____________ refers to cutting width of a saw blade. **(Kerf)**

185. A table showing the estimated or calculated amount of lumber which may be sawed from logs or given length and diameter is called_____________. **(Log rule)**

186. The tax on owner pays after the timber land is sold is called_____________. **(Capital gains tax)**

187. The two most important uses for the nation's timber are_____________ and_____________. **(Sawlogs and pulpwood)**

188. _____________ involves encircling a tree stem with downward, overlapping axe cuts. **(Frilling)**

189. The most common method of measuring pulpwood today at the wood buying centres is in terms of_____________. **(Gross weight or tons)**

190. The leading country in growing softwood lumber in the world is_____________. **(Russia)**

191. Charcoal is an example of a____________wood product. **(Converted)**

192. Limbing a tree is generally performed prior to____________ and____________. **(Bucking and Skidding)**

193. The part of the cut that controls the tree's direction of fall is the____________. **(Undercut)**

194. A cant hook is used to____________the log. **(Roll)**

195. When using a chain saw to remove a large limb on a standing tree the first cut should be made on the____________ of the limb. **(Bottom)**

196. When using a chain saw, a 'kick back' is caused by____________. **(Wood binding)**

197. The area of the end of a log is called the____________ area. **(Basal)**

198. ____________ refers to the value of trees in standing condition. **(Stumpage)**

199. When a logger bids one total price for all the wood to be cut is known as ____________ Sale. **(Lump sum)**

200. To avoid pinching when removing a tree branch with a chain saw, the operator should make a____________ and____________ beyond the undercut. **(Limbing cut , top down)**

201. Stumps should be cut____________ to avoid waste. **(Close to ground)**

202. When felling a tree with a chainsaw the back cut should be____________ with the bottom of the open face notch. **(Level)**

203. A snag is a____________. **(Standing dead tree)**

204. A source of revenue for a logging operation is____________. **(Log sales)**

205. Any chemical purchased for commercial use will be accompanied by details relating to chemical make-up, first-aid instructions, proper disposal and health concerns, These information is contained in the____________. **(Material Safety Data Sheets, MSD Sheet)**

206. Workplace safety is overseen by OSHA which stands for____________. (Occupational Safety and Health Administration)

207. A ____________ is most likely used to handle pulpwood stick. **(Hookeroon)**

208. On hilly ground the tree should be felled in____________ side. **(Uphill)**

209. Reducing a felled tree into marketable size logs is called____________. **(Bulking)**

210. Oriented Strand Board (OSB) is made from____________. **(Wood chips)**

211. Occurrence of forest fire is common in ________type of forest. **(Dry deciduous)**

212. ____________ is used for dragging, lifting and rolling of logs. **(Log hook)**

213. From seasoning point of view felling is not carried out in____________ season. **(Summer)**

214. On hilly ground tree should be felled in____________ direction. **(Uphill)**

215. Estimated loss from the high stump height in logging is about____________ per cent. **(2-5)**

216. The rounded logs split with axe are called____________. **(Hakries)**

217. Square and rectangular, generally 15 cm in section is called____________. **(Scantlings)**

218. Minor transportation is also called____________ transportation. **(Off road)**

219. ____________ are used in felling trees and prevent jamming of the saw in cross cutting and longitudinal splitting. **(Wedges)**

220. Most common toothing is used in saw is____________. **(Peg toothing)**

221. The standard length of the axe handle is____________. **(90 cm)**

222. On the recommendation of____________ and____________ logging branch opened in 2 five year plan scheme in FRI, Dehradun. **(Dr. A. Huber, Dr. A. Koroleff)**

223. Most of the industrial timber comes from____________ forest. **(Conifer)**

224. Processed log (sown timber) is called____________. **(Lumber)**

225. ____________ species of grass is yielding palmarosa oil. **(*Cymbopogon martinii*)**

226. The botanical name of white dammer is____________. **(*Vateria indica*)**

227. ________ is the economical part of climbing lily. **(Tuber)**

228. The scientific name of muga silk is_____________. (***Antheraea assamica***)

229. The most important character for efficiency of fuel wood is_________________________. (**Biomass per unit area**)

230. *Pinus ruxburghii* produces mainly ________. (**Resins**)

231. Guggal is obtained from_____________. (***Boswelia serrate***)

232. _____________ is the technology provides conservation of raw dust into smoke less fuel. (**Briquetting**)

233. Aspirin was first developed from the bark of_____________. (**Willows**)

234. The capacity of motor truck is generally_____________ tons. (**1.5-5**)

235. Long inter node bamboo used for kite production is _____________. (**Dullooa bamboo**)

236. Dullooa bamboo is scientifically known as ________. (***Schizostachyum dullooa***)

237. The weight of an ordinary felling axe head varies from _________ to __________ kg. (**0.7 kg to 1.8 kg**)

238. _____________ is used for felling Bamboo. (**Bill-hook**)

239. _____________ type of ripping saw used for conversion of scantling and planks for Himalaya forests. (**The Frame saw**)

240. _____________ type of ripping saw used for conversion of scantling and planks for cutting sal in Utter Pradesh. (**The Delhi saw**)

241. _____________ type of ripping saw used for conversion of large logs into beams and used in South India. (**The pit saw**)

242. The angle between the face of a tooth and the line passing through points of the teeth is know as _____________. (**Pitch**)

243. Peg-toothed saw consists of all teeth in the form of__________ triangles. (**Isosceles**)

244. Baw saw is very useful for a single man cutting upto diameter __________. (**20 cm**)

245. __________ is used for splitting firewood and also driving is wedges. (**Cleaver**)

246. More than 4 meter length having end cross section more than 15 cm is called __________. (**Beams**)

247. Wooden slabs having maximum thickness 5 cm is called ___________. **(Planks)**

248. A place where wood or other forest produce is stored, pending its disposal is called ___________. **(Depot)**

249. Security deposit for auction is ___________ % of total contract value or money. **(10)**

250. For colouring of pulp, in single stage bleaching ___________ is used and in multi stage bleaching ________ & ___________ is used for bleaching. **($CaOcl_2$, Cl_2 & ClO_2)**

251. Physical characteristics of paper depends on __________. **(Beating)**

252. Mechanical pulp differs from chemical pulp having all lignin of original wood and wood fibres in the form of ________ & ___________. **(Bundles & Fragments)**

253. Live continuous tapping method of resin is otherwise called ___________. **(Living method of tapping)**

254. Heavy tapping method of resin is otherwise called ___________. **(Tapping of death)**

255. Rill method of resin tapping method was developed by __________. **(FRI)**

256. Rill methods gives _________% more resin yield as compare to Cup & lip method without loss of timber. **(22)**

257. Crude pine resin consists of two principal constituents a liquid known as _________ and a solid known as ___________. **(Turpentine oil and Rosin)**

258. The process of separation or extracting fibres from one another is called ___________. **(Ratting)**

259. Oils from plants are obtained by __________ & __________ method. **(Distillation & extraction)**

260. Linaloe oil is obtained from the heart wood of _____________. (***Bursera delpechiana***)

261. The only commercial wood distilling plant is situated in ___________. **(Bhadravati, Karnataka)**

262. The acacia gums largely known as ____________. **(gum Arabic)**

263. The gum and resins that exude from cracks or cuts and solidify with air exposure called ____________. (**dammar**)

264. Gum which is not soluble in water but absorb water and swell up into a mucilaginous mass called as _____________. (**Gum-tetracanth**)

265. Combustibility of wood signifies ___________. (**Readiness**)

266. The quantity of heat emitted by a given weight of wood during combustion process is called ___________. (**heating power/ Calorific value of wood**)

267. Teakwood is not suitable fuel wood for cooking due to its ____________. (**Smokiness**)

268. Defect caused by fungal attack and decay in wood is called as _____________. (**unsoundness**)

269. The most serious fungal defects of wood are ________ & _________. (**Rots & Stains**)

270. When the wood grain twisted spirally making an angle __________ with vertical axis is called twisted fibre. (**40°**)

271. When wood fibres shows a wavy appearance instead of true vertical straight line with main axis it is called ___________. (**Waviness**)

272. Separation of fibres forming a cracks or fissure in the peice of wood and extending to other face or end of wood piece is called ____________. (**Split**)

273. A crack starting from the pith region and extending radially outwards is called ________. (**Heart or star shakes**)

274. Cracks which start from the periphery and extend radially inward towards the centre is called ___________. (**Radial shakes**)

275. Main food of termites is ___________. (**Cellulose**)

276. _______________ is the ability of cell wall to reflect the light. (**Lustre**)

277. The density of any material is defined as weight of unit volume of the material and expressed as __________. (**kg/m^3**)

278. Specific gravity of water is __________. (**1 gm/cm^3**)

279. ____________ is the capacity of wood to bend out of shape freely without rupture. (**Flexibility**)

280. ___________ is the capacity of wood to regain its original shape after the stress has produced deformation. (**Elasticity**)

281. __________ is the capacity of wood to split along the fibre during conversion of wood with a axe or a wedge. (**Fissility/Fissibility**)

282. The process of removal of excess moisture present in timber in its green state is known as ____________. (**Wood seasoning**)

283. Wood preservation on scientific and modern basis was introduced in India by __________. (**Sir Ralph Pearson**)

284. __________ process of preservation combines both seasoning and preservation. (**Boulton**)

285. The floating method is used to transport the woods for a distance upto __________ KM. (**100-200 km**)

286. The average number of bamboos carried in a raft is __________. (**10,000**)

287. Boom is a Dutch word which means ___________. (**Floating obstruction**)

288. Single arm or one way boom is attached permanently at its upper ends and guy ropes always attached at angle ________ to the boom. (**90°**)

289. The arms of two way or V-shaped boom incline at an angle _______ at the apex on the downstream side. (**45°**)

290. Plywood is glued wood made up of veneers in such a manner that the grain of each veneer is at ________ to the adjacent veneer in the assembly. (**Right angle**)

291. The grain direction in laminated wood is __________ to the adjacent veneer. (**Parallel**)

292. The glue laminated wood is called as ________. (**Glulam**)

293. A core board having width of 7.5 cm is called __________. (**Batten board**)

294. A core board not more than 2.5 cm wide is called __________. (**Block-board**)

295. A core board having width less than 7.0 mm is called _________. (**Lamin board**)

296. Veneers treated with a thermosetting fibre penetrating resin is called as ________. (**Impreg**)

297. ____________wood is an excellent materials for aeroplane propeller blades. **(Compregnated wood)**

298. Santalal is derived from __________. (***Santalum album***)

299. Keora oil is obtained from the species ____________. (***Pandanus tectorius***)

300. The flowers yield the 'cassie' perfume of commerce if extracted from species ___________. (***Acacia farnesiana***)

6

Wood Science and Technology

1. Physiologically heart wood is____________. **(Dead)**
2. Sapwood is rich in____________ material. **(Starch)**
3. Growth rings are also called as____________. **(Annual rings)**
4. Earlywood and late wood are referred as and____________ respectively. **(Spring wood and Summer wood)**
5. ____________ is the light coloured outer portion of a tree log consisting of the youngest growth rings. **(Sapwood)**
6. The deep coloured central portion of wood is____________. **(Heartwood)**
7. ____________ and____________ are function of sapwood. **(Conduction of sap and Food storage)**
8. The function of the heartwood is____________. **(Mechanical strength)**
9. Vessels and fibres are absent in____________ wood. **(Non-porous)**
10. The abnormal patches of parenchyma as a result of injury to the cambium due to insect attack are known as____________. **(Pith flecks)**
11. The separation of fibres which forms a crack not extending through the piece from one face to another face is known as____________. **(Checks)**
12. ____________ is the crack extends from face to face. **(Split)**
13. Grain orientation may be classified in to____________, ____________, ____________ and____________. **(Straight, Spiral, Interlocked and Wavy)**
14. Parenchyma distribution is of two classes such as and____________. **(Apo tracheal and Para tracheal)**
15. The basic structural and physiological unit of a plant is____________. **(Cell)**

16. A group of similar cells performing the same function is called as_____________. (**Tissue**)
17. The ability of cell wall to reflect light is termed as_____________. (**Lusture**)
18. The free water is held by_____________ inside the free spaces in the cells and fibres. (**Capillary action**)
19. The development of cell occurs in_____________ stages. (**Three**)
20. Pith may be classified into_____________ and_____________. (**Simple and bordered**)
21. The dark colour sap wood, which appears to be in a transitional stage is called. (**Kutcha-pukka**)
22. The formula for estimating the moisture content of wood is_____________ **{ (Green wt.- OD wt./OD wt.) * 100}**
23. The wood which are liable to crack, split and warp are called as_____________. (**Refractory wood**)
24. The timber which is capable of withstanding rapid seasoning is known as_____________. (**Non-refractory wood**)
25. The wood which are liable to splitting and cracking during seasoning are known as _____________. (**Moderately refractory wood**)
26. Majority of Indian hardwoods are_____________type of wood. (**Diffuse porous**)
27. Thin places in the wall which serves as a mean of intercommunication between two adjacent cells are known as_____________. (**Pits**)
28. The small soft mass of tissue in the centre of the plant is_____________. (**Pith**)
29. In broad leaved species the mechanical function is performed by_____________. (**Fibres**)
30. In softwood, conduction and mechanical functions are performed by_____________. (**Tracheids**)
31. In broad leaved species the conduction function is performed by_____________. (**Vessels**)
32. Deposition of_____________ and_____________ are the reason for susceptible to fungal attack in the sapwood. (**Carbohydrate and starch**)

33. During seasoning process the_____________ evaporates away at first till a point reached where most of the free moisture is removed and only the is left, this point is called as fibre saturation point. (**Free moisture and absorbed moisture**)

34. The greater durability of heart wood has often been attributed to_____________. (**Tylose**)

35. The lignin content of hard woods range from_____________per cent. (**18 to 25%**)

36. _____________ combines the function of vessels as well as fibres of porous wood. (**Tracheids**)

37. The fibres saturation point for tropical timbers ranged between_____________ per cent. (**15 to 30%**)

38. _____________ are formed from branches being incorporated in the main stemby occlusion. (**Knots**)

39. _____________ and_____________ kinds of knots present in the timber. (**Live and Dead knots**)

40. The fibre saturation point is found to vary from_____________ on dry weight basis. (**20 to 30%**)

41. Tracheid cells are minute, narrow thick walled with closed tapering ends, closed packed together resembles a_____________. (**Honey-comb**)

42. _____________in wood is caused by unequal shrinkage across the grain. (**Warping**)

43. In_____________ woods, the conduction and mechanical functions are performed by a single type of cells known as tracheid. (**Softwood**)

44. _____________are indicative function of durability of timber. (**Tylose**)

45. The rays are grouped as broad, high and conspicuous based on the_____________. (**Size**)

46. The abnormal conditions and features of wood which permanently reduce the economic value of wood are termed as_____________. (**Defects**)

47. _____________ is the most common commercial chemical agents used for bleaching. (**Chlorine compounds**)

48. Intercellular canals are commonly referred as_____________ or_____________. (**Resin canals or Gum ducts**)

49. ____________ and____________ are the two main defects in the wood due to fungal attack. (**Stain and Decay**)

50. The outer rough bark is traceable to activity of a second cambium called. (**Phellogen**)

51. ____________ is the first step in forest harvesting. (**Felling**)

52. Cutting down of trees into log a length is known as____________. (**Bucking**)

53. The term seasoning is meant for removal of____________. (**Moisture**)

54. ____________of timber is indicated by natural colour. (**Soundness**)

55. ____________wood is rich in gums, resins and tannin. (**Heartwood**)

56. The discolouration of sapwood is caused by____________. (**Sap-stain**)

57. ____________ is the major difference between the late wood and early wood. (**Annual rings**)

58. ____________is the food material for termite. (**Cellulose**)

59. ____________ ____________ is a chemical pulping process, commonly used in India. (**Sulphate process**)

60. The group of horizontally arranged cells which is running radially from bark towards the centre of the tree is____________. (**Medullary rays**)

61. The defects caused in timber due to____________ with the wood of local wounds that have healed up is known as rind galls. (**Inclusion**)

62. Shakes is a crack starting from the pith and extending radially towards one way or both ways is called as____________. (**Star shakes**)

63. The cracks which start from the periphery of the standing tree and extend radially inwards towards the centre is____________. (**Radial shakes**)

64. ____________ ____________ process is used for treatment of the sapwood of almost all green timbers. (**Boucher's process**)

65. Composite wood are built up____________ products. (**Bonded**)

66. Cup and ring shakes, the cracks follow the direction of____________ and may extend for a distance varying from a part. (**Annual rings**)

67. ____________ type kiln is suitable for the seasoning of hard woods. (**Direct heating**)

68. The durability of wood can_____________ by impregnating the wood with the preservative. (**Increased**)

69. Wood preservation on scientific and modern basis was introduced in India by_____________. (**Sir Ralph Pearson**)

70. The addition of non-fibrous mineral matter to stock prior to sheet formation is known as _____________. (**loading**)

71. Paper was first made from ground mulberry bark by the_____________. (**Chinese**)

72. The pulp should contain more_____________for the manufacture of rayon. (**Cellulose**)

73. In ply wood construction, veneers are in such a manner that the grain of each veneer is at _____________ to the adjacent veneer. (**Right angle**)

74. Plywood consist of________ number of plies. (**Odd**)

75. A built of product made of wood layers is known as_____________. (**Laminated wood**)

76. _____________is a composite board built up of a core composed of strips of wood of various dimensions glued together. (**Core board**)

77. _____________is a built up boards having a core of light material faced on both sides with a relatively thin layer of material having high strength properties. (**Sandwich board**)

78. A sheet of material made from fibres of wood is_____________. (**Fibreboard**)

79. _____________is a board constituted from fragments of wood and other ligno cellulosic materials bonded with organic binders. (**Particle board**)

80. The localised discoloration of timber in the form of streaks is known as_____________. (**Mineral streaks**)

81. Treatable timber should be dried below the moisture content of_____________ per cent. (**25%**)

82. The profession dealing with maintenance of ornamental trees is called_____________. (**Arboriculture**)

83. _____________ is the most common method of stacking timber for air seasoning. (**Horizontal stacking**)

84. _____________method of stacking is used for non-refractory wood. **(Vertical)**

85. The stack is used to separate one layer of seasoning timber from another is known as _____________. **(Crossers)**

86. Cross cutting of felled trees is called as_____________. **(Bucking)**

87. Rough lumber accounts for_____________ per cent of wastage in conversion. **(30%)**

88. The surface treatments in which the preservative is applied by brush/ spraying is referred as_____________. **(Superficial)**

89. Wood that has been specially treated in various ways to increase its strength is known as _____________. **(Improved wood)**

90. The wood is impregnated with all possible impregments is referred as_____________. **(Impregnated wood)**

91. The wood is compressed without impregnating it with any synthetic resins is known as _____________. **(Compressed wood)**

92. _____________ is a special type of improved wood that is both impregnated with synthetic resin and compressed. **(Compregnated wood)**

93. Water contained within the cell wall of wood is called_____________ water. **(Bound water)**

94. Treatment wood in a closed cylinder under pressure is the best and most effective form of preservative treatment which is called as_____________ process. **(Pressure)**

95. The stain caused fungal defects in_____________ part of wood. **(Sapwood)**

96. A novel method of treating green poles was developed by_____________. **(Dr. Boucherie)**

97. The three "E'S" associated with the forest Stewardship programme are_____________, _____________ and_____________. **(Environmental, Esthetics and Economics)**

98. The object of wood preservation is to_____________ the life of perishable timbers. **(Prolong)**

99. The modern science of wood preservation was developed at_____________. **(Europe)**

100. Earthen slides are also called as_____________. **(Chutes)**

101. _____________ and_____________ are the major transportation. **(Land and Water)**

102. _____________ instruments is used to prevent jamming of the saw in cross cutting and for longitudinal splitting. **(Wedges)**

103. The saw used in the saw mill are generally grouped into_____________ and. **(Head saw and re-saw)**

104. The entire opening between two adjacent teeth is known as_____________. **(Gullet)**

105. The cost of felling, conversion and transport is approximately_____________ per cent of total cost of delivery of timber in the market. **(50%)**

106. The edge of the tooth which faces the cutting direction is known as_____________ and the opposite edge of tooth is known as_____________. **(Face and Back)**

107. The angle between the face of a tooth and the line passing through the points of the teeth is called as_____________. **(Pitch)**

108. The width of the saw cut is known as_____________. **(Kerf)**

109. Rays are arranged_____________ to the stem axis. **(Perpendicular)**

110. Tylose are present only_____________wood. **(Hard wood)**

111. Aspiratory pit present only in_____________. **(Soft wood)**

112. _____________ process aims at maximum penetration of preservative with minimum absorption. **(Empty cell)**

113. Institute of wood science and technology is located at_____________. **(Bangalore)**

114. The modern synthetic resin adhesives used in manufacture of composite wood is _____________ and_____________. **(UF and PF)**

115. Acid process in pulp and paper production is otherwise known as_____________. **(Sulphite process)**

116. _____________ is the main advantages of plywood over the use of solid wood. **(Dimensional stability)**

117. Logging costs typically represent_____________per cent of the total cost of saw timber delivered to a mill. **(33%)**

118. The minimum thickness which can be peeled for manufacture of plywood may be as low as______________. **(0.05 mm)**

119. The chemical composition of ACCU is______________ ,______________ and. **(Arsenic pentoxide, Copper sulphate and Sodium dichromate)**

120. For treatment of Bamboo poles commonly used wood preservation process is ______________ process. **(Boucherie)**

121. Wood preservative ACCU was developed in______________ by______________. **(1993, Kamasan)**

122. ______________ is the reciprocal of resistivity and is evaluated in units mhos per centimetre. **(Conductivity)**

123. A tight knot free from decay and cracks in which the fibres are completely intergrown with those of the surrounding wood is referred as______________. **(Live knot)**

124. Crack is the separation of the fibres in the______________ direction. **(Longitudinal)**

125. A separation of fibres along the grain which extends through the piece from one surface to another is called as______________. **(Crack)**

126. ______________is a natural defect in the growth of a tree indicating its departure from the cylindrical shape. **(Flute)**

127. A log which is deviations from its straightness is referred as______________. **(Crookedness)**

128. The floating consists in construction of telescopic chutes is known as______________. **(Pathru)**

129. The floating obstruction is known as______________. **(Boom)**

130. A quantity of wood firmly bound together is termed as______________. **(Rafting)**

131. ______________ is a depots where the timber or other forest produce is stored pending its disposal sale. **(Sale depots)**

132. A place wood or other forest produce is stored, pending its disposal is known as ______________. **(Depots)**

133. The practice of measuring the volume of logs or bolts after they are harvested is called ______________. **(Scaling)**

134. A table showing the estimated or calculated amount of lumber which may be sawed from logs of given length and diameter is called____________. **(Log rule)**

135. A source of revenue for a logging operation is____________. **(Log sales)**

136. ACCU is a____________ type preservative. **(Fixed water soluble)**

137. Based on the seasoning behaviour, the woods can be classified into ____________ categories. **(Three)**

138. ____________ is the commonly used fillers in the pulp and paper manufacturing process. **(Talc)**

139. Radial growth in trees due to the activity of____________. **(Rays)**

140. The vascular bundles of monocots are____________. **(Scattered)**

141. Decay or wood rot is caused by____________ organisms. **(Fungi)**

142. The class of trees angiosperms is made up of____________ trees. **(Hardwood)**

143. Arrangement, colour, texture, type, size and shape refer to____________ characteristics used in tree identification. **(Leaf)**

144. Roots which grow vertically downward and anchor the tree are____________. **(Tap roots)**

145. The trunk as well as the branches and roots grow in diameter by means of cell division in the____________. **(Cambium layer)**

146. The inner core of a woody stem, wholly composed of non-living cells, darker in colour than the outer layer of the stem is the____________. **(Heartwood)**

147. Radial sawed lumber is also called____________. **(Quarter sawn)**

148. __________ and____________ are the main constituents of wood. **(Cellulose & Lignin)**

149. In a mature forest ecosystem, the large trees grows at a _____rate. **(Slow)**

150. In the temperate zone summer growth is often completed____________ high maximum summer temperatures are reached. **(Before)**

151. ____________ or____________ influences diameter growth in trees. **(Day length /Photoperiod)**

152. Periods of long daylight in spring and early summer result in the production of_____________

153. _____________ ,_____________ ,_____________ cells. (**large diameter, thin walled, spring wood**)

154. Growth at the end of the stem is called_____________. (**Terminal growth**) _____________ provided strength to the wood. (**Fibre**)

155. Total usable rough timber from the average tree is_____________ per cent. (**37-52**)

156. _____________is a branch of science which determining past climate from the trees primarily on the basis of annual rings. (**Dendroclimatology**)

157. _____________ method of stacking practiced for stacking of railway sleepers in case of refractory wood. (**Close rib method**)

158. Fibre saturation point for most tropical timbers ranged between_____________. (**12 to 20**)

159. Difference in girth at the butt and top is called_____________. (**Taper**)

160. A defect cause in timber due to inclusion with the wood of local wounds that have healed up is termed as_____________. (**Rindgalls**)

161. Shakes that are formed rupture of tissues in a circular direction is called_____________. (**Cup and Ring shakes**)

162. Complex knot formed at point where dormant buds show abnormal vigour is called_____________. (**Bur**)

163. All cracks appearing along the radii is called_____________ shake. (**Radial**)

164. A crack originating from centre and extending radially in one or opposite direction is called_____________ shake. (**Heart**)

165. Several splits or cracks originate from the centre of the timber is called shake. (**Star**)

166. The partial or complete removal of pith with or without removal of heartwoods is called _____________. (**Hollow centre**)

167. _____________ is a knot in which fibres are completely inter grown with surrounding wood. (**Live knot**)

168. A knot in which fibres are not completely inter grown with surrounding wood is called _____________. (**Dead knot**)

169. A portion of branch that embedded in the woody by natural growth is called. **(Knot)**

170. ____________ is a small soft core occurring in the structural centre. **(Centre heart)**

171. Separation of fibres along the grain which extends from one surface to another is called ____________. **(Split)**

172. The cambium which is self perpetuating layer consists of two types of cells viz. ____________ and____________. **(Fusiform and Ray initial)**

173. Snag means a____________. **(Standing dead tree)**

174. ____________snag are usually marketable and they are free from decay. **(Hard)**

175. Soft snags show stages of____________. **(Deterioration)**

176. Density of trees in a forest stand is normally expressed as____________. **(Basal area of the stand in square feet)**

177. A wire cable or chain used to move logs is called __________. **(Choker)**

178. Nutrients and water absorbed from the soil move up through a tree in the __________. **(Xylem)**

179. Treating wood posts and lumber with penta, creosote or CCA improves its __________. **(Resistance to decay and insects)**

180. A Pulaski is used in firefighting for __________ and __________. **(Grubbing & Chopping)**

181. The function of the cambium of the tree is growth of ________ and __________. **(Wood and Bark cells)**

182. Chromated zinc chloride is superior to zinc chloride as a preservative because of its ____________. **(Resistance to leaching)**

183. The part of the logging operation in which logs are dragged from where they were cut to a central location for loading is called ___________. **(Skidding)**

184. ___________ part of tree carries food made in the leaves down to branches, trunk and roots. **(Phloem)**

185. Both the inner bark and sapwood are formed by cell divisions of the ____________ layer. **(Cambium)**

186. ______________ is a direct product of photosynthesis. **(Carbohydrates)**

187. The only living part of the tree trunk is the ____________. (**Cambium layer**)

188. ___________ carries the sap from the roots to the leaves. (**Xylem**)

189. Closely spaced annual rings indicates _________growth. (**Slow**)

190. ___________ helps to protect the tree trunk from injury. (**Outer Bark**)

191. Debris left after a logging operation is called __________. (**Slash**)

192. The cross banded construction of wood is called as ___________. (**Ply wood**)

193. The caustic soda treated cellulose is called as ____________. (**Cellulose**)

194. ASCU is _____________ type of preservative. (**Fixed water soluble**)

195. Heating of wood in the absence of oxygen to produce methanol is called as _________. (**Wood distillation**)

196. Paper which is more or less impervious to the penetration of inks, is said to be _____________. (**Sizing**)

197. _____________ glue is commonly used in ply wood industries. (**Phenolic glue**)

198. Mechanical treatment given to fibres to increase tensile strength, bursting strength etc is _____________. (**Beating**)

199. Glues once cured are not softened by heat are called as ___________. (**Thermosetting glues**)

200. Estimating the amount of board feet, cords, or pieces in a timber stand is called ___________. (**Cruising**)

201. When timber buyers openly bid against one another at a timber sale it is called a/an ____________. (**Auction sale**)

202. The tree which has more than one leader stem is called _____________. (**Forked**)

203. ___________ causes root rot in sal. (**Polyporus**)

204. The capacity of wood to bend out the shape without rupture is called __________. (**Flexibility**)

205. The capacity of wood to split along with fibres is termed as ___________. (**Fissility**)

206. The layer that produced between xylem and phloem is ___________. (**Cambium**)

207. Secondary thickening in wood is caused by ____________. **(Cambium)**

208. The region in which cells divide repeatedly and forms new cells is termed as ____________. **(Meristematic region)**

209. The defects which show longitudinal distoration of wood in which one of the wood is concave is called __________. **(Bow)**

210. Epiblema of roots is equivalent to _________. **(Stele)**

211. The liquids moving in tree pass mainly through the __________. **(Pits)**

212. The durability of heartwood attributed to the presence of __________. **(Chemical deposits)**

213. The concentric rings usually present the wood formed during one growing season is called as ___________. **(Growth rings)**

214. In slow growing tree the width of growth rings is _________. **(8 rings/cm)**

215. A ring is cut all around the tree near the base usually from 10 cm to 15 cm broad and all the sap and bark is removed is known as __________. **(Girdling)**

216. Splitting the butt end of the leaning tree can be prevented by using ____________. **(Stem tightner)**

217. The distance from tooth to tooth measured at points of two adjacent teeth is referred as _____________. **(Space)**

218. The extent to which the teeth are bent to either side of the plane of the blade is _______. **(Set)**

219. Separations of fibres in the longitudinal direction is called as _________. **(Crack)**

220. __________ method of stacking timber is used for rapid surface drying. **(Vertical)**

221. The heating power of wood is known as ___________. **(Calorific value)**

222. The property of wood to remain sound is called __________. **(Durability)**

7

Forest Protection

1. Life history of Indian termite was first published by_____________. (**J.G. Koeing**)
2. Plant disease written by_____________. (**R.S. Singh**)
3. J. W. Helfer seems to have the first to collect insects because these were pest of trees during the year_____________. (**1810**)
4. The largest insect order is_____________. (**Coleoptera**)
5. Project directorate of biological control is located at_____________. (**Bangalore**)
6. Main skeletal structure to which most of the muscle attached is_____________. (**Thorax**)
7. The outmost waxy layer of insect body is called_____________. (**Cuticle**)
8. Insect body consists of series of segment designated as_____________. (**Sclerite**)
9. _____________ is the author of the book entitled "The ecology of forest insect". (**CFC Beesom**)
10. Service journal Indian forester first published during the year_____________. (**1875**)
11. Insect which feeds on plants or matter of vegetable origin is called_____________. (**Herbivore**)
12. Largest order in animal kingdom contains nearly quarter of million insect species is _____________. (**Coleoptera**)
13. Red rust in Eucalyptus is caused by_____________. (**Algae**)
14. Borers of felled bamboo are known as_____________. (**Ghoon**)
15. Seed borer of albizia species is_____________. (***Bruchus bilineatopygus***)

16. *Eutectona machaeralis* is a major pest of____________. **(Teak)**
17. *Hyblaea puera* is a____________. **(Defoliater)**
18. Teak skelitonizer belongs to family____________. **(Pyralidae)**
19. Termites belongs to order____________. **(Isoptera)**
20. Grass hopper feeds on____________. **(Shoots and leaves)**
21. Black root rot of forest trees is caused by____________. **(Macrophomina)**
22. Cricket laid eggs during the month of____________. **(September-October)**
23. *Selepa celtis* is a major pest of____________. **(Aonla)**
24. *Hoplocerambyx spnicornis* is a major pest of____________. **(Sal)**
25. Newly hatched larva of teak skeletonizer is about____________ mm long. **(2)**
26. *Plecoptere reflexa* is a pest of____________. **(Imli)**
27. Casuarina root nodules are formed in association with____________. **(Frankia)**
28. Female moth of shisham defoliator lay eggs singly on host tree within ____________ days. **(6)**
29. *Celosterna scabrator* is a____________ insect. **(Defoliator)**
30. Aphids damage forest trees by____________. **(Sucking)**
31. Most insect damage is done while the insects are in the____________ stage. **(Larva)**
32. Tea rust is caused by____________. **(Algae)**
33. *Olethreutes paragramma* belongs to ____________ order. **(Lepidoptera)**
34. *Estigmena chinensis* lay eggs on the surface of internodes of bamboo culms during the month____________. **(July)**
35. Tendu gall forming insect feeds on____________. **(Leaves)**
36. Sinoxylon species belongs to____________ order. **(Coleoptera)**
37. Ghun belongs to family____________. **(Plateypodidae)**
38. *Calo termintidae* termite are known as____________. **(Wood dweller termite)**

39. Peak infestation period of inflorescences feeder is during. (**August-September**)
40. Damage stage of white grub is____________. (**Larval/Grub stage**)
41. The interior region of insect head forms the____________. (**Librium**)
42. Locomotion in insect is preferred entirely by _____ and. (**Legs and wings**)
43. Antennae of insect function as____________. (**Sensory**)
44. Body of arthropod are____________. (**Bilaterally symmetrical**)
45. Aphids and white flies are effectively controlled by____________ insecticide. (**Systemic**)
46. The branch of entomology which deals with role of insect in the economy of man is known as____________ entomology. (**Economic entomology**)
47. Third and posterior region of three made division of insect body is known as ____________. (**Metathorax**)
48. A sexually matured terminal stage in insect is____________. (**Adult**)
49. In____________ insect, ovipositor is modified into spine. (**Wasp**)
50. Development of insect without metamorphosis is____________. (**Apterygota**)
51. A sub class of wingless insect comprising four orders are known as. (**Apterygota**)
52. Paired sensory segmented appendages present on the head of the insect is called ____________. (**Antennae**)
53. A compact of artificial control involving use of organisms such as predator and parasite to control pest species is known as____________. (**Biological control**)
54. Insects feeding on animal are known as____________. (**Herbivorous insects**)
55. ____________ state of India rank first in raw silk production. (**Karnataka**)
56. A silk encase constructed for the protection of pupa is known as ____________. (**Cocoon**)

57. A hierarchical division in animal kingdom between a phylum and an order is known as_____________. (**Class**)

58. Immature stage in butterfly/moth is called_____________. (**Pupa**)

59. Insect passes through four distinct stages i.e. egg, larva, pupa and adult is known as _____________. (**Complete metamorphosis**)

60. The number of generation an insect completes in one year in its natural condition is known as_____________. (**Voltinism**)

61. Chemical compound which kill insect by contact with external body surface is known as_____________ insecticide. (**Contact**)

62. Parasites that lives inside the body of its host is known as_____________. (**Endoparasite**)

63. An order of class insect comprising true flies of_____________. (**Diptera**)

64. Population of insects attaining of high levels of incidence is_____________. (**Epidemic**)

65. A stage of insect dormancy during winter is_____________. (**Aestivation**)

66. Study of form and structure of the bodies of insect is known as_____________. (**Morphology**)

67. Study of insect in relation to environment is known as. (**Insect ecology**)

68. Every insect during its growth, shed its skin one or more time which is known as _________. (**Moulting**)

69. Resting phase of inactive instars in all holometabolous insect is known as _____________. (**Pupa**)

70. Manipulation in which involve change in temperature and humidity to destroy pest is known as_____________. (**Physical control**)

71. Chemical or other stimuli that cause insect to make oriental movement towards the source is known as_____________. (**Insect attractant**)

72. Complete abolition of pest species from an area is known as_____________. (**Eradication**)

73. Integrated pest management (IPM) aims at keeping pest population _____________ injurious levels. (**Below**)

74. A member of class insecta having three body regions three pairs of legs and mostly with wings are known as_____________. (**Insect**)

75. Pest management in which all techniques are consolidated into a unified programme is known as_____________. **(IPM)**

76. A parasite whose host is another parasite is_____________. **(Hyperparasite)**

77. An adult stage of insect is known as_____________. **(Imago)**

78. An insecticidal formulation in which the insecticide is impregnated is known as _____________. **(Granule)**

79. A term is applied to legless immature stages of Diptera known as. **(Maggot)**

80. Insect feeding on one kind of food is called _____________. **(Monophagous)**

81. The middle or second segment of thorax in insect is called_____________. **(Mesothorax)**

82. Regulation of insect population or other living organisms by action of natural forces is called_____________ control. **(Natural control)**

83. The inherent power of an organisms to survive, reproduce and multiply apart from destructive factor of its environment is termed as_____________. **(Abiotic potential)**

84. Pheromone trap attracts_____________ moths. **(Male)**

85. Barrier, screens, tree bands, traps for control of insect pest is known as _____________. **(Mechanical control)**

86. The granular insecticide which is having long residual effect is. **(Aldicarb)**

87. *Bacillus thuringiensis* was discovered by_____________. **(Shigetane Ishiwatari)**

88. The dwelling place of termite is also known as_____________. **(Mound)**

89. The bacterial colony is known as_____________. **(Ooze)**

90. Certain secretion attract the opposite sex of the same species which are termed as _____________. **(Pheromones)**

91. The fatal minimum temperature for insect is_____________. **(10 °C)**

92. Chemical when applied in a gaseous form by exposing a volatile liquid of solid for control of insect is known as_____________. **(Contact poison)**

93. An elementary manual of zoology was published during the year_____________. **(1893)**
94. Destructive Insects Pest Act (DIPA) was passed on_____________. **(1914)**
95. Insect which are active in night are called_____________. **(Nocturnal)**
96. Bark eating caterpillar belong to family_____________. **(Indarbelidae)**
97. Felling of the bamboo in the cold weather for checks the breeding of_____________. **(Borer)**
98. Trap tree operations is suitable for control of_____________. **(Sal borer)**
99. Peak infestation period of teak defoliator is_____________. **(June-July)**
100. Ailanthus defoliator belongs to order_____________. **(Coleoptera)**
101. Babul root and shoot borer has_____________ life cycle in a year. **(Two)**
102. Specialized laying organ in female insect is known as_____________. **(Ovipositor)**
103. In honey bee pollen basket is present in_____________. **(Hind leg)**
104. The process of leaving off the colony by honey bee queen is known as _____________. **(Swarming)**
105. *Apis cerana indica* is known as. **(Indian honey bee)**
106. *Apis dorsata* is Known as_____________.**(Rock honey bee)**
107. Scientific name of Dammer bee is_____________.**(*Melipona iridipennis*)**
108. Scientific name of little honey bee is_____________. **(*Apis florea*)**
109. Insects are the most abundant and diversified of all animals due to their _____________ capacity. **(*Flying*)**
110. Reproduction from egg laid by female of a species is known as_____________. **(*Oviparous*)**
111. The phenomenon of producing or giving birth to living young ones instead of laying eggs is known as_____________. **(Viviparous)**
112. Peak infestation period of teak skelitonizer is_____________.**(October-November)**
113. Root rot in sal is caused by_____________. **(Polyporous)**

114. ____________ insecticide is recommended for the control of termite. **(Chlorpyrifos)**

115. The non-preference plant resistance to insects is also known as. (**Antixenosis**)

116. The major predator of lac insect is____________. (***Eublemma amabilis***)

117. ____________ is an insecticide which is used as seed treatment for sucking pest management. (***Imidacloprid***)

118. Sting in worker honey bee is modified part of____________. (***Ovipositor***)

119. Extremely toxic class of pesticides have colour of lower triangle. **(Red)**

120. Directorate of Plant Protection Quarantine and Storage (**DPPQS**) was established in 1946 at____________. **(Faridabad)**

121. Central Insecticides Laboratory (**CIL**) is located at____________. **(Faridabad)**

122. NCIPM (National Centre for Integrated Pest Management) is located at ____________. **(New Delhi)**

123. National nematode collection of India (NNCI) is located at____________. **(New Delhi)**

124. Father of host plant resistance is____________. **(R.H. Painter)**

125. ____________ is the father of insect physiology. (**Wigglesworth**)

126. ____________ is the first Indian scientist who collect and identified fungi in India. (**K.R. Kirtikar**)

127. Father of plant pathology is____________. (**Heinrich Anton de Bary**)

128. ____________ is the founder of virology. (**Martinus Beijerinck)**

129. Plant pathology written by____________. (**Agrios**)

130. Father of Indian Mycology____________. (**E.J. Butler**)

131. Fungi and disease in plants was written in 1918 by____________. **(E.J. Butler)**

132. Fungi and plant disease was written by____________. **(B. B. Mundakur)**

133. Father of Indian Nematology is____________. **(N.M. Alam)**

134. ____________is father of microbiology. (**Antonie Van Leewenhoek**)

135. Father of Nematology is_____________. **(N.A. Cobb)**

136. First plant parasitic bacteria was reported by_____________. **(T. J. Burrill)**

137. _____________ is the father of host plant resistance. **(R.H. Painter)**

138. _____________is the father plant pathology in India. **(E.J. Buttler)**

139. MLO was first discovered by_____________. **(Doi)**

140. _____________proposed a classification of fungi in 1973. **(Ainworth)**

141. Study of animal behaviour is called_____________. **(Ethology)**

142. First plant parasitic nematode was discovered by_____________ and reported by_____________. **(Anguina and Needham)**

143. In honey bees, the queen substance is secreted by_____________. **(Mandibular gland)**

144. _____________ and_____________ type of mouth parts present in honey bee. **(Chewing and Lapping)**

145. Lac insect is scientifically known as_____________. ***(Laccifer lacca)***

146. Termite is_____________type of pest. **(Polyphagous)**

147. _____________ energy is utilized for making fog, smoke and vapour sprays. **(Thermal)**

148. _____________have monopoly in Muga silk. **(India)**

149. _____________ is an association between algae and fungus. **(Lichen)**

150. In sandal, spike disease is caused by_____________. **(MLO)**

151. The rust fungi completing their life cycle on one host are called. **(Autoecious)**

152. _____________ and_____________ pathogens caused the disease damping off in seedling in nursery stage. ***(Pythium aphanidermatum and Pythium debaryamum)***

153. Pine wood nematode,_____________ causes pine wilt in association with the cerambycid insect Monochamus. ***(Bursaphelencus xylophilus)***

154. _____________ type of leg is present in honey bee. **(Foragial)**

155. _____________is used to produce plants free from virus. **(Meristem culture)**

156. Stem wilt of Casuarina is caused by____________. (**Trichosporium**)

157. The bacteria generally used in genetic engineering for transfer of character from one organism to another are____________. (***Agrobacterium tumefaciens***)

158. ____________ ____________is most widely used bacterium for insect control. (***Bacillus thuringiensis***)

159. The term____________ is given for juvenile hormone. (**3rd generation pesticides**)

160. Pest control using natural agents referred as____________. (**Biological control**)

161. The most destructive group of forest insects are____________. (**Borers**)

162. In India, lac is mainly produced in____________. (**Bihar**)

163. India's place in the production of mulberry silk in the world is____________. (**Third**)

164. Fossil records indicate that insect evolved during____________ period. (**Devonian**)

165. Severe defoliation is most harmful to trees that occur in____________ ____________. (**As the spring foliage is unfolding**)

166. The damage done by the Gypsy moth is when it is in the____________ stage. (**Egg**)

167. Pheromones are also called as____________. (**Ectohormones**)

168. ____________is the author of "Indian insect life". (**Mami**)

169. The Indian lac research Institute is located at____________. (**Ranchi**)

170. ____________ is used for the inoculation of host tree for lac cultivation. (**Brood lac**)

171. Honey bee queen produces____________ thousand eggs per day. (**2.0-3.0**)

172. The first abdominal segment when fused with metathorax in bees is known as ____________ ____________. (**Propodeum**)

173. Application of____________ fertilizer increases the disease resistance in plants. (**Potash**)

174. The plant mycoplasmal disease are transmitted by____________. (**Leaf hoppers**)

175. Salt damage to a tree is identified by_____________. **(Foliage necrosis)**

176. A symptom of winter injury in conifers may be foliage turns_____________ and dies. **(Red)**

177. Abnormal tissue growth on trees that is caused by insects/mites is called. **(Gall)**

178. _____________is the vector for spike disease in sandal. **(Jassid)**

179. Death of tissue in virus infected plant described as_____________. **(Mosaic)**

180. Aldrix is a_____________. **(Fungicide)**

181. The disease damping off is caused by_____________. **(Fungi)**

182. _____________ is the most effective method for preventing forest fire. **(fire lines)**

183. The chemical produced by host plant to attract insects is called as. **(Kairomones)**

184. Wilting of sisoo is caused by_____________. **(*Fusarium solani*)**

185. Spike is a common disease of_____________. **(Sandalwood)**

186. Pine bark borer is_____________. **(*Hypsiphyla robusta*)**

187. _____________ is an ecofriendly method to control insect pests and weeds. **(Biological control)**

188. Accidental forest fires are_____________ to trees and animals. **(Harmful)**

189. The life cycle of insects generally encompasses_____________ stages. **(Four)**

190. Mycorrhiza is associated with_____________ parts of plant. **(Roots)**

191. Sal tree are epidemically affected by_____________ insects. **(Borer)**

192. A fire line deprives fire of_____________. **(Fuel)**

193. A fire set to destroy fuel in order to check a fire's progress is called as_____________. **(Back fire)**

194. Insects that attack young twigs, stems or buds are_____________. **(Tip feeders)**

195. The cause of heart rot and dutch elm disease is_____________. **(Fungus)**

196. A fungus disease which may cause death or damage to newly emerging seedling is called_____________. **(Damping off)**

197. A non grass like herbaceous plant is called a_____________. **(Forb)**

198. A fire which runs through the tops of living tree, bushes is called. **(Crown fire)**

199. The study of disease of forest trees is called_____________. **(Forest pathology)**

200. A drip torch is used in_____________. **(Prescribed burns)**

201. A fire that has been started intentionally by a forester to be used as a management tool is a_____________. **(Prescribed burn)**

202. The burning of peat and humus which produces intense heat below the surface and down ward is a_____________. **(Ground fire)**

203. The science that pertains to disease of forest trees or stands and to the deterioration of forest products by organisms is_____________. **(Forest pathology)**

204. Kevlar is a material used in_____________. **(Safety clothing)**

205. The final stages of fire suppression are called_____________. **(Mop-up)**

206. The use of a counter (Back) fire is_____________ type of forest fire control. **(Indirect)**

207. Herbicide damage to trees may be indicated by_____________ and_____________. **(Twisting of leaves/needles and stunted growth)**

208. Other than fires,_____________ and_____________ are the most damaging forest enemies. **(Disease and Insects)**

209. A process which removes injured, disease or insect-infested trees is called_____________. **(Sanitation cutting)**

210. An economical way to remove excess logging debris and undesirable vegetation prior to planting is_____________. **(Prescribed fire)**

211. A herbicide that will not move or translocate within a plant is called _____________ herbicide. **(Contact)**

212. A forest fire which is out of control is called a_____________. **(Wild fire)**

213. A fire set along the inner edge of a fire control line to stop a spreading fire is called as_____________. **(Backfire)**

214. A fire away from the main fire which started by flying sparks is called as _________. (**Spot fire**)

215. The discovery, confinement and extinguishing of a forest fire is called_____________. (**Fire suppression**)

216. The application of fire to gain beneficial results is called_____________. (**Prescription burn**)

217. A buffer zone in and around forest stands which is used to potentially stop the spread of fire is called a_____________. (**Fire break**)

218. Insect causes_____________ damage to forests than does forest fires. (**More**)

219. The disease fire blight is caused by_____________. (**Bacteria**)

220. Aphids and scales are examples of_____________ with penetrating mouth parts. (**Sap suckers**)

221. When frass is found at the base of a tree the probable cause is_____________. (**Boring insect**)

222. Periodic_____________ fires are valuable in reducing the amount of fuel available in the forest. (**low-intensity**)

223. A Pulaski is used in forest fire fighting for_____________ and_____________. (**Grubbing & chopping**)

224. A hypo-hatchet is used for_____________ a tree with herbicide. (**Injecting**)

225. A_____________ is the type of control burn best used for hazard reduction in young forest stands with a heavy layer of litter. (**Back fire**)

226. Severe defoliation that occurs as the_____________ is most harmful to trees. (**Spring foliage unfolding**)

227. The gypsy moth over winters in the_____________ stage. (**Egg**)

228. The damage done by the gypsy moth is when it is in the_____________ stage. (**Larva**)

229. The male gypsy moth locates the female because she secretes a powerful scent called a_____________. (**Pheromone**)

230. Adult insects have 6 legs, adult mites have_____________ legs. (**8**)

231. Many aphids secrete a clear, sugary liquid called as_____________. (**Honeydew**)

232. Young aphids which hatch in spring and feed on the tender new growth are all ____________. (**Females**)

233. The first consideration in using chemical pesticides should always be. (**Safety**)

234. The life span of the adult gypsy moth is usually about days. (**7**)

235. The part of the logging operation in which logs are dragged from where they were cut to a central location for loading is called____________. (**Skidding**) S to uti

236. Scientific name of teak defoliator is____________. (***Hyblea purea***)

237. The major pest of *Leuceana leucocephala* is____________. (**Psyllids**)

238. The symbol of diseased tree is____________ in regular forest. (**m**)

239. A forest fire spreading slowly over the ground with low flame is termed as ____________ ____________ fire. (**Creeping**)

240. Female moth of mulberry silk worm lays about____________ eggs. (**100 – 200**)

241. VAM fungi are____________symbiont. (**Obligate**)

242. ____________ acts as storage organ for VAM fungi. (**Vesicle**)

243. The VAM fungi with spores regularly arranged on a____________are grouped as sclerocystis. (**Central core**)

244. The VAM fungi with spores bearing bulbous stalks are grouped as____________. (**Gigaspora**)

245. ____________fungi develop characteristic structures called vesicle and arbuscules. (**VAM**)

246. Symbiotic association of____________ with____________ is referred to as mycorrhiza. (**Fungal hyphae, plant roots**)

247. The microbes associated with seed surface as termed as____________. (**Spermosphere**)

248. Biofertilizer inoculation technique was started during____________. (**1895**)

249. A group of microbes with symbiotic relationship between____________ and____________ is termed as lichens. (**Fungi and Algae**)

250. ____________ is the first organic fungicide. (**Thiram**)

251. The micorrhizal infection is normally____________ in 'P' rich soil. **(High)**

252. A definite relatively localized necrotic lesion primarily of the bark and cambium is known as____________. **(Canker)**

253. Isolation trenches are dug-up to prevent____________ diseases. **(Root)**

254. Wilting of Sissoo is caused by____________. (***Fusarium solani***)

255. ____________can be controlled by application broad leaf weeds. **(2, 4-D)**

256. ____________ is an insecticide which is having fumigant and penetrant action. **(DDVP)**

257. The light producing substance in glow worm or fire fly is __________. **(Luciferin)**

258. The organism which can not synthesize protein by own enzymes is __________. **(Virus)**

259. Decomposition of organic matter in submerged soil is carried out by ____________. **(Bacteria)**

260. The root knot nematode generally have interaction with ____________. (***Pseudomonas solanacearum***)

261. Most widely used biocontrol agent is ____________. (***Pseudomonas florescence***)

262. When plant show the partial resistance against all the races of pathogen then it is type of____________. **(Horizontal resistance)**

263. NPV stands for ________________. **(Nuclear Polyhedrosis Virus)**

264. ____________ fungicide is used against downey mildew. **(Metalaxyl)**

265. Nuclear polyhedrosis visus (NPV) is the most effective for control of ____________. (***Helicoverpa armigera***)

266. Founder of Modern bacteriology is ____________. **(Pasteur)**

267. Eucalyptus gall wasp belongs to the order____________. **(Hymenoptera)**

268. Basal culm rot in bamboo is caused by ____________. (***Fusarium moniliforme***)

269. ASCU is a chemical used as ____________. **(Preservative)**

270. IPM relies heavily on ____________. (**Economic threshold level**)

271. Biological agent for lantana weed (*Lantana camera*) is __________. (***Ophiomyia lantanae***)

272. Soft cambium and newly formed wood is damaged by __________, which leave tunnels as a sign of damage. (**Bark borers**)

273. The central insecticide Act came into force from __________. (**1971**)

274. ____________ is the author of "Indian insect Life". (**Mani**)

275. The damaging stage of cut worm is ____________. (**Caterpillar**)

276. The density of pest population at which control measures should be applied ____________. (**ETL**)

277. The first Indian imperial Entomologist is ____________. (**Hem singh pruthi**)

278. The grinding of food in grasshopper is done by __________. (**Mandibles**)

279. ____________ is the main characteristics feature of an insect. (**Three pairs of legs**)

280. __________ is the host tree of eri-silk worm. (**Castor**)

281. The reproductive system of insect is derived from _________. (**Endoderm**)

282. The father of animal classification is ___________. (**Carlous Linnaeus**)

283. Inorganic salts are toxic to insects by __________. (**Ingestion**)

284. Insects feeding on plants of several genera within a family are called ___________. (**Oligophagous**)

285. Micrografting is used to produce plants free from __________. (**Virus**)

286. Corticium salmonicolor is a disease which attacks _________ plant. (**Eucalyptus**)

287. Fungal roots are otherwise called as ___________. (**Mycorrhiza**)

288. An organism on or within which another organism feeds and develops is called _____________. (**Host**)

289. A virus particle with both nucleic acid and protein coat is known as ___________. (**Capsid**)

290. Bacterial cell surrounded by flagella is called ________________. **(Peritrichous)**

291. In fermentation industry, pasteurization is done at ____________ temperature. **(62.8 °C)**

292. Nematodes are also known as ____________. **(Tapeworms)**

293. Powdery mildew disease is caused by ______________. **(Fungus)**

294. Powdery mildews are controlled by spray of _________ fungicides. **(Sulphur)**

295. ______________ nutrient is supplied by micorrhiza to the plant. **(Phosphorus)**

296. ________________ is known as downey mildew fungus. **(Sclerospora)**

8

Tree Improvement

1. Tree improvement consists of____________ and____________. **(Silviculture and tree breading)**
2. Variation within the local progenies is due to____________. **(Environment)**
3. Larger units of similar vegetation are referred as____________. **(Biome)**
4. ____________ is the only gaseous hormone. **(Ethylene)**
5. More genetic gain can be achieved in the____________ seed orchard. **(Clonal)**
6. ____________ is known as father of genetics. **(Mendel)**
7. Pure line theory was proposed by____________. **(Johannsen)**
8. Self-incompatibility hampers the development of____________. **(Inbreds)**
9. Progeny having both parent identities is known as____________. **(Fullsib)**
10. The presence of male and female flowers in different plants is called____________. **(Dioceious)**
11. Maturation of anthers and stigma of a flower at the same time is called____________. **(Homogamy)**
12. Pollination occurs through water is known as____________. **(Hydrophylly)**
13. The increase in performance of F1 hybrid over its parent is called____________. **(Heterosis)**
14. Emasculation by suction method is more suitable for____________. **(Small flower)**
15. The variety of species within a region is referred as____________. **(Species diversity)**

16. In bamboo genetic resource India is the_____________ rich country in the world. (**Second**)
17. Head quarter of WTO situated at_____________. (**Geneva**)
18. IPGRI headquarter is situated in_____________. (**Rome**)
19. Gradual reduction in genetic variability due to elimination of various genotypes is known as_____________. (**Genetic erosion**)
20. TRIPS denote_____________. (**Trade Related Intellectual Property Rights**)
21. IBPGR stands for_____________. (**International Bureau of Plant Genetic Resources**)
22. NBPGR established in_____________. (**1976**)
23. An individual having more than two basic sets of chromosomes is called. (**Polyploid**)
24. The seed act was enacted in the year_____________. (**1966**)
25. Ability of a gene to affect more than one phenotypic characteristic is_____________. (**Pleiotrophism**)
26. Development of seed without sexual process is_____________. (**Apomixis**)
27. A plant with functional pollen fails to set seed when self-pollinated due to. (**Incompatibility**)
28. Usually seed production areas should contain a minimum area of_____________ hectare. (**Four**)
29. The original geographic area from which seed or other propagules were obtained is known as_____________. (**Provenance**)
30. Trees that have proven to be genetically superior by means of progeny testing are _____________. (**Elite tree**)
31. Specification occurs through physical separation of population is called_____________. (**Allopatry**)
32. The plant part used in regeneration in tissue culture is known as_____________. (**Explant**)
33. _____________ hormone is primarily responsible for enhancing apical dominance. (**Auxins**)

34. Promoting seed dormancy is responsible for_____________ hormone. **(Abscisic acid)**
35. Variation that is observed among the plants which are regenerated from callus cultures of somatic explants such as meristem is called_____________. **(Somaclonal variation)**
36. Regeneration of whole plant from tissues of an actively dividing plant part such as stem tip is referred as_____________. **(Meristem culture)**
37. Shortest phase in mitosis is_____________. **(Anaphase)**
38. Crossing over takes place at_____________. **(Meiosis I)**
39. _____________ and_____________ proposed the concept of hormonal control of organ formation. **(Skoog and Miller)**
40. The virus free plant was first produced by_____________ and. **(Cocking and White)**
41. _____________is a major carbon source in plant tissue culture medium. **(Glucose)**
42. Autoclaving is a method of sterilizing with_____________ under high pressure. **(Water vapour)**
43. _____________mediated transformation is an important indirect gene transfer technique. **(Agrobacterium)**
44. The unorganised mass of tissues is called as_____________. **(Callus)**
45. The donor tree from which parts to be propagated are taken, called as_____________. **(Ortet)**
46. In tree improvement programme, many cycle of selection and breeding refers to _____________. **(Recurrent selection)**
47. The trees produced from the seed of a parent are called_____________. **(Progeny)**
48. Area where superior phenotypes or genotypes are established and manage intensively and entirely for seed production is referred as_____________. **(Seed orchard)**
49. Narrow sense heritability is_____________. **(V_A/V_P)**
50. Broad sense heritability is ___________. **(V_G/V_P)**
51. The pollen sterility is caused by nuclear genes is termed as_____________. **(Cytoplasmic male sterility)**

52. PPV and FRA refers to_____________ and_____________ respectively. **(Protection of Plant Varieties and Farmer's Right Act)**
53. Double cross plan was suggested by_____________. **(Jones)**
54. Concept of epigenesist was proposed by_____________. **(Wolf)**
55. The term Germplasm was first used by_____________. **(Weisman)**
56. Double helical structure of DNA was shown by_____________. **(Watson and crick)**
57. The term genetic resource was coined by_____________. **(Frankel)**
58. The term parthenogenesis was coined by_____________. **(Owen)**
59. The word mutation was first used by_____________. **(Morgan)**
60. _____________ is the father of plant tissue culture. **(Haberlandt)**
61. The term gene coined by_____________. **(Johannsen)**
62. Single cell culture was first developed by_____________. **(Muir)**
63. Theory of Gemmules was proposed by _____________. **(Charles Darwin)**
64. Sex in plant was discovered by_____________. **(Camerarious)**
65. Chromosomal theory of inheritance was given by_____________ and_____________. **(Sutton and Boveri)**
66. Cryopreservation technique was first developed by_____________. **(Scibert)**
67. Chromosome was discovered by_____________. **(Strasburger)**
68. _____________ gave the concept of circular chromosome in *E. coli*. **(Cairns)**
69. The term genetics was given by_____________. **(Betson)**
70. The phenomenon of linkage was first observed by_____________ and_____________. **(Bateson and Punnet)**
71. Law of homologus series is given by_____________. **(Vavilov)**
72. Pangenesis theory was proposed by_____________. **(Charles Darwin)**
73. The first artificial hybrid was made by_____________. **(Thomas Fairchild)**

74. Cell linkage theory was proposed by____________. (**Virchow**)

75. Concept of pureline theory was developed by____________. (**Johansen**)

76. Wobble hypothesis was given by____________. (**Crick**)

77. The 'One gene one enzyme' hypothesis was proposed by____________. (**Beadle and Tatum**)

78. Coupling and Repulsion hypothesis was given by____________. (**Bateson**)

79. Incompartibility is very common in plant kingdom was reported by____________. (**Lewis**)

80. Transposable elements were discovered by____________. (**McClintock**)

81. Term Heterosis was introduced by____________. (**Shull**)

82. Anther culture technique was developed by____________. (**Guha and Maheswari**)

83. Gametophytic system is described by____________. (**East and Manglsdorf**)

84. DNA polymerase was discovered by____________. (**Kornberg**)

85. Largest organelle in Eukaryotic cell____________. (**Nucleus**)

86. Quantification of genetic diversity parameters and genome analysis of forest trees can be made through____________ ____________ technique. (**Molecular marker**)

87. The group of related individuals when only one parent is common is called____________. (**Half-sib family**)

88. Development of embryo without fertilization is____________. (**Apomix**)

89. Interaction between alleles of two different loci is called as____________. (**Epistasis**)

90. Occurrence of more than one embryo in seed is known as____________. (**Polyembryony**)

91. ____________ and____________ are the two most important factors influencing the life span of seeds under storage. (**Relative humidity and Temperature**)

92. _____________ is very important to maintain its viability and vigour. **(Seed drying)**

93. Development and formation of pollen grains in anther of the stamen is known as _____________. **(Microsporogensis)**

94. Development of a fruit without fertilization is known as_____________. **(Parthenogenesis)**

95. The most common plant growth regulators compound used in rooting of tree species is _____________ _____________. **(IBA)**

96. Systematic removal of individuals in orchards is called _________. **(Roughing)**

97. Two lines different for a single locus called_____________. **(Isogenic line)**

98. Sum total of genes in a reproductive gamets of a population is called_____________. **(Gene pool)**

99. The existence within the population of disadvantageous allele in a heterozygous genotype is known as _____________. **(Genetic load)**

100. The test cross ratio in complementary gene action is_____________. **(1:3)**

101. The classical test cross ratio in dihybrid is_____________. **(1:1:1:1)**

102. The type of histone proteins present in eukaryotic chromosome are_____________. **(Five)**

103. Xanthamine is deaminated product of_____________. **(Guanine)**

104. A tree that has been selected grading because of its desirable character is. **(Candidate tree)**

105. Genetic gain in tree improvement programme is estimated by the formulae. **($G = S * h^2$)**

106. Functional unit of inheritance is_____________. **(Gene)**

107. Migration of alleles from one population into another is called as_____________. **(Gene flow)**

108. In mitosis, longest phase is _____________. **(Prophase)**

109. Seed certification involves _____________ phases. **(Five)**

110. Highest uniformity is present in_____________ cross. (**Single**)

111. Transfer of pollen grains from the anther to the stigma of same flower is. (**Autogamy**)

112. The staminate and pistilate flowers occurs in the same plant is called_____________. (**Monoecy**)

113. The stamens mature before pistil is called_____________. (**Protandry**)

114. The pistil mature before anthers is called_____________. (**Protogyny**)

115. The development of embryo sac from a megaspore is known as_____________. (**Megagametogenesis**)

116. Chromosome other than sex chromosomes are called_____________. (**Autosomes**)

117. The molecular markers which are codominant in nature are_____________ and_____________. (**RFLP and SSR**)

118. The performance of tree over a full rotation in the new environment is referred as _____________. (**Adaptation**)

119. Removal of male flowers/ anthers is known as_____________. (**Emasculation**)

120. In cryopreservation, cells are stored at temperature of_____________. (**-196° C**)

121. Lack of seed set on self-pollination is_____________. (**Self-incompatibility**)

122. The individuals of a clone are called_____________. (**Ramet**)

123. The situation in which a vegetative propagule does not assume tree form but continues to grow like a branch is called as_____________. (**Plagiotropic**)

124. When on F1 is crossed with the recessive homozygous parent, the cross is_____________. (**Test cross**)

125. An exchange of genetic material between homologous chromosome is called as _____________. (**Crossing over**)

126. Superiority of F1 hybrid over a standard commercial variety is called as_____________. (**Economic heterosis**)

127. The selection method used for improvement of traits for resistance and cold hardiness is _____________. (**Tandem selectio**n)

128. Physical basis of life is_____________. (**Protoplasm**)

129. ISTA was established in the year_____________. (**1924**)

130. NBPGR is located at_____________. (**New Delhi**)

131. A cross involving two pairs of gene each pair affecting a different character is called _____________. (**Dihybrid**)

132. A condition in which pollination and fertilization takes place in an unopened flower is termed as_____________. (**Cleistogamy**)

133. Polyploids are common in_____________species. (**Hardwood**)

134. Epistasis is_____________ type of interaction. (**Inter allelic**)

135. The adaption of whole genotype or gene complex to specific environments is the key characteristic of_____________. (**Ecotype**)

136. The concept of ecotype was suggested by __________. (**Turreson**)

137. The nitrogenous base that is present in RNA but not in DNA is_____________. (**Uracil**)

138. _____________ is used to induced plolyploidy in plants. (**Colchicine**)

139. Nullisomy can be represented by_____________. (**2n-2**)

140. Out crossing breeding system maintains a high degree of_____________. (**Genetic variation**)

141. Dominance variation is due to the interaction of specific alleles at_____________. (**Gene locus**)

142. The increased performance of F1 hybrid over its parents is called_____________. (**Heterosis**)

143. Quantitative traits are controlled by_____________ and_____________ genes. (**Minor and poly**)

144. _____________is the process of making RNA from DNA. (**Transcription**)

145. The common form of DNA present in the living organism is the_____________ form. (**B**)

146. The process of making proteins from RNA is_____________. (**Translation**)

147. _____________ is the best tool for a tree breeder to increase the genetic gain. (**Selection**)

148. Vertical antenna like branches often occurring in conifers outside the normal branch whorl is called as____________. (**Ramicorn**)

149. Minimum number of clones to be planted in the first generation seed orchard is. (**25 – 40**)

150. ____________ orchard has broader genetic base and less selection differential. (**Clonal seed orchard**)

151. ____________ is the most important factor which determines stability of seed. (**Moisture**)

152. The optimum temperature and humidity in the culture room is normally and ____________ respectively. (**25 °C and 80%**)

153. When germination of all seeds occurs over a relatively brief period it is called. (**Quasi simultaneous**)

154. ____________ is the systematic removal of individuals in orchards. (**Roughing**)

155. ____________is a fertilized mature ovule which possesses an embryonic plant, food reserve and productive coating in a plant species. (**Seed**)

156. The visible scar observed on a seed at the place of the detachment from seed stack is known as____________. (**Hillum**)

157. DNA may be described functionally as a source for____________ production. (**RNA**)

158. Ratio of the genetic variation to the total variance which determines the relationship between phenotype and genotype is known as____________. (**Heritability**)

159. Some plants take advantage of the sex drive of certain insects, it is called as. (**Plant mimicry**)

160. The smallest unit of genetic material that cannot be divided but is interchangeable among units refers to__________. (**Recon**)

161. Amino acids are joined together in proteins by____________ bonds. (**Peptide**)

162. Cells without cell wall are known as____________. (**Protoplasts**)

163. The role of mRNA is to carry the genetic instructions to the____________. (**Cytoplasm**)

164. In cross pollinated species, a true breeding line obtained by continuous inbreeding is called as____________. (**Inbred**)

165. The coding of amino acids by multiple sets of nucleotides is referred to as ____________ coding. (**Triplet**)

166. The individual with____________ chromosomes are commonly known as Aneuploid. (**2n ±1**)

167. The influence of genetic drift on allele frequencies increases as decreases. (**Population size**)

168. Progeny of a single plant obtained by asexual reproduction is____________. (**Clone**)

169. The progeny of a single homozygous self-pollinated plant is known as____________. (**Pure line**)

170. ____________ formula is useful in determining the extent to which a sexually reproducing population is evolving. (**Hardy-Weinberg**)

171. The genetic code is made up of units consisting of____________ nucleotides. (**Three**)

172. The immediate effect of pollen on the character of endosperm is called as____________. (**Xenia**)

173. Fungi and bacteria usually enter through the____________ region of tree. (**Stomata**)

174. The rudimentary root of the seed or seedling that forms the primary root of the young plant____________. (**Radicle**)

175. Naked seeds are generally found in____________. (**Gymnosperm**)

176. Multiple effect of a single gene is known as____________. (**Pleiotropy**)

177. ____________ is the responsible replacing RNA primers with new DNA in Okazaki fragments. (**DNA polymerase I**)

178. An organisim which has acquired a heritable variation as a result of sudden change in the hereditary material is called as____________. (**Mutation**)

179. A polyploidy having chromosome sets from different sources such as different species is known as____________. (**Allopolypoid**)

180. Heterosis over the better parent is called as____________. (**Heterobeltiosis**)

181. The inherent capacity of an organism to increase in numbers under ideal condition is known as____________. (**Biotic potential**)

182. The original geographic area from which seeds or other propagules were obtained is known as_____________. (**Provenance**)

183. The cross of an inbred line with an open pollinated variety is known as_____________. (**Top cross**)

184. The average performance of an individual in a particular series of crosses is termed as _____________. (**General combining ability**)

185. DNA synthesis is takes place in_____________ phase of cell division. (**S phase**)

186. The number of bases in the single turn of the DNA helix is_____________. (**Ten**)

187. _____________ form of RNA that carries the code from the DNA to the site where the protein is assembled. (**mRNA**)

188. The relationship between RNA and the DNA that served as the template is. (**Complementary**)

189. The nitrogenous base found in DNA but not in RNA is_____________. (**Thymine**)

190. _____________ substances is found in RNA but not in DNA. (**Ribose**)

191. Messenger RNA synthesized_____________ of a cell. (**Nucleus**)

192. The codon is found in_____________ RNA. (**Messenger**)

193. _____________is a type of mutation in which mutation in acodon leads to the substitution of one amino acid with another. (**Missense mutation**)

194. Eye colour in drosophila is an example of_____________ inheritance. (**Sex linked**)

195. In a population sex linked genes are carried by homogametric sex is_____________ of the total sex linked genes in the population. (**Half**)

196. In the second generation, hybrid vigour will be_____________ of the first. (**Half**)

197. _____________ is the practice of rapidly multiplying stock plant material to produce a large number of progeny plants. (**Micropropagation**)

198. Substances that cause cells to increase production of an enzyme are known as. (**Inducers**)

199. _____________ form of out breeding is out crossing. (**Mildest**)

200. The formation of callus from explants is called____________. **(Dedifferentiation)**

201. The respective stigma of *Eucalyptus teriticornis* shows exudation and change in colour from blue to____________. **(Cremish white)**

202. Physical separation of____________ and____________ is called hergogamy. **(Anther and stigma)**

203. An amino acid which is related to drought tolerance in plants is____________. **(Proline)**

204. The situation in which one gene hides the effect of a second gene when both are present in a chromosome is called____________. **(Epistasis)**

205. The smallest unit of genetic material that cannot be divided but is interchangeable among units refers to____________. **(Recon)**

206. The capacity of a cell to develop into a complete plant is called____________. **(Totipotency)**

207. The growth of tissue of living plants on a suitable culture medium in aseptic condtion is called____________. **(Plant tissue culture)**

208. ____________ is the most effective method for transfer of an oligogenic character in an agronomically superior genotype. **(Back cross)**

209. ____________is a major carbon source in plant tissue culture medium. **(Glucose)**

210. The two polynucleotide strands of the DNA double helix are____________ and ____________. **(Parallel and Complementary)**

211. ____________ hormone promotes ripening of fruit. **(Ethylene)**

212. ____________hormones regulates cell division and differentiation. **(Cytokinins)**

213. ____________hormone has anti-ageing effects. **(Cytokinins)**

214. ____________ hormone is responsible for stem elongation. **(Gibberellins)**

215. IAA is____________. **(Naturally occurring auxin)**

216. ____________ hormone is most useful in causing the rooting of plant cuttings. **(Auxins)**

217. The application of____________ hormones would be a logical first choice in an attempt to produce normal growth in mutant dwarf plants. **(Gibberlin)**

218. _____________ is the best method to break the unfavourable linkage. **(Diallel selective mating)**

219. The condition in which the pollen of a flower has no fertilizing effect on the stigma of the same flower is called_____________. **(Self-sterility)**

220. When the male gametes are shed into embryo sac through micropyle is called. **(Porogamy)**

221. The Shannon-Wiener index is defined as_____________. **(H= - sum [(P_i) × In (P_i)])**

222. Pollen was stored for periods ranging from ______to_____________. **(1 day to 3 years)**

223. The methods by which trees are combined to produce progeny is called. **(Mating design)**

224. TMV stands for_____________. **(Tobacco Mosaic Virus)**

225. Genetically pure tree stand is a group of individuals of a_____________ species. (Single)

226. Plus tree are with superior_____________. **(Phenotypes)**

227. Clonal seed orchards of forest trees are raised by_____________. **(Vegetative materials)**

228. In forest the seed production areas are meant for production of_____________ seeds. **(Quality)**

229. Tissue culture plants are developed from_____________. (V**egetative tissue**)

230. In forestry the letters 'tsi' stand for the words_____________. **(Timber stand improvement)**

231. Hybridization is a_____________ between two different species. **(Cross)**

232. Trees produce seed in_____________ quantity from year to year. **(Varying)**

233. The term_____________ is used to describe the physical expression of an inherited tree trait that can be measured and observed. **(Phenotype)**

234. The term_____________ is used to describe the genetic makeup of an individual tree. **(Genotype)**

235. Minimum area needed for seed orchard is_____________ acres. **(10)**

236. _____________ refers to same quantity of yield obtained for perpetuity. **(Sustained yield)**

237. The genetic purity of foundation seeds is_____________. **(99.5)**

238. Progeny of a homozygous single self pollinated plant is called_____________. **(Pure line)**

239. Phenotypically superior tree but not yet progeny tested is called_____________. **(Plus tree)**

240. _____________ means application of principles of trees breeding and silviculture in plantation forestry. **(Tree improvement)**

241. The term provenance means_____________ source. **(Geographic)**

242. Clonal seed orchards means orchards established with_____________ clones. **(Genetically superior)**

243. Quantitative traits exhibit continuous variation as they are controlled by. **(Polygenes)**

244. The chemical basis of heredity is_____________. **(DNA)**

245. The enzyme that replicates DNA by taking instruction from DNA template is_____________. **(DNA-helicase)**

246. The purines base of DNA is_____________. **(Adenine)**

247. The monomeric units of nucleic acids are_____________. **(Nucleotides)**

248. Piece of plant tissue used to initiate a culture is called_____________. **(Explant)**

249. A plus tree whose genetic worth has been established is called_____________. **(Elite tree)**

250. International teak provenances trials were initiated by_____________. **(FRI)**

251. Genetics is the collective study of _________ and __________. **(Heredity and Variations)**

252. The transmission of genetic characters from parents to offsprings is known as __________. **(Heredity)**

253. _____________ is the alternative form of a gene which are located on same position on the homologous chromosomes. **(Allele)**

254. Gene which controls more than one character is called ____________ gene. **(Pleiotropic)**

255. When a gene prevents the expression of another non-allelic gene then it is known as __________ gene. (**Epistatic**)

256. When F_1 individuals are crossed with any of their parents (either recessive or dominant) is called _________. (**Back Cross**)

257. When F_1 individuals is crossed with dominant parents then it is called __________. (**Out cross**)

258. When F_1 progeny is crossed with recessive parents then it is called ____________. (**Test cross**)

259. More than two alternative forms (due to mutation) of same gene are called as ___________. (**Multiple allele**)

260. Gene which causes death of an individual when it comes in homozygous condition is called ___________. (**Lethal gene**)

261. Collective inheritance of character is called __________. (**Linkage**)

262. Total gene component inherited haploid sets of chromoses in an individual is called as __________. (**Genome**)

263. ____________ was the first plant in which complete genome was sequenced. (***Arabidopsis thaliana***)

264. Formation of a variety after a sudden change in genetic material of a cell is called ___________. (**Mutant**)

265. Hardest material in animal kingdom is __________ while in plant kingdom is ___________. (**Enamel & Sporopollenin**)

266. DNA replication takes place in __________ direction with the help of DNA polymerase. (**5'-3'**)

267. The formation of m-RNA from DNA and then synthesis of protein from it, is known as __________. (**Central dogma**)

268. Rough Endoplasmic Reticulum helps in ___________ synthesis while Smooth Endoplasmic Reticulum helps in __________ synthesis. (**Protein & Lipid**)

269. Golgi body termed as _________________. (**Director of macromolecular traffic in cell/ middle man of cell**)

270. _____________ is called as suicidal bags of cell. (**Lysosomes**)

271. _______________ is called power house of the cell. (**Mitochondria**)

272. _______________ are also called protein factory of cell. **(Ribosomes)**

273. _____________ is considered as controller or director of cell. **(Necleus)**

274. Father of tree improvement is _____________. **(Bruce zobel)**

275. Group of population that generally interbreed with one another and that intergrade more or less continuously are referred to as _________. **(Race)**

276. Individuals that are more closely related to each other than to other individuals in a population are called __________. **(Family)**

277. A group of individuals within a family are referred to as _________. **(Siblings)**

278. Variation in a population is ___________ by mutation & gene flow and _________ by natural selection & genetic drift. **(Increased and reduced)**

279. The average performance of the progeny of an individual when it is mated to a number of the other individuals in the population is know as ___________. **(General Combining Ability (GCA))**

280. The average performance of the progeny of a cross between two specific parents that are different from what would be expected on the basis of their GCA alone is known as ___________. **(Specific Combining Ability (SCA))**

281. Tree that are located in the same stand having nearly same age and growing on the same or better site as the select tree and against which the select tree is graded is called _____________. **(Comparision/ check tree)**

282. Mating of individuals more closely related by ancestry is called ___________. **(Inbreeding)**

283. Meiosis takes place in the ____________. **(Reproductive cells)**

284. Genetic superiority of an individual is tested through ___________. **(Progeny testing)**

285. The most important statistical parameter for a breeder is _____________. **(Variance)**

286. ____________ is the method for extraction of cones. **(Tumbling)**

287. Factorial and diallel designs are commonly used in the seed orchard of __________. **(Conifers)**

288. A group of similar plants in structural features and performance is known as _______________. **(Variety)**

289. Antibiotic that cause misincorporation of amino acids in protein synthesis is called _____________. **(Puromycin)**

290. rDNA analysis provides information on ___________ relationships. **(Phylogenetic)**

291. In seed testing, the grow out test is conducted to determine the _____________. **(Genetic purity)**

292. Inability of a hybrid to produce viable offspring is known as ______________. **(Hybrid sterility)**

293. Reverse transcription is also known as ____________. **(Teminism)**

294. The tissue which nourishes microspores is ____________. **(Tapatum)**

295. Laws of heredity were first discovered by ____________. **(Mandel)**

296. IUCN stands for _________________________. **(International Union for Conservation of Nature and Natural Resources)**

297. India is one of the ________ of biological diversity in the world. **(12 centres)**

298. _____________ is used for area rich in endemic plants and containing high diversity of species. **(Biodiversity hotspots)**

299. GATT established in the year ______________. **(1984)**

300. The prophylactic measure that is used to prevent the entry of new diseases, insects and weeds from other countries is called as ___________. **(Quarantine)**

301. ___________ types of RNA molecules are needed to make protein. **(Three)**

302. The protein of the DNA molecules that is translated is composed of ___________. **(Exons)**

303. The protein of the DNA molecules that is not translated and is a noncoding portion of DNA is composed of ___________. **(Introns)**

304. Cytosine will form a base pair only with _____________. **(Guanine)**

305. ______________ developed first electron microscope. **(Knoll & Ruska)**

306. The stage of meiosis in which chromosomes pair and cross over is ____________. **(Prophase I)**

307. In translocation, the exchange of chromosomal segments occurs within the ___________. **(Same chromosomes)**

308. Xanthine is deaminated product of ______________. **(Guanine)**

309. The most effective method of selection is ___________. **(Progeny testing)**

310. Quantitative traits show __________ variations. **(Continuous)**

311. Range of h^2 and fitness is ___________. **(0 to 1)**

312. Crossing of two inbread lines of the same breed is ___________. **(Incrossing)**

313. Self incompatibility uses first reported by ____________. (Koelreuter)

314. The respective stigma of *Eucalyptus tereticornis* shows exudation and change in colour blue to _____________. **(Creamish white)**

315. Tropical trees are predominantly ___________ trees. **(Outcross)**

316. Emasculation of flower is necessary for hybridization in _________ plants. **(Self pollinated)**

317. Polyploid derived from two different species is called ____________. **(Allopolyploid)**

318. The number of bases in the single turn of the DNA helix is ________. **(Ten)**

319. The common form of DNA present in the living organisms is the ________________. **(B form)**

320. Highest uniformity is present in _____________ cross. **(Double)**

321. An individual with gametic chromosome number is ___________. **(Haploid)**

322. DNA can be denatured by heating at a temperature of _____________. **(94°C)**

323. The best method for developing disease resistant plant is ___________. **(Back crossing)**

324. For a mind pollinated species, the width of the pollen dilution zone is ___________. **(200-400m)**

325. ______________ is the unique functional position of organism. (**Niche**)

326. The book on applied tree improvement was written by ______________. (**Bruce Zobel and John Talbert**)

327. Selection differential can be calculated using ______________ and ____________. (**Selection intensity and Phenotypic variance**)

328. The average phenotypic value of the selected individuals, expressed as a deviation from the population mean is __________. (**Selection differential**)

329. Synthetic seed in sandal wood is attributable to ______&______. (**Bapat & Rao**)

330. More genetic gain can be achieved in the ____________ seed orchard. (**Clonal**)

331. Self incompatibility hampers the development of ___________. (**Inbreds**)

332. Selection can be resorted for ___________ trait. (**Heritable**)

333. ___________ seed orchard is used for hybridising diverse and compatible provenance. (**Provenance hybrid**)

334. The presence of male and female flowers in different plants is called __________. (**Dioecious**)

335. The unorganised mass of tissues are called as __________. (**Callus**)

336. The tree produced from the seed of a parent called _____________. (**Progeny**)

337. An individual having more than two basic sets of chromosomes is known as ____________. (**Polyploids**)

338. Usually, seed production areas should contain a minimum area of ___________. (**4 ha**)

339. Breeding value is ___________of the TA. (**Twice**)

340. Repeatability value set in ___________ of heritability (**Upper limit of h^2**)

341. Wide crossing is also known as ___________ hybridization. (**Distant**)

342. Character showing continuous variation and considerable effect of the environment is ______________. (**Quantitative character**)

343. Non additive variation accounts for __________ percent phenotypic variation. (**25**)

344. Lack of seed set on self pollination is ___________. (**Self incompatibility**)

345. If all variation was due to genetics, the broad sense heritability would be equal to ___________. (**One**)

346. _______________ studies are performed in order to describe genetic variation within and between population. (**Population variation**)

347. The best quality seeds would be obtained for producing quality planting stock from _____________. (**Clonal orchard**)

348. If the coefficient of selection is 0.25, then the fitness is ___________. (**0.75**)

349. The application of _____________ hormones would be a logical first choice in an attempt to produce normal growth in mutant dwarf plants. (**Gibberellin**)

350. While responses to plant hormones are slow, ____________ hormones has been shown to be involved in the rapid opening and closing of stomata. (**Abscisic acid**)

351. The period from when the flower open to when seed is set called as _____________. (**Anthesis**)

352. _______________ is the best method to carry forward the maximum number of genotypes/ genes to the advanced generations against the force of natural selection. (**Single seed descent method**)

353. Single seed descent method is the modified form of ___________ method. (**Bulk method**)

354. Pollen grains falling on the stigma of another flower of the same kind but on a different plant is called as ____________. (**Cross pollination**)

355. A measure of the possible choices of information provided by a gene is called as _____________. (**Genetic diversity**)

356. Assessment of Genetic Diversity can be done by __________. (**Species richness**)

357. ____________ is a measure that accounts for both richness and proportion of each species. (**Simpson diversity index**)

358. The current method of choice to infer phylogenetic trees is __________. (**Cladistics**)

359. ___________ is a broad term for the diagrammatic presentation of a phylogenetic tree. (**Dendrogram**)

360. _________ in which only one parent is known for any given progeny. (**Incomplete pedigree design**)

361. Groups of parents of one se are mated to members of the other se is known as __________. (**Nested design**)

362. The selection method predominantly followed in plantation is ___________. (**Comparison tree method**)

363. The exotic tree which has natural hybrids in India is __________. (**Eucalyptus**)

364. ___________ mating system is the mating between phynotypic dissimilar individuals belonging to same populations. (**Phynotypic disassortative**)

365. _________ mating system reduces frequency of other alleles. (**Random mating system**)

366. EDTA stands for _________. (**Ethylene diamine tetra acetic acid**)

367. RAPD stands for _________. (**Randomly Amplified Polymorphic DNA**)

368. NOA stands for _________. (**Napthoxy acetic acid**)

369. The membrane filter used in tissue culture should contain pore size of __________. (**Less than 0.45 micro meter**)

370. From a single cell, development of reproducible non zygotic embryos capable of germinating to form complete plants is called __________. (**Somatic embryogenesis**)

371. The formation of callus from explants is called __________. (**De differentiation**)

372. Rhizogenesis in plant tissue culture involves ___________. (**Root formation**)

373. Aseptic isolation and growth of sexually produced embryos in vitro to produce viable plant is called ____________. (**Embryo culture**)

374. National chemical laboratory is located at ____________. (**Pune**)

375. In recurrent mass selection the genetic gain for the traits of medium heritability is _________. (**12.5%**)

9

Forest Biotechnology

1. Term biotechnology was coined by_____________. (**Karl Ereky**)
2. Applied use of molecular biology and recombinant DNA technology is _______. (**Biotechnology**)
3. Bacteria are present everywhere except in_____________. (**Distil water**)
4. The branch of science dealing with the study of microbes is known as _______. (**Microbiology**)
5. The unique properties of each amino acid are determined by its particular_____________. (**R-groups**)
6. Optical isomerism arises from the presence of a/ an_____________ atom. (**Asymmetric carbon**)
7. In the breakdown of glucose to pyruvic acid,_____________ enzyme catalyses the successive steps. (**Eleven**)
8. _____________ plants are used to accumulate pollutants from soil thus clears soil from pollutants. (**Phytoremediation**)
9. _____________ contains only one type of nucleic acid. (**Virus**)
10. The molecular scissors used to cut DNA into specific genes of interest is called _____________. (**Restriction endonucleases**)
11. When pyrimidine is substituted by purines and vice versa, this type of mutation is known as_____________. (**Transversion**)
12. A set of disease resistance mechanisms that are not specific to a particular pathogen comes under_____________ immunity. (**Innate**)
13. Microbes have a relatively_____________ area that allows the chemical conversions possible. (**Large surface**)
14. Products will be applicable for patenting only if they are____________, _____________ and _____________. (**Novel, Inventive and Useful**)

15. The roots of a plant are converted into drug-producing structure in a process called as ______________. (**Microcosm establishment**)
16. The medium is added continuously to the ferment or tank to replace that, which has been fermented. This is true in case of______________ technique. (**Continuous flow**)
17. Removal of tropical strain by including the negative super coiling is carried out by ______________. (**DNA gyrases**)
18. Chromosomal aberration in which two breaks occur in a chromosome and the intercalary segment reunites in a reverse order is categorized as______________. (**Inversion**)
19. The term 'rhizosphere' was coined by______________. (**Hiltner**)
20. Structure involved in the transfer of genetic material during conjugation of bacteria is ______________. (**Pilus**)
21. Opine genes are present on a Ti plasmid, but not on any component of a/ an ______________ system. (**Binary vector**)
22. The quickest way to produce homozygous breeding lines from heterozygous parents is through______________. (**Double haploids**)
23. Virions are naked______________ viruses. (**RNA**)
24. ______________ bond in the DNA provides the backbone. (**Phosphodiester**)
25. Haploid production cannot be done through______________ culture. (**Callus**)
26. The rhizobium bacterial genes responsible for nodule formation in legumes by symbiosis are termed as______________. (**Sym genes**)
27. DNA replication occurs in______________ stage of cell division. (**S phase**)
28. During meiosis actual reduction of chromosomes occurs in______________ stagc. (**Anaphase**)
29. For clearing oil spills______________micro organism is used. (***Pseudomonas putida***)
30. RAPD molecular markers are______________ in nature. (**Dominant**)
31. AFLP molecular markers are______________ in nature. (**Co-dominant**)

32. STMS stands for_____________. (**Sequence Tagged Microsatellite Sites**)
33. RAPD stands for_____________. (**Random Amplified Polymorphic DNA**)
34. _____________ technique is used for DNA fingerprinting. (**RFLP**)
35. The nitrogen fixation of a nodulating plant can be assessed by_____________. (**Acetylene reduction assay**)
36. The process of production of endospores from a vegetative cell is called as_____________. (**Sporogenesis**)
37. The technique is used for amplification of DNA is_____________. (**PCR**)
38. _____________ techniques uses bacterial enzymes for *in vitro* amplification of DNA. (**PCR**)
39. Azotobacter fixes nitrogen in_____________ soil. (**Rich in organic matter**)
40. Transgenic expression study can be done by_____________. (RT-PCR)
41. DNA is fit for making tools in case of nanotechnology and due to_____________ nature of DNA,_____________in DNA conformation and_____________ of DNA. (**Branching, Flexibility and Small size**)
42. Cytoplasim hybrids are known as_____________. (**Cybrids**)
43. _____________ phyto hormone is used for development of synthetic seeds. (**ABA**)
44. During stress_____________ amino acid is accumulated. (**Proline**)
45. The bond between sugar and nitrogeneous base in case of DNA is known as. (**Glycosidic bond**)
46. In tissue culture regeneration of shoot and root occurs by manipulating the balance of_____________ and_____________. (**Auxin and Cytokinin**)
47. Fermentation carried out in a limited volume of closed container is termed as_____________. (**Batch**)
48. _____________DNA structure forms left hand helix. (**Z DNA**)
49. _____________and_____________was given the double helical structure of DNA. (**Watson and Crick**)

50. ____________ amino acid participate in urea cycle. (**Arginine**)
51. The ability of Plant cell to develop into a complete plant is known as.____________(**Totipotency**)
52. In the breakdown of xylan, the action of exo enzyme yields____________. (**Xylose**)
53. After transformation of plant cells through agrobacterium,____________ antibiotic is used to carry out agrobacterium free selection of cells. (**Carbencillin**)
54. Sieve tubes of phloem tissue lack in____________. (**Nucleus**)
55. Transfer of genes in between different species is known as____________. (**Horizontal gene flow**)
56. During DNA isolation____________ chemical is used for breaking of plasma membrane. (**CTAB**)
57. Humus is primarily formed by____________. (**Bacteria**)
58. In the breakdown of xylan, the action of endo enzyme yields____________. (**Xylobiose**)
59. In a/ an____________ protocol, bacteria with engineered abilities to detoxify pollutants are intentionally released in an area. (**Bioremediation**)
60. ____________purification can be done by affinity chromatography, ion exchange chromatography and sephadex column. (**Protein**)
61. Development of high density molecular linkage map of a particular mapping population can be used for____________ ,____________ and____________. (**MAS, map based cloning and species characterization**)
62. During RNA isolation____________chemical is used to prevent RNA from contamination. (**DEPC**)
63. L. T. Plasmid in *E. coli* produces____________. (**Enterotoxin**)
64. ____________is the acid preferring nitrogen fixing bacteria that dominates forest floor. (**Thiobacillus**)
65. Aerobic bacteria generally convert cellulose into____________ and____________. (**CO_2 and cell substance**)
66. In lac operon, allolactose act as____________ inducer. (**Gratituous**)

67. In tissue culture disease resistance can be obtained by_____________ variation. (**Somaclonal**)

68. Plant_____________ disease can be controlled by antisense RNA technology, by using ribozymes, RNAi technology and satellite RNA. (**Viral**)

69. For patenting a plant variety it should be_____________. ({**DUS, Distinctness, Uniformity and Stability**})

70. DNA chip is a wafer of_____________ and carries a large number of DNA probes. (**Silicon**)

71. The process of conversion of nitrate into element nitrogen is termed as_____________. (**Denitrification**)

72. DNA-synthesizing enzyme required specifically for replication is called as ________._____________ (**DNA replicase**)

73. _____________ are used as purified enzymes. (**Biosensers**)

74. _____________ enzyme is used in brewing. (**β-amylase**)

75. Selection of protoplast fusion hybrids can be done by_____________ selection. (**Visual**)

76. _____________ plants can be obtained by meristem culture. (**Virus free**)

77. Complete set of cloned DNA fragments of an organism is known as_____________. (**Genomic library**)

78. Bacteria possessing restriction enzymes get not affected because their genome is _____________. (**Methylated**)

79. Complete set of chromosomes found in the gamete of true diploid is_____________. (**Genome**)

80. _____________hormone related to drought tolerance. (**Absicis acid**)

81. In woody perennials, the bulk of the organic matter lost is derived from. (**Cellulose**)

82. _____________ organism is involved in iron transformation. (**Ferrobacillus**)

83. A group of microbes capable of utilizing CO_2 as a carbon source is termed as. (**Heterotrophs**)

84. Cryopreservation is done to store cells or cultures etc. for a long period at _____________ temperature. (**- 196 °C**)

85. Organelles such as mitochondria and chloroplasts in eukaryotic cells evolved from prokaryotic organisms by the engulfment by primitive nucleated cells is known as ____________ theory. (**Endosymbiotic**)

86. The techniques to clone the first gene were developed in the year____________. (**1972**)

87. Translation of histone proteins occur during____________ phase of cell division only. (**S phase**)

88. High density molecular map can be developed by____________. (**SNPs**)

89. ____________ is dominant insoluble form of phosphorus present in acidic soil. (**Iron phosphate**)

90. Methanogenic bacteria are____________ in nature. (**Anaerobic**)

91. Bioinformatics is the combination of____________ and____________. (**Biology and IT**)

92. Commonly used host in gene cloning is____________. (***Escherichia coli***)

93. The kind of microflora dominate in acidic environment is____________. (**Fungi**)

94. Restriction site for cloning should be____________ in nature. (**Palindromic**)

95. For cloning eukaryotic gene in prokaryotes, genes should be isolated from____________. (**cDNA library**)

96. Soil organisms depending on dead tissue of both plants and animals as food source is called as____________. (**Detritovores**)

97. ____________ enzyme plays important role in reverse transcription. (**Reverse transcriptase**)

98. The group of enzyme always produced in microbial cell irrespective of presence or absence of a substrate is called____________ enzyme. (**Constitutive**)

99. During the process of reverse transcription DNA is synthesized from____________. (mRNA)

100. ____________ amino acid provide twists in tertiary structure of proteins. (**Proline**)

101. Serine amino acid plays role in splicing of____________. (**Proteins**)

102. The undefined constituents of root exudates required for growth of ectomycorrhizae are called____________. (**M-factor**)

103. *In vitro* multiplication of plants from a small tissue explants is called as____________. (**Micropropagation**)

104. Cytokinin is mostly synthesized in____________parts of plants. (**Root tips**)

105. A technique uses electric discharge to produce pores on cell membrane for intake of recombinant DNA is called as____________. (**Electroporation**)

106. Bacteria which grow at a temperature around____________ are called as physhrophites. (**10 °C**)

107. Extra chromosomal genetic element found within bacterial cells and replicates independently of chromosomal DNA is known as____________. (**Plasmids**)

108. ____________ bacteria oxidizes molecular hydrogen into water in soil. (**Bacillus pantotrophus**)

109. ____________ is the main building block of pectin. (**Galactouronic acid**)

110. The mycorrhizae are____________ microflora. (**Endorhizosphere**)

111. Nucleus was discovered by____________. (**Robert Brown**)

112. Crossing of plants through fusion of____________ is called as somatic hybridization. (**Somatic cell**)

113. Transplanting a cell, tissue or organ from one nutrient medium to another is called as ____________. (**Subculture**)

114. The Azospirillum was first isolated and developed by____________. (**Dobereiner**)

115. ____________ is a process of formation of somatic embryos from callus. (**Embryogenesis**)

116. Isolating a gene and process of producing identical copies is called as____________. (**Gene cloning**)

117. An organism that can grow either in the presence or in the absence of O_2 is called ______. (**Facultative anaerobe**)

118. Quality control laboratories of National Biofertilizer Development Corporation are located at____________. **(Bangalore)**

119. For the growth of mesophilic microbes, the optimum temperature ranges between ____________. **(30 to 40 °C)**

120. ____________ is the father of genetic engineering. **(Paul Berg)**

121. Crossing of two genotypically different plants and genes from both plants are brought together is called____________. **(Hybridization)**

122. The process of production of endospores from a vegetative cell is called as. **(Sporogenesis)**

123. The microbial process by which contaminants are removed from soil is termed as ____________. **(Microbial remediation)**

124. The group of genes or cloned DNA fragments is called as____________. **(Gene bank)**

125. The DNA fragments used as finger print in identification of any organism is____________. **(Molecular marker)**

126. PCR stands for____________. **(Polymerase Chain Reaction)**

127. PCR was developed by _______. **(Karl Mullis)**

128. m-RNA is primarily used for transcription of____________. **(DNA)**

129. PCR techniques provide several copies of specific____________ sequence. **(DNA)**

130. Science of identifying the sequence of DNA in species is called as____________. **(Genomics)**

131. ____________is the carrier material normally used for multiplication of VAM. **(Vermiculite)**

132. Technique of genetic engineering by which a gene sequence with many identical copies is replicated is known as____________. **(Gene cloning)**

133. The distance between two genes that recombine with a frequency of____________ per cent map unit. **(One)**

134. Bacteria multiplying in the Dead sea are mostly likely to be____________. **(Halophiles)**

135. YEMA medium was introduced for rhizobium by____________. **(Fred)**

136. Mycorrhizae help in mobilizing____________ element. (**Phosphorus**)

137. Process by which a cell monolayer or a plant explants is transferred without subdivision into a fresh medium is called as____________. (**Reculture**)

138. ____________ is the basic unit of lignin. (**Phenylpropanoid**)

139. Transgenic plants are____________ organisms. (**Genetic modified**)

140. A process by which genetic material carried by a species cell is altered by incorporation of exogenous DNA in genome is called as____________. (**Transformation**)

141. Process during which the information in a sequence of DNA is used to construct mRNA is called as____________. (**Transcription**)

142. Replication of chloroplast DNA is controlled by____________. (**Nuclear DNA**)

143. The evidence for semi-conservative replication of DNA was first given by____________. (**Stahl and Meselson**)

144. The substances produces by microbes which can kill other microbes at low concentration is known as____________. (**Antibiotics**)

145. The rhizosphere effect can be quantitatively represented by____________ ratio. (**R:S**)

146. ____________ is one of a number of different forms of a gene. (**Allele**)

147. Enzyme used in polymerase chain reaction is____________. (**Taq-polymerase**)

148. A sequence of bases at a unique physical location in genome is called as____________. (**Marker**)

149. Microorganism was first time observed by____________. (**A. V. Leeuwenhoek**)

150. A compound that regulates entry of O_2 into the site of nitrogen fixation in leguminous plants is____________. (**Leghaemoglobin**)

151. Power house of cell is called as____________. (**Mitochondria**)

152. In methemoglobinemia, the haemoglobin is converted to____________. (**Methemoglobin**)

153. Apical bud dominance is caused by____________hormone. (**Auxin**)

154. Dormancy breaking hormone is____________. (**Cytokinin**)

155. Stress hardening in plants can be active by_____________. **(ABA)**

156. Fruit ripening hormone is_____________. **(Ethylene)**

157. Rhizobium from leguminous root nodules was first isolated by_____________. **(Beijerinck)**

158. Acetylene Reduction Essay is done to determine_____________activity. **(Nitrogenase)**

159. The most common genera of actinomycetes in soil are_____________. **(Streptomyces)**

160. Number of units of inorganic nitrogen immobilized for each hundred units is referred as_____________. **(Nitrogen factor)**

161. Bacteria thriving well within a temperature of_____________ ^{0}C are called mesophillic. **(15 – 45)**

162. The rhizosphere zone extends from root surface within_____________ mm. **(1-2)**

163. Individual fungal filaments are called_____________. **(Hyphae)**

164. Mycorrhiza is a symbiotic association between_____________ and_____________. **(Fungi and roots of higher plants)**

165. Aseptic technique and isolated bacteria in pure culture was developed by_____________. **(Joseph Lister)**

166. Rhizobium from root nodules of a non legume was first isolated by_____________. **(Trinick)**

167. The anaerobic breakdown of protein rich materials is termed as_____________. **(Putrefaction)**

168. Study of rhizophere was initiated by_____________. **(Hiltner)**

169. The enzyme responsible for breakdown of chitin are_____________ and_____________. **(Chitinase and Chitobiase)**

170. Halophiles grow in conditions of_____________salt. **(High)**

171. _____________ first time obtained pure culture of bacteria by serial dilution in liquid media. **(Joseph Lister)**

172. The principle of immunization was first time demonstrated by_____________. **(Louis Pasteur)**

173. N_2 fixing bacteria in soil were discovered by_____________. **(S. N. Winogradsky)**

174. Germ theory of disease was first proposed by____________. **(Fracastorius)**

175. Bacteria from root nodules of legume were isolated for the first time by____________. **(M. W. Beijerinck)**

176. Five kingdom scheme of classification was proposed by____________. **(R. H. Whittaker)**

177. Legume can obtained nitrogen from air, which was first time showed by____________. **(J. B. Boussingault)**

178. Distinction between prokaryotic and eukaryotic cells was made on the basis of absence or presence of____________. **(Nuclear membrane)**

179. ____________ is the building block of chitin. **(N-acetylglucosamine)**

180. Siderophores show strongest affinity to bind____________element. **(Iron)**

181. ____________ is the father of field experiment. **(J. B. Boussingault)**

182. Thermophiles grow in conditions of____________ temperature. **(High)**

183. The ideal pH range for the growth of fungi in the soil is____________. **(4.5 to 6.5)**

184. The ideal pH range for the growth of bacteria in soil is____________. **(6.5 to 8.0)**

185. The optimum C : N ratio for organic matter is____________. **(10:1)**

186. The term biochemistry was introduced by ______________. **(Carl Neuberg)**

187. The plant virus was first isolated in crystalline form of __________. **(TMV)**

188. Ribosomes are synthesized in ____________. **(Nucleolus)**

189. Taq polymerase used in PCR is _____________. **(Thermotolerant)**

190. The bacterial genes lack ___________. **(Introns)**

191. In cell culture medium commonly used carbon source is __________. **(Sucrose)**

192. __________ factor is responsible for detoxification of drugs in case of humans. **(P450)**

193. RNA interference is used to control different viral diseases in case of plants and animals. The active fragment involving in interference is ___________ & _________. (**Si RNA & Micro RNA**)

194. __________ is the targets genes in the chrroloplast. (**Transplastomics**)

195. _________ propagation can be done by through shoot bud proliferation, through adventitious buds and meristem tip culture. (**Micro**)

196. Urokinase enzyme is produced by ____________. (***Baccillus fastidious***)

197. The organism used to accumulate copper from factory waste is _________. (***Zoogloea ramigera***)

198. During splicing of proteins, __________ part is removed. (**Inteins**)

199. Antibiotic resistance is used as a selectable marker for selection of _____________ cells. (**Transformed**)

200. Marker free selection can be done by __________. (**Co-transformation**)

201. Histone proteins are __________ in nature. (**Basic**)

202. Stable transformation of genes in case of plants can be done by ____________. (**Agrobacterium mediated transformation**)

203. __________ enzyme is used to overcome oxidative stress. (**SOD**)

204. Genetic background of a species is known as __________. (**Germplsm**)

205. A group of adjacent genes that are identical or related are termed as ___________. (**Gene cluster**)

206. A virus that live in bacteria is ___________. (**Bacteriophage**)

207. An enzyme that closes nicks or discontinuities in one strand of double stranded DNA by creating a bond is ___________. (**DNA ligase**)

208. Precursor of IAA is ____________. (**Tryptophan**)

209. Root promoting hormone is __________. (**IBA**)

210. Up and down mutations occurs in __________ region. (**Promoter**)

211. The endosymbiont responsible for nodulation in actinorhizal plants is a/ an ___________. (**Actinomycetes**)

212. The plant gene responsible for nodule formation are termed as ______________. (**Nodulins**)

213. The site of nitrogen fixation in Frankia actinorhizal plants is __________. (**Vescicles**)

214. The quantum of nitrogen fixed by *Leucaena leucocephala* is __________________. (**110 kg/ha/yr**)

215. The microbes associated with seed surface is termed as __________. (**Spermosphere**)

216. The enzyme involved in lignin degradation is _________. (**Laccase**)

217. ___________ number of ATP molecules required for nitrogen fixation. (**16**)

218. The nitrogenise of Azotobacter overcomes oxygen toxicity by ___________. (**Conformational production**)

219. The expary period of any carrier based bioinculants is ________ months. (**Six**)

220. The rate of organic matter decomposistion under tropical to subtropical condition over temperate is _________. (**Faster**)

221. The native microflora are termed as __________. (**Autochthonous**)

222. The total microbial load of soil can be assessed by __________. (**Dehydrogenase**)

223. Volume of soil under the influence of roots of growing plants is referred as ___________. (**Rhizosphere**)

224. _____________ act as a storage organ for VAM fungi. (**Vesicle**)

225. ___________ is used as food source by microphytic feeders. (**Microflora**)

226. The size range of soil inhabiting protozoa is _________. (**5-100 μm**)

227. Mycotoxin is produced by __________. (**Fungi**)

228. The nitrogen fixing organism living inside the leaves of the aquatic fern azolla is a ____________. (**Cyanobacteria**)

229. The smallest inhabitants of soil is ____________. (**Bacteriophages**)

230. In cellulose molecule, individual glucose units are bound together by ____________. (**β-1-4**)

231. _____________ is a well known synthetic chelator. (**EDTA**)

232. The first organic fungicide is ____________. (**Thiram**)

233. ____________ represents the liason tissue between the fungal sheath and the host cells. (**Hartig** net)

234. __________ structures of VAMF connect the fungal ramifications inside roots with the mycelium of the fungus outside roots. (**Appressoria**)

235. The VAM fungi with spores bearing straight or angular stalks are grouped as __________. (**Glomus**)

236. The VAM fungi with spores bearing bulbous stalks are grouped as __________. (**Gigaspora**)

237. THE VAM fungi with sessile spores are grouped as __________. (**Acaulospora**)

238. The VAM fungi with spores regularly arranged on a central core are grouped as ____________. (**Sclerocystis**)

239. The phenomenon of enhancement in native humus decomposition owing to external addition of organic matter is termed as _________. (**Priming**)

240. The most abundant organic compound in nature is __________. (**Cellulose**)

241. The most soil moisture conditions favour nitrogen mineralization is __________. (**Alternate drying and wetting**)

242. Denitrification may also be termed as ____________. (**Nitrate respiration**)

243. ____________ represents a dominant anaerobic N2 fixer. (**Clostridium**)

244. Nif gene is responsible for ____________. (**Synthesis and regulation of Nitrogenase**)

245. The bacterial population per gram of rhizosphere of soil is ____________. (**6778 × 10^6**)

246. The temperature range of mesophiles is ____________. (**20 – 40°C**)

247. Solidifying agent in a medium is called as ___________. (**Agar agar**)

248. Extra chromosomal genetic materials is ____________. (**Plasmid**)

249. Bacteria which grow at a temperature around 10° are called as ____________. (**Phychrophites**)

250. Thermophiles grow in conditions of high ____________. (**Temperature**)

251. Halophiles grow in conditions of high ____________. (**Salt**)

252. An organism that can grow either in the presence or in the absence of mol. O_2 is called a/an ____________. (**Acidophiles**)

253. Germ theory of disease was first proposed by ____________. (**Jacob Manle**)

254. The production of CAMP from ATP is catalysed by ____________ membrane bound enzyme. (**Adenylate Cyclase**)

255. ____________ bacteria oxidizes ferrous compounds into ferric ones. (**Leptothrix ochracae**)

10

Tree Seed Technology

1. _____________ is a fertilize ovule consisting of intact embryo stored food and seed coat, which is viable and has got the capacity to germinate. **(Seed)**
2. The content of other than seeds such as straw, soil particles, small stones and seeds of other species is called_____________. **(Inert material)**
3. Initial seed obtained from the selected individual plants of a particular variety or parental line, used to produce breeder seed and forms the basis of the total seed production chain is_____________ seed. **(Nucleus)**
4. The loss of weight of a sample when it is dried due to decrease in_____________. **(Moisture)**
5. A specified quantity of seed, physically identifiable in respect of which an international analysis certificate may be issued is called_____________. **(Seed lot)**
6. The blowers are used for_____________. **(Purity of seeds)**
7. A type of seed that is not certified but the labelling is done to indicate the quality or standard in a truthful manner is called_____________ seed. **(Truthfully labelled seed)**
8. The reagent used for biochemical test for determining seed viability is_____________. **(Tetrazolium chloride or Bromide)**
9. The food material contained in_____________that nourishes the embryo during its development. **(Endosperm)**
10. A large proportion of seed used in the forest regeneration is obtained from_____________. **(Natural stand)**
11. The condition of a viable seed which presents it form germinating when supplied with suitable environmental condition is called_____________. **(Dormancy)**
12. The disruption of seed coat by chemical treatment in a strong acid to increase their permeability to water and gases is called_____________. **(Scarification)**

13. The capacity of the seed to remain capable of germination for some specific period of time is known as_____________. (**Seed viability**)

14. Tetrazolium chloride test of seed viability is also known as_____________ test. (**Biological**)

15. Rainfall during_____________ is the most limited factor that affect the fruit setting and seed production. (**Flowering**)

16. The reduction in_____________ considerably slow down the metabolic process which in term retard respiration process. (**Moisture content**)

17. A group of genetically identical plants desired asexually from a single mother tree is called_____________. (**Clone**)

18. _____________ ensures the supply of reproductive material for scientific trials and introduction of species in exotic areas. (**Evaluation**)

19. The genetically superior seeds for afforestation programme can be produced in _____________. (**Seed orchards**)

20. The colour changes in fruit is a suitable criteria for judging the_____________. (**Seed maturity**)

21. The propagators or protectors of their own kind are known as_____________. (**Seed**)

22. _____________ denotes the defective floral structure such as short style and long filament and long style and short filament. (**Heterostyle**)

23. International Seed Testing Association was founded in 1924 at_____________. (**Switzerland**)

24. The full name of ICIA is_____________. (**International Crop Improvement Association**)

25. National seed corporation was established in the year_____________. (**1963**)

26. The safe moisture content is storage in_____________ per cent. (**4-8%**)

27. To meet the continuous short and long term supply of reproductive material for a programme is called_____________. (**Utilization**)

28. _____________ is the supply of seeds for establishment of gene bank, botanical gardens, expansion of arboretums, seed herbaria. (**Conservation**)

29. _____________is mature floral ovary which may contain one or more seeds and includes accessory floral parts. (**Fruit**)

30. _____________refers to a condition in viable seed which prevents it from germinating when supplied with the factors which are adequate for germination. (**Dormancy**)

31. _____________is defined as one which can germinate under favourable conditions, providing any dormancy may be present is removed. (**Viable seeds**)

32. Method through which the genetic and physical characteristics of the seeds could be improved is called_____________. (**Seed technology**)

33. _____________ is the practice of burying seeds in moist medium, in alternate layers to overcome dormancy which keeps seeds in a cold and moist environment. (**Stratification**)

34. During fatty acid synthesis, the growing fatty acid chain is attached to_____________. (**ACP**)

35. _____________denotes seeds properties which determine the potential of rapid uniform emergence and development of normal seedling under a wide range of field condition. (**Vigor**)

36. _____________ plays an important role in fruit setting and seed formulation. (**Temperature**)

37. _____________ at the time of flowering is the most limiting factor that affects the fruit setting and seed production. (**Rainfall**)

38. In deciduous forest_____________plays an important role in pollination. (**Wind**)

39. Excess supply of_____________ and_____________ may result in fast vegetative growth, poor flowering and fruit bearing. (**Nutrient supply and Water**)

40. The seeds produced by over matured and immature trees are ________. (**Infertile**)

41. Impurity percentage in real value of seed is called as_____________. (**Dockage**)

42. Real value of seed is calculated by_____________. (**(Purity % × Germination % / 100)**)

43. The plants raised by____________ means flowers early as compared to the plants raised by seeds. (**Vegetative**)

44. ____________ is the cause of self-unfruitfulness in which the pollens of some varieties are in capable of fertilizing the flowers of other varieties or of the same varieties. (**Incompatibility**)

45. High temperature is injurious to____________ in which blossom drip occurs in summer due to low atmospheric humidity. (**Pollination**)

46. The process and development of fruit and seed will not be normal, the young fruits turn yellow, shrivel and fall off, if there is not adequate supply of____________. (**Food material/Nutrient**)

47. When seeds for____________ can be sown immediately after collection, no storage of seed is required. (**Afforestation**)

48. The seeds which can be dried own to a low moisture content of around 8-10% and can be stored at low temperature for long period of time are called____________. (**Orthodox seeds**)

49. The seeds which cannot survive drying below high moisture content (20-50%) and cannot be stored for long time is called____________. (**Recalcitrant seeds**)

50. When seeds are stored in____________ heating occurs which is caused by the respiration of seeds. (**High moisture content**)

51. When seeds are stored in____________ the seeds survive for longer period in storage without loss of germination capacity. (**Low moisture content**)

52. A reduction in moisture content considerably slows down the____________ process which interm retards the respiration process. (**Metabolic process**)

53. If the seeds are stored at lower temperature the rate of respiration will be____________ thus higher will be the life span of seeds. (**Lower**)

54. ____________ light is the photosynthetic active radiation. (**Red light**)

55. Fully ripped seeds return viability longer than seeds collected when immature because of formation of____________ in matured seeds which conserved viability. (**Biochemical compounds**)

56. Seeds damaged mechanically during extraction and cleaning rapidly lose its. (**Viability**)

57. The fungicidal and insecticidal treatments should not be used during_____________ as it can be harmful to the seeds. (**Storage**)
58. The low moisture content, high temperature and high O_2 pressure provide optimum storage condition for most of the_____________ seeds. (**Orthodox seeds**)
59. The high moisture content, high temperature and high O_2 pressure provide optimum storage condition for most of the_____________ seeds. (**Recalcitrant seeds**)
60. A number of changes occur in cell tissue with ageing of the seeds, these changes are called_____________. (**Physical changes**)
61. When a tree species seeds abundantly in a particular year is said as_____________. (**Good seed year**)
62. _____________ refers to the area where the mother trees of the seeds are growing. (**Provenance**)
63. The techniques used for vegetative propagation of a clone for mass production is called _____________. (**Grafting**)
64. The outstanding individuals of a particular species which were selected and reserved for improvement of the species by selection is called_____________. (**Plus tree**)
65. _____________ is a plantation of genetically superior trees, isolated in order to minimize pollination from inferior individuals. (**Seed orchards**)
66. A crop of vigorous, health and ideal growth that has been selected and reserved for seed collection for raising seed orchards or plantation is called_____________. (**Seed stand**)
67. The orchards which are raised by grafting clones in the form of bud of plus trees is called _____________. (**Clonal seed orchards**)
68. The orchards which are raised from the seeds of plus tree are called_____________. (**Seedling seed orchards**)
69. The specific gravity of seeds/ fruits decreases as the moisture content of the seeds/ fruits _____________with maturity. (**Decreases**)
70. The method of seed extraction from fleshy fruits is called_____________. (**Dipulping**)
71. _____________ can be used to separate out impurities having different specific gravity than that of the seeds. (**Floatation**)

72. The methods lowering down of the seeds moisture known as____________. **(Seed drying)**

73. The objective of seed certification is to ensure and____________ of the seed to the purchasers. (**Genuineness and quality**)

74. ____________is a legally sanctioned system for quality control of seed multiplication and production which consist of different committee members. (**Seed certification**)

75. The____________ of standing seed crops is an essential step in varying conformity of seed crops to prescribed certification stand orders. (**Field inspection**)

76. The methods prescribed for the determination of____________ are designed to reduce the oxidation, decomposition or loss of other volatile substances. (**Moisture content**)

77. High temperature is injurious to pollination in which blossom drop occurs in summer due to____________. (**Low atmospheric humidity**)

78. The seed should not be stored in high moisture content because heating occurs due to ____________ of seeds. (**Respiration**)

79. The seed lot which loses its germination capacity may be subjected to____________. (**Genetic drift**)

80. ____________ light is harmful to the seeds. (**Ultra violet**)

81. The number of physiological changes occurs in cell tissue will____________ the seeds. (**Ageing**)

82. The seed should be collected from____________ to trees. (**Middle aged to mature**)

83. The seed from young and over mature trees have____________ viability. (**Low**)

84. The collection of seeds should be _____ from isolated trees. (**Avoided**)

85. Inferior trees should be removed from the seed stand to improve the____________ of the seed. (**Genetic quality**)

86. Removal of under growth of trees from the seed stand is very important to facilitate ____________and____________. (**Seed production and Collection**)

87. The_____________ and_____________ should be highly productive for high seed production from selecting location of seed orchards. (**Climate and soil**)
88. The seed orchards should be located within the_____________ zone of the species. (**Natural geographical**)
89. The attack of insect, pest and fungi is the most_____________ on the forest floor. (**Rapid**)
90. Seed quality index is the vigour of________. (**Seed germination**)
91. Unable to viable seed to germinate even under optimal conditions provided for germination is known as_____________. (**Seed dormancy**)
92. Process of breaking or softening the seed coat to make it permeable to water and gases is known as_____________. (**Scarification**)
93. The application of proportional mixture of powders around the seed is known as ____________ (**Pelleting**)
94. Indian society of seed technology was founded in the year_____________. (**1971**)
95. Progeny of a nucleus seed is referred to as_____________ seed. (**Breeder**)
96. A legal system which ensures production of high quality seed in terms of purity and germination is called_____________. (**Seed certification**)
97. _____________ is considered as father of Seed Testing. (**Frederick Nobbe**)
98. Indian seed certification rules are based on the rules proposed by_____________. (**OECD**)
99. Natural inhibitor present in the seed is_____________. (**ABA**)
100. Seed dormancy is caused by_____________. (**ABA**)
101. Seed lot certification colour is_____________ and_____________. (**Orange and green**)
102. _____________ year was celebrated as the seed year. (**1993**)
103. Seed sample colour certificate colour is_____________. (**Blue**)
104. A condition of mature viable seed in which germination is considerably delayed, even though external conditions favour germination is due to_____________. (**Seed dormancy**)

105. The potential capacity of a seed to germinate is called as____________. **(Viability)**

106. The percentage by number of seeds in a given sample that actually germinate irrespective of time is known as____________. **(Germinative capacity)**

107. The percentage by number of seeds in a given sample that have germinated upto the time when the rate of germination reaches its peak is known as____________. **(Germinative energy)**

108. Percentage of the number of the seeds in a sample that develop into seedling at the end of the growing season is referred as____________. **(Plant per cent)**

109. ____________ and __________ can improve the physical quality of a seedlot. **(Cleaning & Grading)**

110. Morphological, physiological and functional changes that occurs from the time of fertilization until the matured ovules (seeds) are ready for harvest is known as ___________. **(Seed maturation)**

111. *Tectona grandis* seed mature in ____________ days. **(120-200 days)**

112. Indian seed Act was formed in ____________ and it has ___________ section. **(1966 & 25)**

113. Seed rule was formed in __________ and seed order in _________. **(1968 & 1983)**

114. Tag colour of the breeder seed is ____________.**(Yellow)**

115. Tag colour of Foundation seed is ___________. **(White)**

116. Tag colour of Registered seed is ____________. **(Purple)**

117. Tag colour of Certified seed is ____________. **(Azar blue)**

118. Systematic removal of individuals not desired for the perpetuation of the population is called ______________. **(Roughing)**

119. ______________ is the first step during seed germination. **(Imbibition)**

120. Maximum seed germination in _________ light and minimum in __________ light. **(Red & far-red)**

121. Branches developed after removal of surrounding trees are called ______________. **(Epicormics branches)**

122. Mostly vegetative orchards have been established by ____________. **(Grafting)**

123. Seed are formed, but the embryos developed without fertilization is called ____________. **(Apomixis)**

124. Embryos develop from the embryo sac without pollination is ____________. **(Parthenogenesis)**

125. Chromosome number of *Jatropha curcus* is ____________. **(2n = 22)**

126. ____________ is the most important factor which determines stability of seed. **(Moisture)**

127. In seed testing, the grow out test is conducted to determine ____________. **(Genetic purity)**

128. The stalk of the ovules is called ____________. **(Funicle)**

129. A fertilized mature ovule, which possess an embryonic plant, food reserve and protective coating in a plant species is ____________. **(Embryonic callus)**

130. Usually seed production areas should contain a minimum area of ____________ hectare. **(Four)**

131. The visible scar observed on a seed at the place of the detachment from seed is called ____________. **(Hillum)**

132. Genetically most pure seed is ____________. **(Nucleus seed)**

133. The growth hormones used for shoot multiplication under in vitro condition are collectively called as ____________. **(Cytokinins)**

134. Tetrazolium test for testing viability of seeds is attributable to ____________. **(Lakon)**

135. Biopiracy means ____________. **(Exploitation of bioresources without authentic permission)**

136. The resumption of active growth of embryo is called as ____________. **(Germination)**

137. In most of monocot plants, germination is ____________. **(hypogeal)**

138. A set standards used to regulate own or community activity in relation to biological world is ____________. **(Bioethics)**

139. Seed drying is very important to maintain its ____________ & ____________. (**Viability & Vigour**)

140. The rudimentary root of the seed or seedling that forms the primary root of the young plant is known as ____________. (**Radicle**)

141. The stage t which the maximum dry matter accumulation in seed is ____________. (**Physiological maturity**)

142. Seeds of tree species can be graded based on its density or weight by using ____________. (**Specific gravity separator**)

143. Keeping the fruits at moist condition for prolonged period before extraction is called ____________. (**Pre-curing**)

144. Bulk reduction in fleshy fruits can be achieved by ____________. (**Removing the fruit pulp**)

145. Removal of wings from seeds or fruits during processing is called ____________. (**Dewinging**)

146. The original tree from which parts to be propagated are taken as called as ____________. (**Ortet**)

147. The most common plant growth regulators compound used in rooting of tree species is ____________. (**IBA**)

148. Seed lot certificate colour is ____________. (**Orange and green**)

149. Embryo eased test ____________ is used for surface disinfestations. (**NaOCl**)

150. A seed which is just waiting for favourable environment condition to germinate is called ____________. (**Quiescent seed**)

151. The principle mode of pollination in Gymnosperm is by ____________. (**Wind**)

152. The "Embryon" is published by ____________. (**ASEAN forest tree seed centre**)

153. In the great majority of plant species, seed germination begins with ____________. (**Radicle**)

11

Forest Ecology, Biodiversity and Environment

1. _____________ is the main source of energy for the earth. (**Sun**)
2. The major cause of 'Minamata' disease in Japan was due to pollution by_____________. (**Mercury**)
3. _____________ method in water analysis is used for the quantitative estimation of dissolved oxygen. (**Modified Winkler's method**)
4. Systematic scientific study of fresh water ecosystem is called as_____________. (**Limnology**)
5. The Earth Summit held in 1992, was organised at_____________. (**Rio de Janeiro**)
6. The altitude of troposphere in atmosphere is about_____________ kilometre. (**Upto 11 km**)
7. Bhopal gas tragedy took place in the year_____________. (**1984**)
8. The unit of noise pollution is measured in_____________. (**Decibel**)
9. Most important mitigation strategy for reducing global warming and climate change is _____________. (**Carbon sequestration**)
10. _____________ is an agreement to mitigate emission of green house gases. (**Kyoto Protocol**)
11. _____________ radiations are injurious to living beings because they cause dehydration. (**Ultra-violet**)
12. The World Environment day is celebrated on_____________. (**5th June**)
13. The term 'Ecosystem' was first coined by_____________. (**Transley**)
14. Transition zone between two structurally different ecosystems is called. (**Ecotone**)
15. _____________ and_____________ is known as potential bio-indicator of air pollution. (**Bryophytes and Lichens**)

16. F.S Rowland, M. Molina and P. Crutzen were awarded the Nobel prize in chemistry in 1995, for_____________. (**Explaining the formation and depletion of ozone layer**)

17. The country which is famous for the use of wind energy as a non-conventional energy sources is_____________. (**Netherlands**)

18. _____________ country is popularly known as the "land of the windmills". (**Netherlands**)

19. Biological diversity act was passed by parliament in the year_____________. (**2002**)

20. Environment protection act was passed by parliament in the year_____________. (**1986**)

21. Green-house effect is due to increase in the atmospheric concentration of. (**CO_2**)

22. _____________per cent of nitrogen is present in atmosphere. (**78**)

23. Ozone depletion is caused by the increase in the level of_____________. (**Chloroflurocarbon**)

24. _____________ first used and defined "Ecology" in 1866. (**Hackel**)

25. _____________ plants may be found in a primary state of succession. (**Lichen**)

26. An abandoned field changes into a mature forest over an extended period of time through a process called_____________. (**Succession**)

27. The identification and classification of trees refers to_____________. (**Dendrology**)

28. The part of a tree that grows into a seed is called a/an_____________. (**Ovule**)

29. The separation between lobes on a leaf is a/an_____________. (Sinus)

30. The science of classification of trees or any other living matter is_____________. (**Taxonomy**)

31. The development of the system of scientific name for living organisms, the system of binomial nomenclature, is credited to_____________. (**Carolus Linnaeus**)

32. The class of trees, Gymnosperm is made up of_____________ trees. (**Coniferous**)

33. Roots which grow horizontally and help keep the tree upright are_____________. (**Lateral roots**)
34. Dendrology defined as the_____________ of all woody plants. (**Taxonomy**)
35. The system used to name all living organisms is properly called_____________. (**Binomial classification**)
36. An ovate leaf is_____________shaped. (**Egg**)
37. _____________gas released from paddy field. (**Methane**)
38. The Kyoto conference was held in the year_____________. (**1997**)
39. Places which receive equal rainfall are called_____________. (**Isohytes**)
40. Species become locally adopted population along the environmental gradient, such local population of given species is known as_____________. (**Ecotype**)
41. _____________is used to measure infiltration. (**Lysimeter**)
42. All India forecasts of rainfall is prepared at_____________. (**Pune**)
43. Humidity is measured by_____________. (**Hygrometer**)
44. Hydrology is the science which deals with_____________ as well as water. (**Surface and ground**)
45. Carnivores are usually_____________consumers. (**Secondary or Tertiary**)
46. The importance of ecosystem lies in_____________ and_____________. (**Cycling of materials and flow of energy**)
47. Ecosystem has two components *i.e.* and. (**Biotic and Abiotic**)
48. Biotic components includes_____________, and_____________. (**Producers, consumer and decomposers**)
49. Pyramid of number in grassland or a crop ecosystem is. (**Upright**)
50. In tree ecosystem, pyramid of number is_____________. (**Inverted**)
51. In grassland and tree ecosystem pyramid of biomass is always_____________. (**Inverted**)
52. In pond ecosystem, pyramid of biomass is_____________. (**Inverted**)
53. Pyramidofenergyingrasslandorpondecosystemisalways_____________ only. (**Upright**)

54. An ecosystem consisting of hawk, mice, grasses and snakes is at dynamic equilibrium; so_____________ organism would be found at the bottom of an energy pyramid is _______. (**Grasses**)

55. In most food webs, green plants are of prime importance because they provide_____________ for animals. (**Protection**)

56. Competition for food, light and space etc., is probably most severe between two closely related species occupying the_____________. (**Same niche/habitat**)

57. In an ecosystem, the population of primary_____________ is more than that of primary _____________. (**Producers, Consumers**)

58. The food chain, in which microorganisms break down the energy rich compounds synthesized by producers is_____________ food chain. (**Detritus**)

59. The cycling of elements in an ecosystem is called_____________. (**Biogeochemical cycle**)

60. The first link in any food chain is always_____________ , because they only can fix atmospheric CO_2 in the presence of_____________. (**Green plants, Sunlight**)

61. Trophic levels are formed by_____________linked in food chain. (**Organisms**)

62. _____________ fallout is most harmful for life on the globe. (**Nuclear**)

63. The direct entry of gasses pollutant may takes place into_____________ parts of the plants. (**Stomata**)

64. _____________ part of living cell is affected by SO. (**Plasmodesmata**)

65. _____________ is most pollutant country of Asia. (**China**)

66. _____________is the phenomenon of increase in the concentration of non-biodegradable substances with passage through food chain within the body of organism. (**Biomagnification**)

67. _____________ is an indicator of air quality. (**Lichen**)

68. Increase in temperature due to increase in CO_2 concentration in the atmosphere is called _____________. (**Greenhouse effect**)

69. _____________ is main source of water pollution. (**Industrial effluents**)

70. Biosphere is composed of organisms,____________ ,____________ and____________. **(Living organisms, Lithosphere, Atmosphere and Hydrosphere)**

71. The highest layer of atmosphere is____________. **(Thermosphere)**

72. The percentage of oxygen is____________ per cent in the atmosphere. **(20.947)**

73. Lithosphere is composed of____________ and____________ layers. **(Core and Mantle)**

74. Worldwide, the source of maximum consumption of fuel is____________. **(Oil)**

75. Soil is a complex mixture of____________ ,____________ and. **(Sand, Silt and Clay)**

76. ____________layers of atmosphere maintain an almost uniform horizontal temperature. **(Stratosphere)**

77. Coal burning power plants and factories are responsible for about 86% of____________ in atmosphere. **(SO_2)**

78. Fall in temperature with increasing elevation is known as____________. **(Lapse rate)**

79. A powerful eye irritant present in smog is____________. **(Proxy Acetyl Nitrate)**

80. Biological oxygen demand measures____________ needed by microbes. **(Dissolved oxygen)**

81. Biological treatment of water pollution is done with the help of____________. **(Phytoplanktons)**

82. Sulphur dioxide pollution is indicated by____________. **(Lichens)**

83. Checking of re-radiating heat by atmosphere CO_2, methane, ozone, dust is called ____________. **(Greenhouse effect)**

84. Acid rain is caused by high concentration of effect of____________ and____________. **(NO_2 and SO_2)**

85. Ozonosphere is mainly depleted by ____________. **(Chlorofluorocarbons)**

86. Air pollution is highest in____________ state of India. **(Kolkata)**

87. Certain epiphytic lichens indicate changing levels atmospheric pollution, specially of ____________. **(SO_2)**

88. Ozone layer is chiefly disturbed by____________. **(Supersonic jet)**

89. BOD of a river was found to be very low, this means water was____________. **(Clean)**

90. Nuclear power generates____________ pollution. **(Thermal)**

91. When pollution load is more in the lakes they cannot takes place self-purification because there is no____________to sustain microbial activity. **(Oxygen)**

92. The pollutions emitted by jet aeroplanes in outer atmosphere as fluorocarbons are known as____________. **(Aerosols)**

93. UV radiation to disinfect polluted water is not generally recommended because____________and____________ of water reduces it disinfection potential. **(Colour, turbidity)**

94. Decomposition of domestic wastes under natural processes is known as. **(Biodegradable process)**

95. Biochemical oxygen demand means amount of oxygen taken up by the ____________ present in the water. **(Microorganisms)**

96. ____________gas is produced in paddy and is associated with global warming. **(CH_4)**

97. Green house effect is enhanced in the environment by the gas____________ .**(CO)**

98. Taj Mahal is threatened due to the effect of____________. **(Sulphur dioxide)**

99. When huge amount of sewage is dumped into river, the BOD will ____________. **(Increase)**

100. The intensity of sound in the normal conversation is____________. **(10-20 decibel)**

101. At present the most significant cause of dwindling biodiversity is____________. **(Destruction of habitat)**

102. In the year 1984, the Bhopal gas tragedy was caused by the leakage of____________. **(Methyl isocyanate)**

103. Formation of ozone hole is maximum over_____________. (**Antarctica**)

104. _____________ organisms is used as indicator of water quality. (**Biggiatoa**)

105. The major contribution of green house gases to the atmosphere is. (**USA**)

106. Sewage drained into water bodies kills fishes because it increase competition for _____________. (**Dissolved oxygen**)

107. Ozone holes refers to hole in ozone layer in _____________ . (**Stratosphere**)

108. The water of holy Ganga river is pure due to the presence of_____________. (**Bacteriohages**)

109. Leaf curling is caused of_____________. (**O_3**)

110. Air pollution effects are usually found on_____________ parts of the plant. (**Leaves**)

111. Plants are known as purifiers of air due to process of_____________. (Photosynthesis)

112. The major source of methane in India is_____________. (**Rice field**)

113. _____________ is a combination of smoke and fog. (**Smog**)

114. Photochemical smog always contains_____________ and_____________. (**PAN and O_3**)

115. The indiscriminate use of fertilizers causes_____________ pollution. (**Water**)

116. Lichens is an indicator of_____________pollution. (**SO_2**)

117. The UV-rays in atmosphere are checked by_____________. (**O_3**)

118. The effect of SO_2 and its transformation products on plants is_____________. (**Chlorosis**)

119. Sound becomes a hazardous noise pollution if its level is above_____________. (**80 dB**)

120. BOD of a pond is related to_____________ per unit volume of water. (**Microbes**)

121. If the fertilizers are added to water_____________ will occur. (**Eutrophication**)

122. The branch of biology that deals with environment is called _____________. **(Ecology)**

123. An organism with other individual of the same species is known as. **(Population)**

124. Basic unit of Ecological hierarchy is_____________. **(Organism)**

125. A population genetically adapted to its particular environment is known as. **(Ecotype)**

126. Thin life supporting belt on and around earth is_____________. **(Biosphere)**

127. An area constituted under the Indian forest Act is known as_____________. **(Reserved forest)**

128. The average height of a regular crop is determined by_____________ formula. **(Lorey's formula)**

129. The microscopic, photosynthetic organisms that float on water surface are called _____________. **(Phytoplankton)**

130. A unit of land with a natural boundary having a mosaic of patches is known as _____________. **(Landscape)**

131. Solid earth crust is known as_____________. **(Lithosphere)**

132. The extent of biosphere from ocean floor to mountain top is_____________ kilometre. **(22.5 km)**

133. The role of an organism in the ecological system is known as_____________. **(Niche)**

134. The regulatory factor of environment are_____________ and_____________. **(Temperature and light)**

135. _____________ is determined by the temperature and rainfall of an area. **(Climate)**

136. The regions in 20°–40° latitude are known as_____________. **(Subtropical region)**

137. Temperature varies with change in_____________. **(Latitude and Altitude)**

138. As the height of the mountain increases, the temperature_____________. **(Decreases)**

139. The ecological factors related to soil and substratum is called_____________ factors. (**Edaphic**)

140. Capability of an organism to survive and reproduce in a particular environment is known as_____________. (**Adaptation**)

141. The ability of some animals to blend with the surrounding is_____________. (**Camouflage**)

142. The term 'Niche' was first coined by_____________. (**Grinnel**)

143. The sum of the effective environmental conditions under which an organism lives is known as_____________. (**Habitat**)

144. The maximum energy of solar radiation occurs in the region of_____________ and _____________. (**Green and Yellow**)

145. The lower stratum of woody and other vegetation above the ground cover is called as_____________. (**Under growth**)

146. The estimated light used in photosynthesis is less than_____________ per cent. (**2%**)

147. The response of the plant to photoperiod is called as_____________. (**Photoperiodism**)

148. The natural process by which plants become adapted to drought, cold or heat is known as _____________. (**Hardening**)

149. The relation of a site to weather conditions especially sun and wind is_____________. (**Exposure**)

150. The climate which is modified or adjusted by climatic factors is called. (**Bioclimate**)

151. The habitat together with the functions of a species is called_____________. (**Niche**)

152. Groups of species exploiting a common resource are called _____________. (**Guilds**)

153. The organisms in different geographical regions but occupying same niche are called _____________. (**Ecological equivalents**)

154. The animals that consume food for energy are called_____________. (**Consumers**)

155. The environment which has been modified by human activity is called. (**Anthropogenic environment/ Man made environment**)

156. Excessive plantation of trees to save the diminishing forest cover is known as ____________. (**Afforestation**)

157. Growing drought resistant grasses is known as____________. (**Lay farming**)

158. Restoration of soil fertility is a____________ process. (**Slow**)

159. A slope is divided into a number of small flat fields which is known as____________. (**Terrace farming**)

160. Ecological influence on characteristics of the soil brought about by its physical and chemical characteristics is____________. (**Edaphic factors**)

161. Excess nitrate in drinking water may lead to____________ and____________. (**Blue Baby syndrome and Methaemoglobinemia**)

162. The structural and functional unit of biosphere is____________. (Ecosystem)

163. The relationship between biotic and abiotic environment is called____________. (**Holocoenosis**)

164. Anthropogenic ecosystems are due to____________ activity. (**Human**)

165. Herbivores are____________ consumers. (**Primary**)

166. In shallow water the dominant producers are____________ . (**Macrophytes**)

167. In aquatic systems the dominant producers are____________. (**Phytoplankton**)

168. During photosynthesis, solar energy is converted into____________ energy. (**Chemical**)

169. ____________ and are the animals which feed on the dead bodies of other organisms. (**Scavengers and Detritivores**)

170. The amount of abiotic material present in an ecosystem is known as ____________. (**Standing stage**)

171. The cycling of gases in the biotic system is called____________. (**Biogeochemical cycle**)

172. The organisms in an ecosystem form one or more layers is known as ____________. (**Stratification**)

173. _____________ feed upon the organism of all trophic levels. (**Parasites**)

174. The rate at which organic matter is stored after the basic metabolic process is _____________. (**Net primary productivity**)

175. The transfer of energy from one trophic level to the next trophic level is called _____________. (**Food chain**)

176. A network of food chains in an ecosystem is known as_____________. (**Food web**)

177. Ecological pyramids were developed by_____________. (**Charles Elton**)

178. The graphical representation of ecological parameters is called_____________. (**Ecological pyramid**)

179. Ecological pyramids are also called_____________. (**Eltonian pyramids**)

180. A graphical representation of numerical relationship between different trophic levels of the food chain is known as_____________. (**Pyramid of numbers**)

181. The amount of living organic matter present in a particular environment is called _____________. (**Biomass**)

182. The major ecosystems with their groups of climax plants and associated animals are called _____________. (**Biome**)

183. The major forest biomes of India are_____________. (**Tropical rain forest and Coniferous forests**)

184. The gradual replacement of one community by another in the development of vegetation towards a climax is_____________. (**Plant succession**)

185. The culmination stage in plant succession for a given environment is known as. (**Climax**)

186. The change of plant and animals communities in an orderly sequence in an area is _____________. (**Ecological Succession**)

187. The process of change is called_____________. (**Reaction**)

188. The first community which inhabits a bare area is called_____________. (**Pioneer community**)

189. Succession beginning in dry habitat is called_____________. (**Xerach**)

190. The last and stable community in an area is called____________. (**Climax community**)

191. Sequences of successional stages on a bare rock are called____________. (**Lithosere**)

192. The entire series of communities of biotic succession from pioneer to climax community is known as ____________. (**Sere**)

193. Increase in fauna and decrease in flora would be harmful due to increase in. (**CO_2**)

194. A forest community will develop only if the climax is ____________. (**Moist**)

195. A mature soil developed in cool and humid climates by the process of podsolization is called as____________. (**Podsol**)

196. A group of soils having same range in particle size and physical properties based on texture is called as____________. (**Soil class**)

197. The extent to which the gross volume of the soil is unoccupied by solid particles is called as____________. (**Soil porosity**)

198. The ratio of the weight of organic carbon to the weight of total nitrogen in a soil or organic material is called as____________. (**C:N ratio**)

199. The food chain in which microorganisms breakdown dead producers is called ____________food chain. (**Detritus**)

200. Variety and variability among living organisms is known as____________. (**Biodiversity**)

201. The genetic variation in the same genes is known as____________. (**Alleles**)

202. The genetic variation within a species is called____________. (**Genetic diversity**)

203. The evolution of new species is known as____________. (**Speciation**)

204. The species which are confined to a particular country are called____________. (**Endemic species**)

205. The diversity between communities is____________ diversity. (**Beta**)

206. Biodiversity increases when we move from to. (**Poles to Equator**)

207. Red list of threatened species is published by____________. (**IUCN**)

208. IUCN stands for____________. **(International Union for Conservation of Nature and Natural Resources)**

209. Kaziranga National park is famous for____________. **(Rhinoceros)**

210. National park and wildlife sanctuaries are example of____________ biodiversity conservation. **(In situ)**

211. The hot spots concept was developed by____________. **(Norman Myers)**

212. The easiest way to store germplasm of wild plants is by____________. **(Seed gene banks)**

213. Undisturbed and legally protected ecosystem is known as____________. **(Core zone)**

214. The outer most part of the Biosphere resources is____________. **(Buffer zone)**

215. The Biosphere reserve was formulated by____________. **(UNESCO)**

216. ____________ deals with the conservation of ecosystem and the genetic resources. **(Biosphere reserves)**

217. The concept of bio reserves was first time started in the year____________. **(1975)**

218. Keoladeo Ghana Bird sanctuary is situated at____________. **(Bharatpur, Rajastan)**

219. First national park in India to start Tiger reserve is at____________. **(Nainital)**

220. The national park famous for Asiatic lion is____________. **(Sasan Gir)**

221. The national park known for one horned Rhinoceros is____________. **(Kaziranga)**

222. ____________ is a catalogue of Taxa that are facing the risk of extinction. **(IUCN Red list)**

223. The species which are not endangered and vulnerable presently but at risk are called ____________. **(Rare)**

224. When no member of the species remains alive anywhere in the world, this is____________. **(Extinction)**

225. Disappearance of a species from earth due to human activities is known as ____________ extinction. **(Anthropogenic)**

226. The extinction of species in the geological past is known as_____________ extinction. (**Background**)

227. The largest flying bird is_____________. (**California Condor**)

228. Research and educational activities are conducted in_____________ zone of biosphere. (**Buffer**)

229. Totally_____________ number of zoo around the world. (**800**)

230. The diversity of habitat over the total landscape or geographical area is known as_____________diversity. (**Gamma**)

231. The area reserved for the welfare of wildlife is called_____________. (**National park**)

232. Species diversity_____________from lower to higher altitude on a mountain. (**Decreases**)

233. "Origin of species" is the book written by_____________. (**Charles Darwin**)

234. Similipal biosphere reserve situated at_____________. (**Orissa**)

235. A Taxon not critically endangered but facing a high risk of extinction in median term future is known as_____________. (**Vulnerable**)

236. The Man and Biosphere programme was started by_____________. (**UNESCO**)

237. The traits determining particular characteristics of a species are called_____________. (**Genes**)

238. The study of reciprocal relationship between organisms and their environment is called_____________ . (**Ecology**)

239. _____________can be visualized as a thin file supporting belt on and around earth. (**Biosphere**)

240. The microscopic, photosynthetic organisms that float on water surface are called _____________. (**Phytoplankton**)

241. The solid and dry crust of the earth that forms the contents and other land masses is called_____________. (**Lithosphere**)

242. The water mantle of the earth is known as_____________. (**Hydrosphere**)

243. The study of soil and its related aspects is known as_____________. (**Pedology**)

244. _____________ are group of species exploiting a common resource in a similar fashion. (**Guilds**)

245. _____________ showed that closely related birds could occur together due to difference in there nesting and feeding habitat. (**Lack**)

246. Bisons of Asia, Kangaroos of Australia and Zebra of Africa are. (**Ecological Equivalents**)

247. Physical space occupied by an organism is_____________. (**Spacial**)

248. The place occupied by an organism, population or community is called_____________. (**Habitat**)

249. The term 'Niche' was first introduced by in the year. (**Grinnel, 1917**)

250. The first atomic bomb explosion was done in and the second in. (**Nagasaki, Hiroshima**)

251. Prolonged exposure to noise of_____________ or more may lead to impaired hearing. (**80 db**)

252. Degradation of organic wastes by microorganism in presence of oxygen is known as _____________. (**Composting**)

253. The process of controlled burning of the waste at high temperature of 850 ^{0}C is known as_____________. (**Incineration**)

254. The magnitude or intensity of energy released by an earthquake is measured by _____________. (**Richter Scale**)

255. Tropical cyclone in Caribbean Sea is called as_____________. (**Hurricanes**)

256. Property of certain elements to emit protons or electrons is known as. (**Radioactivity**)

257. Ozone is formed in the atmosphere by_____________. (**Photochemical reaction**)

258. Tabacco smoke contains a hydrocarbon called_____________. (**Benzopyrene**)

259. Petroleum used in automobiles contains_____________ as a antiknock agent. (**Tetraethyl lead**)

260. _____________ is the most abundant hydrocarbon in the atmosphere. (**Methane**)

261. Increase in concentration of soluble salts in soil is called_____________. **(Salination)**

262. Deposition of acids (Sulphuric acid/nitric acid) on the earth form from atmosphere is known as_____________. **(Acid rain)**

263. The unit of measurement of sound is_____________. **(Decibel)**

264. The radioactive gases are also called_____________. **(Greenhouse gasses)**

265. The thinning of stratospheric zone layer during spring time is called_____________. **(Ozone hole)**

266. The management of rainfall with resultant runoff is called_____________. **(Watershed management)**

267. The type of vegetation found in temperature forests is_____________. **(Conifers mixed with broad leaves species)**

268. Deep channels worn by water are called_____________. **(Gullies)**

269. Soil on river bank is poor in nutrients because of_____________. **(Leaching)**

270. A land for grazing is known as_____________. **(Pasture)**

271. The three components of environments are_____________, _____________, _____________. **(Productive, protective and waste assimilation system)**

272. The difference between the number of births and that of death is known as_____________. **(Rate of natural increase)**

273. The replacement levels in developed and developing countries are_____________ and_____________. **(2.1 & 2.7)**

274. The total fertility rate in developed countries is closely_____________. **(Replacement level)**

275. Replacement level in developing countries is always higher than 2.0 because some children die before reaching_____________age. **(Reproductive)**

276. The number of children a couple must produce to replace themselves is called _____________. **(Replacement level)**

277. The movement of individuals into a place or country is known as_____________. **(Immigration)**

278. The time required for a population to double itself is known as_____________. **(Doubling time)**

279. The number of babies produced per thousand individuals is known as_____________. **(Natality)**

280. Biological Oxygen Demand measures_____________. **(Dissolved O_2 needed by microbes to decompose organic waste)**

281. _____________ can serve as indicators to detect pollution of air by SO_2. **(Lichens)**

282. Sound becomes hazardous noise pollution at decibels above_____________. **(80)**

283. Preservation of living tissues/parts of plants in liquid nitrogen at a temperature of -196 °C known as_____________ .**(Cryopreservation)**

284. Derivative chemicals synthesized from plant products are called_____________. **(Botano chemicals)**

285. The diversity of communities along the environmental gradient is_____________. **(Beta diversity)**

286. The pioneer plant of xerach is_____________. **(Lichens)**

287. The ecological pyramids were devised by_____________. **(Charles Elton)**

288. The plant species that invade the bare habitat initially are known as_____________. **(Pioneer species)**

289. Graphic presentation of ecological parameters at each trophic level in an ecosystem is _____________. **(Ecological pyramids)**

290. Interconnected matrix of food chain in an ecosystem is known as_____________. **(Food web)**

291. A protected area where multiple use of the land is permitted by dividing it into certain zone is known as_____________. **(Biosphere reserve)**

292. The richest and most threatened reservoir of plant and animal life on earth is known as _____________. **(Hotspots)**

293. Excess fluoride in drinking water causes_____________. **(Fluorosis)**

294. The ability of reproductively active individuals to produce babies is known as _____________. **(Fertility)**

295. If the individuals of a species remain alive only in captivity or other human controlled conditions, the species is said to be____________. (**Extinct in the world**)

296. The increased concentration of greenhouse gasses in the atmosphere leads to ____________. (Global warming)

297. A glass house used for raising delicate plants is called____________. (**Greenhouse**)

298. ____________ is a product of incomplete decomposition caused by a group of bacteria called methanogens. (**Methane**)

299. The response of plants to elevated concentrations of CO_2 is known as____________. (**Carbon dioxide fertilization effect**)

300. Collecting and storage of rain water for future use is known as____________. (**Rain water harvesting**)

301. Most biotic resources are____________. (**Renewable**)

302. The practice of felling and burning of forests followed by cultivation of crop for few years is called____________. (**Shifting cultivation**)

303. The first movement against tree felling was fought by Bishnoi women of Keharili Village named____________. (**Amrita Devi**)

304. The practice of farming with increased organic input to the soil is known as____________. (**Organic farming**)

305. The land that has been destroyed due to mining is known as____________. (**Mine spoil**)

306. Conversation of grassland/forests into deserts is known as____________. (**Desertification**)

307. Raising of plant species having high calorific value and growth rate is known as ____________. (**Energy plantation**)

308. The book list containing a record of threatened species is____________. (**Red Data Book**)

309. The most rich biodiversity zones of India are. (**Western Ghat and North East**)

310. A zero population growth when birth and death rates are equal is known as ____________. (**Demographic Transition**)

311. UNESCO stands for____________. (**United Nations Educational, Scientific and Cultural Organisation**)

312. World wildlife week is celebrated during. (**1th – 7th October**)

313. Lithosphere, Hydrosphere and the Atmosphere will constitute the____________. (**Physical environment**)

314. Competition, mutualism and predation are the interaction occurring between ____________. (**Different population in an area**)

315. The assemblage of population of plants, animals, bacteria and fungi that live in is known as____________. (**Biological community**)

316. ____________ extends to about 22.5 km in thickness from ocean to mountain tops. (**Biosphere**)

317. Because of their small size, decomposers are known as____________. (**Micro consumers**)

318. The minerals and atmospheric gases enter into biotic system and after the death and decay of organisms return to the soil and atmosphere which is known as ____________. (**Biogeochemical cycle**)

319. The rate of increase in the biomass of consumers per unit and time is called. (**Secondary productivity**)

320. The larger carnivores prey upon the primary carnivores is called____________. (**Secondary carnivores**)

321. All type of ecosystem possess two type of food chain, they are____________ and. (**Grazing food chain and detritus food chain**)

322. Ecological pyramids were first devised by a British ecologist. (**Charles Elton**)

323. A semi-closed coastal body of water that has a free connection with sea is known as ____________. (**Estury**)

324. The sequence of succession stages on a bare rock are called as____________. (**Lithosphere**)

325. The last and stable community in the area is called as____________. (**Climax community**)

326. The first arrivals in a bare area are called____________. (**Pioneers**)

327. The sequence of succession stages on sand is known as____________. (**Psammosere**)

328. The sequence of succession stages of secondary succession is called. **(Subsere)**

329. A by-product of paper industry resembling raw dust is known as. **(Bregoli)**

330. The acid water received through rain, frost and snow is known as_____________. **(Wet deposition)**

331. Plants which are non-green, grow on other plants and obtain food from the host are known as_____________. **(Parasites)**

332. An area which is strictly reserved for the welfare of wildlife and activities like forestry, grazing and cultivation are not allowed is known as_____________. **(National Park)**

333. The first National park established in India was_____________. **(Jim Corbett National Park, Nainital)**

334. A species is considered_____________ when no member of the species remains alive. **(Extinct)**

335. The most undisturbed forests without any human impact are known as_____________. **(Pristine forests)**

336. The major drawback of incineration is_____________. **(releases a number of toxic chemicals in the atmosphere)**

337. The major disappearance of species from the face of earth due to human activities is known as_____________. **(Anthropogenic extinction)**

338. _____________ is the most comprehensive inventory of the global conservation status of plant and animal species. **(2000 IUCN Red list)**

339. Growing ornamental and fruit trees in urban areas alongside roads is known as _____________. **(Urban forestry)**

340. Using same land for farming and forestry is known as_____________. **(Agroforestry)**

341. About 3/4 of the earth's surface is occupied by_____________. **(Oceans)**

342. The biomass can be used to obtain fuels like_____________. **(Methanol and Ethanol)**

343. Biogas is a mixture of two gas namely_____________ and_____________. **(Methane & Carbon dioxide)**

344. The study that deals with the origin, formation and geographic distribution of the soil is known as_____________. (**Pedology**)

345. _____________ converts direct solar energy into electricity. (**Photo Voltaic cell**)

346. All materials originating from photosynthesis are known as. (**Biomass**)

347. The energy produced from the kinetic energy of water falling from a height is known as _____________. (**Hydro power**)

348. Breaking down of rock into small particles is known as _______. (**Weathering**)

349. Maturation of soil through development of humus is known as_____________. (**Pedogenesis**)

350. The removal of top soil from its place by various physical agencies is called_____________. (**Soil erosion**)

351. Men folk opposing the large scale felling of forest trees, the movement was known as _____________. (**Chipko movement**)

352. A special programme of tree plantation is held every year in the country known as _____________. (**Van mohotsava**)

353. The measures and practices taken to protect the soil against loss is known as. (**Soil conservation**)

354. Water logging and application of non-degradable chemicals leads to loss of_____________. (**Soil fertility**)

355. Planting of crops in rows or strips to check flow of water is called_____________. (**Strip cropping**)

356. The best soil for healthy and vigorous growth of most plants is_____________. (**Loamy soil**)

357. The successful establishment of a species after reaching to new area, as a result of adjustment with the conditions prevailing there is known as_____________. (**Ecesis**)

358. Successive replacement of communities in an area over a period of time is known as _____________. (**Succession**)

359. Autotrophic plankton is called_____________. (**Phytoplankton**)

360. Tropical forests occur in India in_____________ and. (**Kerala & Assam**)

361. The rate of biomass or organic matter production by the organisms is known as ____________. (**Productivity**)

362. New approach to conservation of environment is the establishment of____________. (**Biosphere conservation**)

363. The totality of genes, species and the ecosystems of a region are known as ____________. (**Biodiversity**)

364. The evenness in the number of individuals of a species is known as____________. (**Equitability**)

365. Chilika lake bird sanctuary is in ____________. (**Orissa**)

366. The two main centres of biological diversity in Western Ghats are &. (**Agastryamalai hills, Silent valley and The new Ammambalam reserve**)

367. Animals having high protection prospects are. (**Himalayan bear and musk deer**)

368. The widely used aerosol propellant and refrigerant is. (**Chlorofluoro-methane**)

369. ____________ dissolves silver and lead paints. (**Hydrogen Sulphide**)

370. A primary succession on bare rock starts with____________. (**Lichens**)

371. The surrounding in which the organism lives is known as____________. (**Environment**)

372. Short term properties of atmosphere is known as____________. (**Weather**)

373. The modified stems developed from double accessory buds in the axils of cotyledons is termed as____________. (**Lignotubers**)

374. The growth in trees is confined only in certain regions is called____________. (**Growing points**)

375. The process of releasing large amount of materials as a result of biogeochemical cycles and flow of energy is known as____________. (**Biological production**)

376. An assemblage of number of organisms usually of different species which occupy the same habitat is known as____________. (**Community**)

377. Conservation in natural habitat is known as____________. (***In-situ***)

378. Important constituent of biogas is____________. (**Methane**)

379. India ratified the Kyoto protocol in the year____________. **(2002)**

380. Biological diversity act was enacted during the year____________. **(2002)**

381. IUCN stands for____________. **(International Union for Conservation of Nature and Natural resources)**

382. CITES stands for____________. **(Convention on International trade in Endangered Species of Wild Fauna and Flora)**

383. The Ramsar convention relates to____________genetic resources. **(Wetland)**

384. Plants grows in extremely cold soils are____________. **(Psychrophytes)**

385. In submerged hydrophytes functional stomata are found in __________. **(No where in the plant)**

386. The first plant to reappear in a badly burned forest area will most probably be ________. **(Ferns)**

387. Energy and nutrients enter a community by way of the __________. **(Producers)**

388. The food chain that starts from plants and goes from smaller to larger animal is ___________. **(Predator food chain)**

389. ____________ is the driving force of ecosystem. **(Solar energy)**

390. In an aqueous environment microscopic animals and plants are collectively known as ___________. **(Planktons)**

391. A plant which tends to be restricted to calcarious soil is __________. **(Calciphile)**

392. The term 'Forest Succession' was first used by __________. **(Thoreau)**

393. The climax stage of estuarine succession is ____________. **(Evergreen Forest)**

394. Study of the biotic community and its abiotic environment is called ____________. **(Ecosystem ecology)**

395. Population ecology is otherwise called as ______________. **(Synecology)**

396. The poly climax theory in plant succession was developed by ____________. **(Tansley)**

397. The concentration of ozone is found maximum in ______________. **(Lower stratosphere)**

398. The __________ elemental cycle has no atmospheric reservoir. **(Phosphorus)**

399. The ministry of environment was set up in __________. **(1980)**

400. The first international conference on environment education held in ___________. **(New Delhi)**

401. Indian Institute of petroleum is situated at ___________. **(Dehradun)**

402. The major nutrient responsible for eutrophication is ___________. **(Phosphorus)**

403. The maximum permissible limit of free residual chlorine in water is ___________. **(0.2 ppm)**

404. The atmospheric layer reflecting radio waves is called ___________. **(Ionosphere)**

405. Biodiversity act was proposed in the year _____________. **(2000)**

406. Headquarter of UNEP is situated at ___________. **(Kenya)**

407. The dominance ofa new genetic form as a result of environment change is called ___________. **(Natural selection)**

408. The Jawaharlal Nehru Community Biodiversity movement lunched in India in the year ____________. **(1989)**

409. Project tiger was launched in India In the year ___________. **(1973)**

410. The most commonly used method for desalination of water is ___________. **(Reverse Osmosis)**

411. Flyash is the environmental pollutant generated by ___________. **(Thermal power plant)**

412. The use of DDT was banned for agricultural purpose in India in the year ___________. **(1985)**

413. The term symbiosis was coined by _____________. **(De Bary)**

414. The main pollutant of London smog was __________. **(SO_2)**

415. ____________ is the largest salt water lake in India. **(Chilka)**

416. ___________ produce food themselves through photosynthesis. **(Phototrophs)**

417. Wind speed is measured by ___________. **(Anemometer)**
418. PAR is the radiation of wavelength is ___________. **(400 to 700 nm)**
419. Blue colour of sky and red colour of sunset is due to ___________. **(Dispersion)**
420. Ultra violet radiations are absorbed by ___________. **(Ozone layer)**
421. A tropical grassland, usually scattered with trees is ___________. **(Savanna)**
422. World water resource day is observed on ___________. **(March 11th every year)**
423. Permissible limit of iron in drinking water is ___________. **(1 ppm)**
424. A disease caused by mercury pollution in Japan in 1953 is ___________. **(Minamata)**
425. Indian institute of tropical meteorology is situated at ___________. **(Pune)**
426. Meteorology is derived from ___________ word meteros which means ___________. **(Greek & high in the sky)**
427. Reflected solar radiations without any change in quality is known as ___________. **(Albedo)**
428. Solar radiations are measured by ___________. **(Pyranometer)**
429. Atmospheric pressure is measured by ___________. **(Aneroid Barometer)**
430. Relative humidity is measured by ___________. **(Pyschrometer)**
431. Rainfall is measured by ___________. **(Rain gauge)**
432. Chemical used for artificial rainfall is ___________. **(Silver Iodide)**
433. Reduction of nitrates and nitrites to nitrogen by microorganisms is known as ___________. **(Denitrification)**
434. A wooded wetland in which water is near or above ground level is known as ___________. **(Swamp)**
435. Rapid growth of algae in surface waters due to increase in inorganic nutrients is called as ___________. **(Algal bloom)**
436. Pollutant free alternative to petrol for automobiles is ___________. **(Propane)**

437. A chemical or physical agent that causes cancer is called __________. (**Carcinogen**)

438. ___________ gas may causes acid rain in an industrial area. (**SO_2**)

439. Agriculture in which several plant species are grown simultaneously to reduce insect infection and disease is called as ___________. (**Heteroculture**)

440. In a polluted lake, the index of population is ___________. (**Daphnia**)

441. Dome of heated air surrounds an urban area and traps a pollutant is known as ____________. (**Dust dome**)

442. Fungi are plant that lack ___________. (**Oxygen**)

443. Most poisonous pollutant in water is ___________. (**Arsenic**)

444. The gas which do not pollute air is ___________. (**Nitrogen Dioxide**)

445. Primary consumers in aquatic system are __________. (**Zooplanktons**)

446. An organic compound which build protein molecules is ___________. (**Amino acid**)

447. Colourless odourless and non-corrosive air pollutant is __________. (**SO_2**)

448. ____________ % of global water is occupied by Hydrosphere. (**70%**)

449. Maximum concentration of ozone is found in ______________. (**Lower stratosphere**)

450. _____________ test has self purification capacity of water body. (**BOD**)

451. Most commonly used disinfectant in water purification is ____________. (**Chlorine**)

452. Instrument used for measuring the concentration of ozone in air is called ____________. (**Ozonometer**)

453. The term ecosystem was coined and developed by _____________. (**A. G. Tansley**)

454. The organisms who eat dead organic matter are called __________. (**Decomposers**)

455. In our country 'van Mahotsav Day' is observed on ___________. (**First of July**)

456. The process of conversion of gas into a liquid is called ____________. **(Condensation)**

457. CNG stand for ________________. **(Compressed Natural Gas)**

458. Green petrol is ____________. **(Unleaded petrol)**

459. Eco labelling is governed by ____________. **(ISO 14001)**

460. Organism living at the bottom of a lake are ____________. **(Benthos)**

461. Kyoto protocol came into force on ____________. **(16 February, 2004)**

462. Succession showing changes in communities at a place is called ____________. **(Biotic succession)**

463. One or more ecotype in a single species is called ____________. **(Ecospecies)**

464. The science of relationship of life to weather is called ____________. **(Biometeorology)**

465. The ration between number of species and number of individuals is called ___________. **(Diversity Index)**

466. The process by which species evolve in response to other living members of their ecosystem is called ____________. **(Coevolution)**

467. The most favoured habitat of a plant is called __________. **(Microhabitat)**

468. Sound limit in industrial area is ____________. **(75 dB)**

469. The permissible Arsenic levels in drinking water is _________. **(20 µg/l)**

470. The first Environmental law in India was enacted in __________. **(1950)**

12

Forest Soil

1. Forest soils are_____________ in nature. (**Acidic**)
2. The term soil originated from latin word 'Solum' which means_____________. (**Floor**)
3. Most commonly used sources of N fertilizer in forestry is_____________. **$(NH_4)_2SO_4$**
4. _____________ is considered as the father of pedology. (**V. V. Dokuchaev**)
5. _____________ is considered as father of soil testing. (**Troug**)
6. Land capability class was given by_____________ and_____________. (**Klingebiel and Montgomery**)
7. The law of the minimum was given by_____________ in 1862. (**Liebig**)
8. Law of restriction was proposed by_____________. (**Liebig**)
9. The soil structural classification followed in India is according to_____________. (**USDA**)
10. One ton is equal to_____________ kilogram. (**907**)
11. _____________ number of categories in the land capability classification. (**Three**)
12. To study the origin, classification and description of soil is known as_____________. (**Pedology**)
13. _____________ is a process of accumulation in soil silica. (**Podzolization**)
14. Laterization is a process of accumulation in soil_____________. (**Iron and aluminium oxides**)
15. The alkaline permanganate method of soil available nitrogen analysis was developed by _____________. (**Subbiah and Asija**)
16. The behaviour of plant in relation to day length is termed as_____________. (**Photoperiodism**)

17. The relative proportion of sand, silt and clay in a soil is known as_____________. **(Soil texture)**
18. Density of soil is measured by_____________. **(Pycnometer)**
19. Bulk density of soil is generally_____________ of particle density. **(Half)**
20. _____________element is consider as energy currency for the plant. **(Phosphrus)**
21. _____________ nutrient element is indirectly related with drought and disease resistance. **(Potassium)**
22. Presences of_____________in soils increase availability of phosphorus. **(Mycorrhiza)**
23. _____________ number of nutrients are termed as essential elements for plants. **(17)**
24. The plants growing in salt water is known as_____________. **(Halophites)**
25. Instrument used to measure soil moisture in the field is_____________. **(Tensiometer)**
26. The plant nutrient element that stimulates root development, flowering and seed formation is_____________. **(Phosphorus)**
27. Application of clay to sandy soil is known as_____________. **(Marling)**
28. Total pore space is highest in_____________ soil. **(Sandy soil)**
29. Most dominant mineral on earth crust is_____________ and_____________ per cent. **(Feldspars and 48 %)**
30. Mechanical analysis of soil separation is done by_____________ method. **(Hydrometric)**
31. _____________ and_____________ are the active soil forming factors. **(Climate and Vegetation)**
32. Parent material, topography and time are the_____________soil forming factors. **(Passive)**
33. Urea contains_____________ per cent of nitrogen. **(46%)**
34. SSP contains_____________ per cent sulphur. **(12%)**
35. Walkley and Blacks method of organic matter determination allow per cent recovery. **(74)**

36. Boron is harmful for plants having_____________concentration. (**More than 3ppm**)
37. _____________ conditions are favourable for podsolization. (Cool and **humid climatic**)
38. Process of eluviation of oxides of iron and aluminium, humus and carbonates from the upper horizons and their accumulation in sub surface horizons is called as_____________. (**Podsolization**)
39. The end product of tropical weathering is_____________. (**Laterite**)
40. _____________ climatic condition are favourable for laterization. (**Warm and humid climate**)
41. Black soil cover approximately_____________ per cent of the geographical area of India. (**22**)
42. Red soils cover approximately_____________ per cent of the geographical area of India. (**20**)
43. The total number of essential nutrient elements required for plant growth is. (**17**)
44. _____________ is the dominant soil order in India. (**Inceptisols**)
45. Total area of saline soils in India is_____________. (**7.2 mha**)
46. Total area of sodic/ alkali soils in India is_____________. (**2.8 mha**)
47. Total area of salt affected soils in India is_____________. (**28 mha**)
48. Physical condition of soil is known as_____________. (**Soil tilth**)
49. _____________ fertilizer is highest nutrient content. (**SSP**)
50. Most of the plant obtains nitrogen from the soil in the form of_____________. (**Nitrate**)
51. _____________ is the essential micronutrient which requires for the growth of symbiotic microorganisms. (**Molybdenum**)
52. _____________ nutrient plays an important role in controlling the rate of transpiration. (**Potassium**)
53. _____________ form of nitrogen is present in urea. (**Amide form**)
54. Indian soil is mainly deficient of_____________ nutrient. (**Nitrogen**)
55. ________ is an element not essential for plant growth. (**Iodine**)
56. Nutrient mobility concept was proposed by_____________. (**Bray**)

57. ____________technique is used to determine age of fossils. (**Carbon dating**)

58. Most of the micronutrient elements are available in the____________ pH. (**Acidic**)

59. The soil microorganisms contain C:N ratio about____________. (**8:1**)

60. The saw dust contain C:N ratio about____________. (**1000:1**)

61. The arrangement of primary and secondary soil particles and their aggregates into a certain definite pattern is called as____________. (**Soil structure**)

62. Height rise of water in a capillary tubes indirectly proportional to____________. (**Surface tension of water**)

63. Available water holding capacity is____________ in clay loam soil. (**Maximum**)

64. The moisture situation in which water drain out of macrospores after saturation is called ____________. (**Field capacity**)

65. Scattering of detached soil particles by the impact of rain drops is called ____________ erosion. (**Splash**)

66. Particle size analysis of soil by____________. (**Picnometer**)

67. Bulk density in soils can be estimated by____________ and____________method. (**Core and Excavation**)

68. Soil bulk density is measured by____________. (**Hydrometer**)

69. Bulk density is high in____________ soil. (**Sandy soil**)

70. Soil aeration limits plants growth when per cent pores filled with water are more than ____________. (**50-60%**)

71. Mycorrhiza is a symbiotic association between____________ and____________. (**Fungi and roots of higher plants**)

72. MOP is chemically called as____________. (**KCl**)

73. Microorganism associated with the symbiotic N_2 fixation in nonlegumes species is ____________. (**Azolla**)

74. Most prominent soil group of soil is ____________ soil. (**Alluvial**)

75. White alkali soils are____________. (**Saline soil**)

76. ____________method is used to for estimation of organic carbon. (**Walkly and Black**)

77. ____________ element is essential for the flocculation of soil particles. **(Calcium)**

78. ____________ element is essential for the deflocculation of soil particle. **(Sodium)**

79. ____________is most widely used for correcting soil acidity. **(Lime)**

80. The soil horizon where maximum leaching of materials is observed is. **(A horizon)**

81. ____________ horizon is generally absent in Black soils. **(B)**

82. ____________ land capability class restricted to recreation, wildlife, water supply or aesthetic purpose. **(VIII)**

83. ____________is most suited phosphatic fertilizer for acid soil. (Rock **phosphate**)

84. ____________type of soil structure is dominant in grasslands. **(Spheroidal)**

85. Free living nitrogen fixing organism is____________. **(Azotobacter)**

86. Wind erosion is calculated by____________equation. **(Chepli and Woodruff)**

87. Erosion created by the activities of man and sometime by animals is called as ____________ soil erosion. **(Accelerated)**

88. Rill with more than____________ cm depth is generally called as gullies. **(30)**

89. Removal of a very thin layer of soil from the entire surface of large area is called ____________ erosion. **(Sheet)**

90. ____________group of microorganisms have highest biomass in soil. **(Fungi)**

91. Zn is deficit in____________soils. **(Alkaline)**

92. ____________ length of diameter of particles that is most susceptible to wind erosion. **(0.1 mm)**

93. Soil moisture retention is measured by____________. **(Permeameter)**

94. The pH of saline sodic soil is____________. **(Less than 8.5)**

95. The electrical conductivity of sodic soil____________ dsm^{-1}. **(Less than 4)**

96. ____________ number of classification in size of groups of soil particles by ISSS. (**Four**)

97. The organic soils commonly contain _______ per cent organic matter by volume and at leastper cent by weight basis. (**more than 50 and 20**)

98. Red colour of the soil is due to the presence of____________. (**Hematite**)

99. The carbohydrate compound of organic matter is____________ and____________. (**Cellulose and Hemicellulose**)

100. Total number of classes in nutrient index is____________. (**Three**)

101. Diagnosis and recommendation integrated system (DRIS) is applicable for____________. (**Long duration crops**)

102. ____________plant element accounts for at least one half of the total number of ions absorbed. (**Nitrogen**)

103. A mature and dry plant tissue contains about____________ per cent proteins. (**1-15**)

104. ____________is responsible for imparting vigour, resistance and regulating water with in the plants. (**Potassium**)

105. The criteria for essentiality was first put forward by____________. (**Arnon & Stout**)

106. A microorganism responsible for oxidation of nitrates is called ____________. (**Nitrobacter**)

107. The universal soil equation is ____________. (**A= RKLSCP**)

108. Application of____________ for the reclamation of sodic soil. (**Gypsum**)

109. Critical limit of electrical conductivity for irrigation water____________ dsm^{-1}. (**3**)

110. Degree of acidity or alkalinity of soil is expressed in____________. (**pH**)

111. ____________ is very slowly decomposing organic compounds in soil. (**Lignin**)

112. Windblown materials, comprised primarily of silt with some fine sand and clay are called as____________. (**Loess**)

113. Weathered products of rocks and rock minerals are called as____________. (**Parent material**)

114. _____________ is the father of soil physics. (**Hillel**)

115. The value of ESP in alkali soils is always_____________. (**More than 15**)

116. Rocks formed by solidification of molten magma at the surface of the earth are called _____________. (**Igneous**)

117. _____________ is the major component of igneous rock. (**Feldspars**)

118. In India,_____________ soil order has negligible area. (**Histosols**)

119. In India,_____________ soil order has highest area. (**Inceptisols**)

120. The power to resist a change in pH by the soil is called its _____________. (**Buffering capacity**)

121. Buffering capacity is_____________ for humus. (**high**)

122. Lowest category of soil classification includes_____________. (**Series**)

123. _____________ is the most dominant mineral in earth crust. (**Feldspar**)

124. Science dealing with the origin of formation of soil from unconsolidated parent material with special reference to factors and process of soil formation is known as _____________. (**Soil genesis**)

125. _____________ number of classes in a group, land suitable for cultivation. (**Four**)

126. Particle density =_____________. (**Weight of dry soil/ volume of soil particles**)

127. Material transported by the action of gravity is called as_____________. (**Colluvium**)

128. _____________ is a process of breakdown or dissolving of rocks and minerals on earth's surface. (**Weathering**)

129. Munsell colour chart contains about_____________ different coloured papers or chips. (**175**)

130. Hydromorphic soils are evolved by_____________ process. (**Gleization**)

131. The process of conversion of inorganic to organic form is called_____________. (**Immobilization**)

132. Silt materials transported by_____________ are called as loess. (**Wind**)

133. Soil water potential always has_____________ sign. (**Negetive**)

134. Application of fertilizer through drip irrigation is known as_____________. **(Fertigation)**

135. End product of denitrification process is_____________ and_____________. **(N_2O and N_2)**

136. Soil order characterized by excess of weathering with dominant iron and aluminium oxides clays, common in tropics is_____________. **(Oxisols)**

137. Wet bulk density=_____________. **(Weight of soil/ volume of soil)**

138. Value of bulk density is_____________than particle density. **(Lower)**

139. Capillary rise of water is due to the force of_____________. **(Cohesion and Adhesion)**

140. The size of capillary pore is_____________. **(more than 6μ)**

141. _____________water in the soil is found in the micro pores. **(Capillary)**

142. The water held between_____________ bar is called plant available water. While capillary water is retained at _____________. **(-0.33 to -15 bar and -1/3 to -15 bar)**

143. The capability of soil to produce plants is called as_____________. **(Soil productivity)**

144. When magma appears above the earth, it is called_____________. **(Lava)**

145. _____________ is the most resistant product to weathering. **(Quartz)**

146. Chief constituent of lime stone is_____________. **(Calcite)**

147. Layering of sedimentary rocks is known as_____________. **(Budding)**

148. _____________ is the study of relationship between plant and soil. **(Edaphology)**

149. Study of rock is called_____________. **(Petrology)**

150. Hygroscopic water is biologically inactive because it moves in_____________ form. **(Vapour)**

151. The inherent capacity of the soil to supply plant nutrients in adequate amount in suitable proportion for the growth of plants is known as_____________. **(Soil fertility)**

152. _____________fertilizer is vital for maintaining growth and helping plant resist diseases. **(Potassium)**

153. A fertile soil may or may not be productive but a productive soil is always. (**Fertile**)

154. Alluvial soils are geologically of two types such as____________ and____________. (**Khadir and Bangar**)

155. Black soils are poor in phosphorus, nitrogen and organic matter but rich in ____________ minerals. (**Kaolinite clay**)

156. The process of removal of soil particles from the parent's material and transportation of such particles by wind and/ or water is known as____________. (**Soil erosion**)

157. The first stage of water erosion is____________. (**Splash erosion**)

158. Rolling or sliding of larger soil particles is known as____________. (**Surface creep**)

159. Size range of particles in surface creep type of wind erosion is ____________. (**0.5-0.2 mm**)

160. The capacity of different agent causing soil erosion is known as____________. (**Soil erosivity**)

161. The susceptibility nature of the soil to erosion is known as____________. (**Soil erodibility**)

162. ____________ fertilizer is most suitable for acidic soil. (**Sodium nitrate**)

163. Fertilizer most suitable for water logged soil is____________. (**Ammonical fertilizers**)

164. ____________ was the first commercial mineral fertilizer and led to the development of the modern plant nutrient industry. (**SSP**)

165. Potassic fertilizer are suitable for acidic and heavy soils but not for____________. (**Alkaline soil**)

166. ____________refers to the decomposed mixture of dung and urine of farm animals along with litter and leaf over material from roughages or fodder fed to the cattle. (**Farm Yard Manure, FYM**)

167. A mass of rotted organic matter made from waste is called____________. (**Compost**)

168. A practice of ploughing or turning into the soil undecomposed fresh green plant tissue for the purpose of improving fertility status and physical structure of the soil is called as____________. (**Green manuring**)

169. ____________ is a group of bacteria that fixes nitrogen in association with the roots of leguminous crops. (***Rhizobium***)

170. ____________ is a renewable biofertilizer and can be mass-produced on the farm like blue-green algae. (**Azolla**)

171. ____________ is essential for seed germination and root development. (**Phosphorus**)

172. Hydrology is the science which deals with____________ and____________water. (**Surface and Underground**)

173. The rocks formed from____________ are called igneous rock. (**Molten magma**)

174. In India, forest resources accounting is mainly based on____________ and ____________ benefits. (**Tangible, Intangible**)

175. The plants which are able to grow in calcium rich soil are called____________. (**Calciphytes**)

176. Alluvial soils are formed as a result of soil deposited by____________. (**Water**)

177. The particle size of silt is____________. (**0.02-0.002mm**)

178. The pH of forest soil is generally less than____________. (7)

179. The pH of soil in teak forest is generally less than____________. (7)

180. Coarse textured soils such as sands are____________ in water holding capacity and nutrient content. (**Low**)

181. Litter from leaves, especially from needles at various stages of decomposition is called____________. (**Duff**)

182. A watershed is a____________ covered with vegetation. (**Drainage basin**)

183. The term alkaline refers to a substance with a pH of____________. (**More than 7.0**)

184. A soil with a pH of 3.5 is____________ in nature. (**Acidic**)

185. Salt damage to a tree is identified by____________. (**Foliage necrosis**)

186. A soil profile is made up of different soil layers which is known as____________. (**Horizon**)

187. Size of the silt particle according to USDA is____________. (**0.02-0.002 mm**)

188. Sand stone is_____________type of rock. (**Sedimentary**)

189. Iron pan is formed in_____________ type of soil. (**Laterite**)

190. The C:N ratio of humus is____________ while for normal soil, it is ____________. (**10:1 and 12:1**)

191. The rate of organic matter decomposition is_____________ , under tropical to subtropical condition. (**Faster**)

192. As the clay content of soil increase, the decay rate ___________. (**Decreases**)

193. The phenomenon of enhancement in native humus decomposition owing to external addition of organic matter is termed as_____________. (**Priming**)

194. density of earth is ____________ while density of rock is ____________. (**5.5 gm/c.c and 2.6-2.7 gm/c.c**)

195. Study of rocks is called _____________. (**Petrology**)

196. Vertical section through a soil is called ____________. (**Soil profile**)

197. A+B horizons collectively called as ___________. (**Solum**)

198. ______________is the most important chemical weathering process. (**Hydrolysis**)

199. The mass per unit volume of soil including the pore space is called as ____________. (**Bulk/ apparent density**)

200. Soil colour chart is given by ___________. (**Munsell**)

201. Movement of water into the soil is called as ___________. (**Infiltration**)

202. Amount of water required to produce a unit quantity of dry weight material is called as _____________. (**Water use efficiency**)

203. Capacity of the soil to change its shape under moist conditions is called as ____________. (**Plasticity**)

204. _______________ element is not found in acidic soil. (**Molybdenum**)

205. The process of removal of silica instead of sesquioxides to concentrate in the solum is called as ____________. (**Laterisation**)

206. Saline soils also called as ______________. (**Solon chalk/ white alkali**)

207. Alkali soil also called as ____________. (**Sodic soil/ solonetz/ non-saline alkali soils**)

208. Particle density and bulk density are expressed in term of____________. **(mg/m^3 or gms/cm^3)**

209. ____________ is called as night soil manure. (**Poudrette**)

210. Specific gravity of soil is _________ and for water it is __________. **(0.2 and 1.0)**

211. One gram of soil contains ____________ of microorganisms. (**14 millions**)

212. The length of engineer chain is ___________ while Gunter chain is ________. **(100 ft and 66 ft)**

213. The transfer of minerals from the top soil to sub soil through water is called as __________. (**Leaching**)

214. __________ fertilizer are fixed immediately after application in the soil. (**Phosphorous**)

215. *Aspergillus niger* is used to test the __________ element status in the soil. (**Potassium**)

216. Amount of exchangeable cations per unit weight of dry soil is called ____________. (**Cation Exchange Capacity**)

217. Cation exchange capacity is expressed in term of __________. (**me/100 g or c mol kg^{-1}**)

218. Capacity of a soil to absorb or release anion under normal soil conditions is called ____________. (**Anion Exchange Capacity**)

219. The process of accumulation insoluble lime salts (Mostly Ca & Mg) in soil horizons is called _____________. (**Calcification**)

220. E-value is used for determination of available _____________ in soils. (**phosphorous**)

221. L-value is used to calculate the ___________ supply to the soil. (**Phosphorus**)

222. _____________ is a process in which colloidal particles tend to attract each other and coagulate like a bigger particle. (**Flocculation**)

223. Formation of inorganic substances from organic substances by the microbial dccomposition is called _____________. (**Mineralization**)

224. The downward movement of water through the soil is called___________. (**Percolation**)

225. The disintegrated and decomposed mass of weathered rocks and soil material on the earth surface in which the soil develops is called ___________. (**Regolith**)

226. The process of movement of soil particles having size ranging from 0.05 to 0.5 mm in a series of bounces or jumps is known as ___________. (**Saltation**)

227. __________ represents the floating of small sized particles in the air stream. (**Suspension**)

228. Size range of particles in suspension type of wind erosion is ___________. (**Less than 0.05 mm**)

229. The rate of soil erosion in India is __________. (**16.4 t/ha**)

230. __________ soils have highest pore space than other soils. (**Clay soil**)

231. Soils of India is broadly classified into _________ groups. (**Eight**)

232. _____________ is the smallest unit of soil classifications. (**Soil series**)

233. Organic matter is found in __________ quantities in grass lands than forest area. (**Highest**)

234. Organic matter = Organic carbon (OC) × 1.724, So 1.724 is known as ___________. (**Bemlen factor**)

235. Highest acidic soils are found in ___________ state of India. (**West Bengal**)

236. When erosion is caused by excessive grazing, deforestation is called _____________. (**Anthropogenic erosion**)

237. Soil having at least _________ percent organic matter is called as organic soils, but it is generally less than _________ percent in Indian soils. (**20 and 0.5**)

238. Zero tillage practice was first started in ____________. (**USA**)

239. K and Na is determined by ____________. (**Flame photometer**)

240. Calcium ammonium nitrate (CAN) is also known as ___________, which contain 26% N. (**Kissan khad**)

241. Land capability classification (LCC) has _________ classes of soils. (**Eight**)

242. The size of fine clay particles in USDA classification of soil texture is _________. (**<0.002 mm**)

243. Particle smaller than ___________ microns cannot be separated by sizing. (<**20**)

244. More than 90 % of gaseous exchange is soil takes place by ___________. (**Diffusion**)

245. Gravitational water in soil move through the __________ pores. (**Macro**)

246. Water held against the force of gravity in microscopes is ______________. (**Capillary water**)

247. ______________ type of water which is available to plants. (**Capillary water**)

248. Soil water suction always has ______________ sign. (**Positive**)

249. Sandstone and limestones are ___________ rocks. (**Sedimentary**)

250. Materials transported and deposited by water are called as _________. (**Alluvium**)

251. Materials transported by water and settled out in lakes are called as _____________. (**Lacustrine**)

252. Sand materials transported by water are called as ___________. (**Aeolian**)

253. B horizon is a ___________. (**Illuvial horizon**)

254. E horizon is a ____________. (**Elluvial horizon**)

255. Most important weathering for soil formation is ____________. (**Chemical weathering**)

256. Soil that are formed from the same kind of parent materials but differing in their topographic positions within the landscape are called as ___________. (**Toposequence**)

257. Zonal soil forming process is ____________. (**Calcification**)

258. The total number of soil order is ____________. (**Twelve**)

259. ____________ percent of K2O in Muriate of Potash. (**58.0**)

260. Under alkaline condition N loss occur as __________. (**Ammonia**)

261. Dominant type of wind erosion in soil is _____________. (**Saltation**)

262. ____________ gave the statement water is the sole nutrient of plant. (**JB Helmont**)

263. Kanker nodules are found in mostly __________ soil. (**Red soil**)

264. Granite is ________ rock. (**Ignite**)

265. ___________ Fertilizer's are completely imported. (**Potassic**)

266. Fixation in soil is a problem in case of ________ fertilizer. (**Phosphatic**)

267. Azotobacter fixes atmospheric nitrogen by __________. (**Non-symbiotically**)

268. __________ micro nutrient deficient in Indian soils. (**Zn**)

269. The readily available form of phosphates in soil solution is ________ and ______. (**H_2PO_4 and HPO_4**)

270. _____________ is the most suitable reclamation measures for heavy metal contamination. (**Phytoremediation**)

271. The minute silicate clay colloid particle is called _________. (**Micelle**)

272. The enzyme responsible for urea hydrolysis is ___________. (**Urease**)

273. Hartig nets in soil produced by ___________. (**Ectomycorrhiza**)

274. Hydrogen bonding takes place in the interlayer of __________. (**Kaolinite**)

275. Journal Plant and soil is published from __________. (**Netherlands**)

276. _________ developed the DAPA tests for available soil micronutrients. (**Lindsay WL**)

277. Fertilizer control order issued in the year __________. (**1957**)

278. A soil profile having two sola, one above the other is known as __________. (**Bisequem profile**)

279. The peeling of the plates from parent rock due to differential expansion and contraction of minerals as a result of changes in temperature, is called as __________. (**Exfoliation**)

280. Flour apatite is a rich source of ____________. (**Phosphorus**)

281. The dominant spectral colour in soil indicates ________. (**Hue**)

282. Inorganic parent material transported by gravity is __________. (**Colluvial**)

283. In primary minerals most resistant to weathering is __________. (**Quartz**)

284. Soil aggregation can be estimated by using ____________. (**Yoders apparatus**)

285. The diagnostic surface horizons are called ____________. (**Epipedons**)

286. For nitrification the optimum pH is __________. (**8.5**)

287. Hill soil are generally _________ in nature. (**Acidic**)

13

Tree Physiology

1. A plant cell constitutes the____________ per cent of water. (**80-90%**)
2. Photosynthesis is the process of conversion of____________ energy into ____________ energy. (**Radiant, Chemical**)
3. Dark reaction of photosynthesis occurs in____________ and light reaction occurs in ____________ region of chloroplast. (**Matrix, Lamellae**)
4. The connecting link between glycolysis and Kreb's cycle is____________. (**Pyruvic acid**)
5. ____________ is responsible for turgidity of plant cells. (**Osmosis**)
6. A diffusion of liquids through a semi-permeable membrane is known as____________. (**Osmosis**)
7. The diffusion of solvent particles into a living cell is known as____________. (**Endo-osmosis**)
8. The diffusion of solvent particles out of a living cell is known as____________. (**Exo-osmosis**)
9. The outward pressure exerted by the cell solution on the cell wall which is developed due to osmotic diffusion of water is called as____________. (**Turgor pressure**)
10. Cohesive force theory of ascent of sap was proposed by____________. (**Dixon**)
11. Formation of male flowers is induced by____________. (**GA_3**)
12. During imbibition energy of water is transformed into____________ energy. (**Kinetic, Heat**)
13. A plant which needs not less than____________ hours of light to flower is called day neutral plant. (**Ten**)
14. ____________ is the site of protein synthesis. (**Ribosome**)

15. Translocation of nutrient and water through xylem is due to____________. **(Transpiration pull)**
16. Water use efficiency higher in plants is____________. **(CAM plants)**
17. ____________ is the first stable product formed in C_3 plants. **(PGA)**
18. The main site for the dark reaction of photosynthesis is____________. **(Stroma)**
19. The most abundant protein present in the world is____________. **(Rubisco)**
20. Photosystem II is absent in____________plants. **(C_4)**
21. RNA synthesis in nucleus participates in protein synthesis in____________. **(Cytosol)**
22. The transpiration pull/ cohesion theory of ascent of sap was given by____________ and ____________. **(Dixon and Jolly)**
23. ____________ scientist worked on drought tolerance. **(RD Asana)**
24. Probably the most useful measurable property in the soil plant air system is ____________ potential. **(Water)**
25. The upward movement of water from base to top of tree is known as____________. **(Ascent of sap)**
26. Effect of low temperature on the initiation and development of flower is called ____________. **(Vernalization)**
27. In vernalization the seeds are allowed to germinate for some time and then are given cold temperature treatments by keeping at____________. **(0 to 5 °C)**
28. The total amount of water present in the soil is called as____________. **(Holard)**
29. The absorption of water by____________ is called as imbibition. **(Hydrophilic colloids)**
30. Vacuoles are surrounded by____________. **(Tonoplast)**
31. The important function of leghaemoglobin in root nodules is____________. **(Oxygen regulation)**
32. The main function of endoplasmic reticulum is____________ synthesis. **(Protein)**

33. ____________,____________ and are the end products of respiration in plants. (**CO_2, H_2O and Energy**)

34. ____________ cell organelles is the site of chemical activity in cells, perhaps over half of the cell's metabolism. (**Mitochondria**)

35. Vanadium as an essential element replaces____________ in some N-fixing bacteria. (**Mo**)

36. The movement of secondary branches of roots and stem growing at right angle is known as____________. (**Diageotropic**)

37. Plasmolysis will occur when the cell is placed in a/an____________ solution. (**Hypertonic**)

38. One gram of fat liberates____________ kcal energy on oxidation. (**9.3**)

39. Growth is mainly affected by two climatic factors such as and____________. (**Light & Temperature**)

40. ____________structures controls the transport of the material into or out of living cells. (**Cell membrane**)

41. First time reported the presence of ribosome in cell by____________. (**Palade**)

42. When dormancy occurs due to____________ conditions is known as imposed dormancy. (**Unfavourable environmental**)

43. The term protoplast was introduced by____________. (**Johannes von Hanstein**)

44. The process by which proteins are synthesized in a cell is called____________. (**Translation**)

45. Peptide linkages are formed among ____________. (**Amino acids**)

46. Wilting of plants results from excessive____________. (**Transpiration**)

47. Opening of stomata is stimulated by____________ hormone. (**Ethylene**)

48. Seeds swell when placed in water is due to____________. (**Imbibition**)

49. Golgi bodies was discovered by____________ in____________. (**Camillio Golgi & 1898**)

50. Off season flowering in plants is positive by giving treatment of____________ and ____________. (**Vernalization and Photoperiodism**)

51. The phenomenon in which germination of seeds is affected by light, such seeds are known as_____________. (**Photoblastic**)
52. During photosynthesis the oxygen in glucose comes from_____________. (**CO_2**)
53. Plant cell supplied with_____________ element have high oil content. (**S**)
54. The externally thin strand of cytoplasm is termed as_____________. (**Plasmadesmata**)
55. In the_____________ climate C_4 plants are more produced. (**Tropical**)
56. Translocation of water and nutrients from roots to above ground parts of plants takes place through_____________. (**Xylem**)
57. Water rises in plant by_____________. (**Transpiration pull**)
58. Photosynthesis is to the chloroplast, as respiration is to the_____________. (**Mitochondria**)
59. The enzyme catalyses a reaction by reducing_____________energy in the system. (**Activation**)
60. _____________ instrument is used to measure stomatal opening. (**Porometer**)
61. Photosynthesis is _____________ process. (**Anabolic**)
62. _____________ type of soil water is most useful for plants. (**Capillary water**)
63. Production of ATP in the presence of light is known as_____________. (**Photophosphorylation**)
64. Growth of plants towards light is known as_____________. (**Phototropism**)
65. Leaf Area Index (LAI) =_____________. (**Leaf area/ Ground area**)
66. _____________ nutrient plays important role in controlling the rate of transpiration. (**Potassium**)
67. _____________ is the end product of glycolysis. (**Pyruvate**)
68. Functions and vital activities of the plants are studied under _________. (**Cytology**)
69. Plant lost water in transpiration upto the extent of _____________ per cent. (**99**)

70. Iron stored in chloroplast as an iron protein complex is called_____________. (**Phytoferritin**)

71. The stage of seeds showing no germination because of internal condition of seed is termed as_____________. (**Dormancy**)

72. During the first phase of glycolysis, ATP is used to form_____________ type of bond with glucose. (**Phosphoester**)

73. Ribosomes are produced in_____________. (**Nucleolus**)

74. Process of endosmosis stops, when the solution become_____________. (**Isotonic**)

75. A formal group replaces_____________ group of chlorophyll-a to form chlorophyll-b. (**CH_3**)

76. Excessive use of fertilizer causes death of plants due to_____________. (**Exosmosis**)

77. Apical dominance is the result of_____________. (**Auxins**)

78. Stomata are regulated by_____________. (**K**)

79. Site of oxidative electron transport is cell is_____________. (**Mitochondria**)

80. The basic element of the cohesion theory for the ascent of sap is_____________. (**Hydration**)

81. The enzyme used to break down fructose 1,6-biphosphate into two trioses is called _____________. (**Aldolase**)

82. The term auxin was first used by_____________ in 1926. (**Frits Went**)

83. Flow of solvent from low concentration to high concentration is known as. (**Osmosis**)

84. _____________ and_____________ are the natural occurring cytokinins. (**Kinetin and Zeatin**)

85. Law of inhibitory factor was given by _____________. (**R.F. Blackman**)

86. The_____________ and_____________ elements in the plants are surrounded by a layer of living cells is called pericycle. (**Xylem and Phloem**)

87. _____________ part of root absorb water and minerals. (**Root hairs**)

88. Maximum free radicals production is takes place in____________. **(Senescence)**
89. The strongest bond is____________ bond. **(Ionic)**
90. The weakest bond is____________ bond. **(Covalent)**
91. ____________ plant hormone is a volatile hormone. **(Ethylene)**
92. Glycolysis occurs in____________ region of the plant cell. **(Cytoplasm)**
93. The young leaves of terminal bud at first typically hooked, finally dying back at tips and margins are the deficiency of____________. **(Calcium)**
94. Growing of plant in soilless nutrient solution is referred as____________. **(Hydroponics)**
95. Growth movement toward or away from the earth's gravitational pull is known as ____________. **(Gravitropism)**
96. The rate of transpiration greatly depends upon____________. **(Position of stomata)**
97. ____________ plant hormone is primary regulator of abscission process. **(ABA)**
98. The pigments which produces the colour of many flowers are generally stored in ____________. **(Vacuole)**
99. Maximum density of water is existing at____________temperature. **(4 °C)**
100. Water potential of soil at field capacity is____________. **(-0.3 bars)**
101. The compound, gibberellin was isolated from was given by Yabuta and Hayashi in 1930. **(Fungus)**
102. Important radiant generated in pentose phosphate pathway of glucose degradation is ____________. **(NADPH)**
103. Transpiration is measured by____________. **(Potometer)**
104. ____________ is the major form of carbon transfer in plants. **(Sucrose)**
105. In a rapidly transpiring plant the water column in xylem will be in____________ pressure. **(Negetive)**
106. ____________ light is the most effective in producing a phototrophic response. **(Blue)**

107. The hormone ABA was first identified and characterized chemically in____________ by ____________. **(1963 & Frederick)**

108. Etiolated plants are formed due to lack of ____________. **(Light)**

109. Oxygenase function of Rubisco was first developed by____________ and .____________ **(Ogren & Bowes)**

110. In the majority of plant species, seed germination begins with____________. **(Radicle)**

111. ____________ and____________ pairs of structures are usually found in both plant and animal cell. **(Nucleolus and Cell membrane)**

112. The plant hormone promote germination of dormant seeds and growth of dormant buds is____________. **(GA)**

113. Denitrification occurs in____________ soils. **(Water logged)**

114. Mineral salt which are absorbed by the roots for the soil are in the form of ____________ solution. **(Concentrated)**

115. Leaves with yellow or colourless patches are termed as____________. **(Variegated)**

116. The rate of transpiration doubles with every rise in temperature by____________. **(20 °C)**

117. ____________ calvin cycles are needed to produce one molecule of glucose. **(Six)**

118. Flowering stimulus is perceived by____________. **(Leaves)**

119. C_3 cycle of carbon fixation takes place in____________ of chloroplast. **(Stroma)**

120. When starch reacts with iodine, produces____________ colour. **(Blue)**

121. Turgidy of a cell is maintained by____________ pressure. **(Turgor)**

122. Mass flow mechanism was proposed by____________. **(Munch)**

123. VAM is mostly used in____________ crops. **(Perennial)**

124. Wilting of leaves is hot weather is due to the presence of____________ and ____________ element. **(Iron and Calcium)**

125. The process by which N_2 is reduced to ammonium is called____________. **(Nitrogen fixation)**

126. The first alkaloid to be isolated and crystalized was the____________. (**Morphine**)

127. Under aerobic condition microbes grow slower but use more sugar and produces more CO_2 and ethanol which is known as____________. (**Pasteur effects**)

128. Antimicrobial compounds synthesized by plants when infected with microbes are ____________. (**Phytoalexins**)

129. When seed is placed in a suitable medium for germination,____________ is the first process water enter into the seed coat. (**Imbibition**)

130. Oxidation in biological system is accompanied by____________. (**Dehydrogenation**)

131. The most striking feature of CAM plants is formation of____________ at night. (**Malic acid**)

132. ____________ element is require for pollen germination. (**B**)

133. Kreb's cycle produces____________ ATP. (**30**)

134. ____________ group of enzymes from double bonds by elimination of a chemical group. (**Lyases**)

135. ____________number of ATPs are required to produce one mole of hexose in photosynthesis. (**18**)

136. In monocots and dicots accumulation of____________hormone causes collapse and lysis of mature cortical cells in the roots, leading to a tissue with large air space. (**Ethylene**)

137. In ammonification, conversion of organic nitrogen to____________ by soil microbes. (**NH_4**)

138. Hydroponics is the growing of plants in____________. (**Water culture**)

139. Electron transport system takes place in____________ part of mitochondria. (**Cristae**)

140. Mature root nodule made largely of ____________ cells. (**Tetraploid**)

141. Nucleus is derives its name from the latin word for____________. (**Kernel**)

142. ____________ instrument is used to measuring stomatal pressure. (**Porometer**)

143. Glycolysis takes place in____________. (**Cytoplasm**)

144. If the accumulation ratio in absorption of nutrients is____________ then it is known as active absorption. (**Greater than one**)

145. ____________ is the most dangerous gas for depletion of ozone layer. (**CFC**)

146. Photochemical reactions of photosynthesis take place in____________. (**Thylakoids**)

147. ____________ is the end product of glycolysis. (**Pyruvic acid**)

148. For C_3 plants, the optimum temperature for photosynthesis is____________. (**15-25 °C**)

149. When a cell is fully turgid, suction pressure will be____________. (**Zero**)

150. Callus is induced to form roots in the medium of more than ____________. (**Auxin, Cytokinins**)

151. ____________ amino acid produced in photorespiration. (**Serine**)

152. Close association of____________, ____________ and____________ in a leaf cell are related with photorespiration. (**Chloroplast, peroxisomes and mitochondria**)

153. Photorespiration is also called as ____________. (**Three cell organelle reaction**)

154. Photorespiration is found in ________ plants and absent in __________ plants. (**C_3 & C_4**)

155. In C_4 plants, photosynthesis inhibited by____________ per cent oxygen. (**21**)

156. Z-scheme of electron transport first proposed by____________ and____________. (**Hil and Bendall**)

157. The break down of large molecules to small molecules, and this process often release energy is known as____________. (**Catabolism**)

158. Photorespiration process is also known as____________. (**Glycolate pathway**)

159. The molecules known as the energy currency of the cell are____________. (**ATP**)

160. The transpiration ratio is____________ for C_3 plants. (**Highest**)

161. ____________ is the main green house gases for global warming. **(CO_2)**

162. ____________ element is for most of dehydrogenase. **(Zn)**

163. The CO_2 concentration, at which photosynthetic fixation just balances respiratory loss is known as____________. **(CO_2 compensation point)**

164. The smooth endospermic reticulum functions in the synthesis of____________. **(Lipids)**

165. Inhibitory effect of high conc. of O_2 on photosynthesis is called as____________. **(Warburg's effect)**

166. ____________ photons are required to produce one molecule of oxygen. **(8)**

167. ____________ ecosystem has highest net primary productivity per unit area. **(Tropical rain forest)**

168. In C_4 plant, the first stable product of photosynthesis is ____________. **(Oxalic acid)**

169. Mycorrihiza increased the availability of____________ nutrient. **(P)**

170. Acetylene reduction to____________is measured as N_2 fixation. **(Ethylene)**

171. Water potential of chemically pure water is____________. **(Zero)**

172. ____________ hormone is associated with acid growth theory. **(Auxin)**

173. In____________ cell organelle, PEP carboxylation is taking place in C_4 plants. **(Mesophyll)**

174. Wavelength of visible light is____________. **(390-760 nm)**

175. To produce____________ and____________ is the primary function of the light reaction of photosynthesis. **(ATP and NADPH)**

176. ____________ enzyme is responsible for the synthesis of malic acid in C_4 plant. **(PEP carboxylase)**

177. Electrophoresis was developed to separate____________. **(Proteins)**

178. RNA-DNA hybridization to quantify gene expression at mRNA level is called as____________. **(Northern blotting)**

179. Acid rain is due to ________ and____________ gases. **(NO_2 and SO_2)**

180. Sulphate reduction in leaves takes place in____________. **(Chloroplast)**

181. Nitrogen fixation is carried out by_____________ enzyme. (**Nitrogenase**)

182. In C_3 plant, rubisco enzyme first reacts with CO_2 to form_____________. (**PGA**)

183. The process of breaking seed dormancy by rupturing the hard seed coat is known as _____________. (**Scarification**)

184. Carbohydrates are present in the plasmalemma in the form of _____________ and _____________. (**Phosphoproteins and phospholipids**)

185. A cell increase in volume when placed in a solution, which is_____________. (**hypertonic**)

186. _____________ element taken up by plant in both cation and anaion form. (**N**)

187. Exogenous application of_____________hormone can induce flowering in long day plants even under short day condition. (**GA**)

188. Water potential concept was formulated by_____________. (**Otto Renner**)

189. The nitrogen base not found in RNA is_____________. (**Thymine**)

190. Transamination involves the amination of a/ an_____________ acid. (**Alpha keto**)

191. The sequential process of fatty acid catabolism to acetyl CoA is called_____________. (**Beta oxidation**)

192. The loss of water in the form of water vapour is termed as transpiration whereas in the form of liquid is known as_____________. (**Guttation**)

193. Normally guttation occurs at_____________. (**Night**)

194. Guttation takes place through aspecial pore like structure called _____________. (**Hydathodes/ water stomata**)

195. The term guttation is given by_____________. (**Burgerstein**)

196. Foliar transpiration is of two types such as_____________ and_____________. (**Stomatal and Cuticular**)

197. Loss of sap from the injured parts of the plant is known as_____________. (**Bleeding/ exudation**)

198. Both guttation and exudation are governed by _____________. (**Root pressure**)

199. Dropping of the soft parts of the plants due to loss of turgidity in their cells is called ____________. **(Wilting)**

200. Photosynthesis is the process of converting____________ energy to____________ energy and storing it in the bonds of sugar. **(Light, Chemical)**

201. The basic reaction of photosynthesis is____________. **($CO_2 + H_2O$ + light energy → $(CH_2O)_n + O_2$)**

202. The reactions leading to the production of____________ and reduction of____________ are called the light reactions because they are initiated by the splitting of water by light energy. **(ATP, NADP+)**

203. The chlorophyll is built into the membranes of the____________. **(Thylakoids)**

204. Light behaves both as a wave phenomenon and as discrete particles of energy called as ____________. **(Photons)**

205. Molecules that have evolved to absorb wavelengths in the visible region of the spectrum very well are called ____________. **(Pigments)**

206. The two primary pigments involved in photosynthesis are ____________ and ____________ in the ratio of____________. **(Chlorophyll a, Chlorophyll b and 3:1)**

207. The first stable product of the celvin cycle is____________. **(Phosphoglycerate, PGA)**

208. Photorespiration is a process in plant metabolism which attempts to ameliorate the consequences of a wasteful oxygenation reaction by the enzyme____________. **(Rubisco)**

209. Photosynthesis active radiation (PAR) =____________. **(400 to 700 nm)**

210. There are two parts of photosynthesis which are light reaction / hill reaction takes place in____________ of chloroplast and Dark reaction of celvin cycle takes place in ____________ of chloroplast. **(Grana and Stroma)**

211. Three types of photosynthetic mechanism are ____________, ____________ and ____________. **(C_3 pathway, C_4 pathway and CAM pathway)**

212. $NADH_2$ will produce____________ ATP while $FADH_2$ produce____________ ATP. **(3, 2)**

213. ATP is a molecule that contains high energy____________bond. **(Phosphate)**

214. In the light phase of photosynthesis, green plants release____________. **(Oxygen)**

215. The most effective colour of visible light spectrum for photosynthesis is ____________. **(Red light)**

216. The site of primary photochemical reaction is____________. **(Quantasome)**

217. The function of ATP in photosynthesis is the transfer of energy from____________ to ____________. **(Light reaction to Dark reaction)**

218. Photosynthesis is a/ an____________ process. **(Oxidation-reduction)**

219. In photorespiration, NAD is reduced to____________. **($NADH_2$)**

220. Photorespiration was discovered by____________ in____________ plants. **(Decker, tobacco)**

221. Respiration in plants consists of____________ ,____________ and____________. **(Glycolysis, Krebs cycle and ETC)**

222. ____________ organelles are concerned with the aerobic respiration supply of chemical energy for the maintenance of life. **(Mitochondria)**

223. Each mitochondria is bounded by two lipoprotein and filled with the colourless fluid called____________. **(Matrix)**

224. Kreb's cycle takes place in the____________. **(Mitochondrial matrix)**

225. One molecule of glucose gives two molecule of____________. **(Pyruvic acid)**

226. The ratio of the volume of CO_2 released to the volume of O_2 absorbed in the respiration is____________. **(Respiratory Quotient/ Respiratory ratio)**

227. Respiratory quotient for organic acid is ____________. **(1.33)**

228. Term photoperiodism was coined by____________. **(T. D. Lysenko)**

229. ____________ group of plants similar to animals in their requirements of oxygen for cellular respiration both in the day and night. **(Fungi)**

230. Aerobic respiration of glucose produces energy of____________kcal. **(673)**

231. The compound released in aerobic respiration is_____________. **(H_2O)**

232. _____________is the preferred substrate in aerobic respiration. **(Glucose)**

233. Glycolysis occurs in the cytoplasm and kreb's cycle and ETC occurs in the _____________. **(Mitochondria)**

234. 1 molecule of ATP is equal to_____________ Kcal. **(7.6)**

235. The first stable product of kreb's cycle is_____________. **(Oxaloacetate)**

236. The first discovery and use of plant growth regulators was with acetylene and ethylene, which enhanced flowering production in ___________. **(Pineapple)**

237. Auxins derive their name from the Greek word *auxein* means_____________. **(To grow/ increase)**

238. The shading, detachment, separation of plant parts is termed as_____________. **(Abscission)**

239. ____________ auxin prevent leaf abscission and fruit drop. **(Synthetic)**

240. Seeds that are difficult to germinate are frequently treated with _______ solution. **(Gibberelic acid)**

241. _____________ plant growth regulator that promote cell division, stimulate shoot proliferation, activate gene expression and metabolic activity. **(Cytokinin)**

242. In 1964, the first naturally occurring cytokinin was isolated from corn called. **(Zeatin)**

243. All cytokinin are chemically similar to_____________. **(Adenine)**

244. Cytokinins are generally used in plant cell culture at a concentration range of _____________. **(0.1 – 10.0 mg/l)**

245. _____________ is a plant growth regulator that induces dormancy, prevents seeds from germinating and causes abscission of leaves, fruits and flowers. **(Abscissic acid)**

246. _____________ is a fatty acid derived plant hormone that is similar in overall structure to physiologically small molecules from animals called prostaglandins. **(Jasmonic acid)**

247. _____________is a phenolic phytohormone which is found in plants with roles in plant growth and development, photosynthesis, transpiration, ion uptake and transport. **(Salicylic acid)**

248. The term phytohormone in plants was suggested by____________. **(Thiamann)**

249. The affect of low temperature on seeds that hastens flowering is called. **(Vernalization)**

250. The physical basis of life is____________. **(Cytoplasm)**

251. Green colour of leaves is due to the presence of____________. **(Chlorophyll)**

252. Nutrients and Water absorbed from the soil move up through a tree in the____________. **(Xylem)**

253. ____________ part of tree carries food made in the leaves down to branches, trunk and roots. **(Phloem)**

254. A flower lacks either a calyx or corolla, it is called ________flower. **(Imperfect)**

255. The male and female flower parts occur in separate flowers on the same tree, the species is said to be____________. **(Monoecious)**

256. Ovate, deltoid, lanceolate, star shaped are example of leaf____________. **(Bases)**

257. Both the inner bark and sapwood are formed by cell divisions of the layer. **(Cambium)**

258. ____________ is a direct product of photosynthesis. **(Carbohydrates)**

259. ____________ carries the sap from the roots to the leaves. **(Xylem)**

260. Closely spaced annual ring indicates____________. **(Slow growth)**

261. The ________ builds wood tissue in the trunk. **(cambium layer)**

262. ____________gives support and strength to the tree. **(Heart wood)**

263. ____________ helps to protect the tree trunk from injury. **(Outer bark)**

264. The manufacture of plant food in the leaves of the plant is called____________. **(Photosynthesis)**

265. The outside edge of a leaf is called the____________. **(Margin)**

266. The leaf stem is also called the____________. **(Petiole)**

267. When a single leaf attached to the twig it is said to be____________. **(Simple)**

268. When any leaflets are attached to the petiole to form a larger leaf, that type of leaf is called_____________. (**Compound**)

269. Deciduous trees are referred to as_____________. (**Hardwood**)

270. Coniferous trees are referred to as_____________. (**Softwood**)

271. Identification of trees according to species is called_____________. (**Dendrology**)

272. A tree sprout arising from the root or stump is called a _____________. (**Sucker**)

273. The function of the xylem in the tree is to conduct_____________ and _____________ to the crown. (**Water and minerals**)

274. The best starter fertilizer for tree has a high concentration of_____________. (**Nitrogen**)

275. The process by which excess water from the tree is allowed to evaporate through opening in the leaves is _____________. (**Transpiration**)

276. In photosynthesis carbon dioxide from the air and water and minerals from the soil into_____________ and_____________ with oxygen as a by-product. (**Sugars & starchs**)

277. The process of consuming carbohydrates and oxygen to obtain energy for the biological process of life support, growth and reproduction is called. (**Respiration**)

278. The loss of water vapour from the leaf of a tree is called_____________. (**Transpiration**)

279. Most of the water vapour given off by a plant escapes through structures called _____________. (**Stomata**)

280. Excess carbohydrates not used in respiration are converted to_____________and _____________, then stored in the tree. (**Cellulose & Lignin**)

281. Length of day light affects the time of flushing or_____________in the spring and the onset of dormancy in the autumn. (**New growth**)

282. Seed formation depends on the amount of reserve_____________ and nitrogenous materials in a tree. (**Carbohydrates**)

283. The term used for dormancy induced by external environment is_____________. (**Quiescence**)

284. During the process of Photosynthesis____________takes place in the absence of light. (**Hill reaction**)

285. The function of root hairs is to____________. (**Absorb water and nutrients**)

286. ____________ is known as father of plant physiology. (**Stephan Hales**)

287. ____________ is known as father of Indian plant physiology ()

288. The movement of molecules or atoms or ions of materials from an area of higher concentration to an area of their lower concentration is called __________. (**Diffusion**)

289. Diffusion pressure of a pure solvent is always ___________ than its solution. (**Higher**)

290. Exchange of gases like CO_2, O_2 and distribution of hormones in plants takes place through ___________. (**Diffusion**)

291. The process of transpiration is a ___________ process. (**Diffusion**)

292. ___________ is defined as the special diffusion of solvent from the solution of lower concentration to the solution of higher concentration through the semi-permeable membrane. (**Osmosis**)

293. Osmosis was disovered by ___________. (**Abbe Nollet**)

294. The exchange of material in and out through the membrane is called as ___________. (**Permeability**)

295. The osmotic pressure of pure water is __________. (**Zero**)

296. The osmotic pressure is highest in __________ and lowest in __________ parts of the plant. (**Leaves & Root**)

297. The highest osmotic pressure is found in __________ while lowest in ___________. (**Halophytes & Hydeophytes**)

298. Osmotic pressure of a solution is measured by ___________. (**Osmometer**)

299. First osmometer was made by ___________. (**Pfeffer**)

300. The formula of vont Hoff for measuring osmotic pressure is ___________. (**OP= mRT**)

301. The opening and closing of stomata is dependent on the process of __________. (**Osmosis**)

302. Turgor pressure regulate plant __________ and _________. (**Movements and maintain the normal shape of cell**)

303. Diffusin pressure deficit (DPD) for _______________ is equal to OP+TP. (**Plasmolysed cell**)

304. The difference between the free energy of molecules of water and free energy of the solution is called as ___________. (**Water potential**)

305. Breaking of seed coat during the seed germination is due to ___________ process. (**Imbibition**)

306. Swelling of wooden door during rainy season due to ___________. (**Imbibition**)

307. Water potential is measured in ______________. (**Bars/Pascal**)

308. Forms of water which reaches at the soil water table due to the gravitational force after the rainfall and not available to plant is called _____________. (**Gravitational water**)

309. Thin film of water which is tightly held by the soil particles is called ___________. (**Hygroscopic water**)

310. Water exits between the soil particles in small capillary pores and most available form to the plants is called as ______________. (**Capillary water**)

311. Total amount of water present in the soil is called as __________. (**Holard**)

312. Maximum absorption of water takes place from _________ region. While maximum absorption of mineral takes place from __________ region. (**Root hair and Meristematic**)

313. The movement of water from cell to cell through plasmodesmata is called __________. (**Symplast / Living path**)

314. About 90% of water in plants is absorbed by _________ way of water absorption. (**Passive**)

315. Term active and passive absorption was proposed by _________. (**Renner**)

316. Translocation of food mainly occurs in the form of __________. (**Sucrose**)

317. Stomatal pore is surrounded by two specialised epidermal cells called as ___________. (**Guard cell**)

318. __________ type of stomata are found in xerophytic plants. (**Sunken stomata**)

319. Active potassium (K+) and hydrogen (H+) theory was given by __________. (**Levitt**)

320. Plant hormone __________ help in closing of stomata. (**Abscisic acid**)

321. Number of stomata present in unit area is called as __________. (**Stomatal index**)

322. Mobile elements (N, P, K & Mg) can be remobilized from senescent organ to young plant parts and deficiency symptoms first appear in __________ plant parts. (**Older**)

323. Immobile elements (B, S, Ca, Co & Fe) can not be remobilized from senescent organ to young plant parts and deficiency symptoms first appear in ________ plant parts. (**Younger**)

324. Plants grown in moistened air with nutrients is called __________. (**Aeroponics**)

325. Growth inhibitor __________ is used in bonsai. (Chlormequat)

326. Term hormone was given by __________ while phytohormone by __________. (**Starling and Thieman**)

327. High auxin and low cytokinin favours __________ formation while high cytokinin and low auxin favours __________ formation. (**Root & Shoot**)

328. ______________ is root initiating hormone in cutting. (**Indole Butyric Acid**)

329. Photoperiodism was first discovered by __________ & __________in Tobaco. (**Garner & Allard**)

330. _____________ pigment is responsible for flowering. (**Phytochrome**)

331. The term phytochrome was given by __________. (**Butler**)

332. Term enzyme was given by __________ and first isolated enzyme is zymase discovered by __________. (**Kuhne & Buchner**)

333. Protein part of conjugated enzyme is called __________. (**Apoenzyme**)

334. Conjugated enzyme (made up of protein + non-protein) is also known as __________. (**holoenzyme**)

335. ATP was discovered by __________. (**Lohman**)

336. Light energy is converted into __________ energy by photosynthesis. **(Chemical)**

337. First true and oxygenic photosynthesis started in __________. **(Cyanobacteria)**

338. Absorption spectrum of photosynthesis is _________ & ________light. **(Blue & Red)**

339. Rate of photosynthesis is higher in _________ wavelength of light but highest in __________ light. **(Red & White)**

340. Dark reaction was discovered by _____________. **(Blackman)**

341. Quantasome was discovered by _____________. **(Park & Biggins)**

342. Quantasome are made up of__________ molecules of various pigments. **(200-400)**

343. Intensity of light at which rate of photosynthesis becomes equal to rate of respiration in plants is known as _______________. **(Light compensation point)**

344. Optimum temperature for photosynthesis in plants is ___________. **(20-35 °C)**

345. The amount of CO_2 in grams absorbed by 1 gram of chlorophyll in 1 hours is called as _____________. **(Photosynthetic number/ Assimilatory number)**

346. Plants which re adapted to grow in high intensity of light are called __________. **(Heliophytes)**

347. Plants which are adapted to grow in shade re called as __________. **(Sciophytes)**

348. Law of minimum was given by __________. **(Liebig)**

349. Law of limiting factors was given by __________. **(Blackman)**

350. Pigment system of bacteria is denoted by ___________. **(B-890/ 870)**

351. Non green plantswhich are depends on dead organic matter are called as ____________. **(Saprophytes)**

352. When respiratory substrates are carbohydrates then respiration is known as ___________ respiration. **(Floating)**

353. If respiratory substrates are proteins then respiration is known as ____________respiration. **(Protoplasmic)**

354. Anaerobic respiration was first reported by ___________. **(Kostytchev)**

355. __________ is the common phase for aerobic & anaerobic respiration. **(Glycolysis)**

356. Substrate level phosphorylation forms ________ number ATP. **(Four)**

357. Glycolysis is known as ___________ pathway. **(EMP)**

358. EMP stands for ___________. **(Embden, Meyerhof, parnas)**

359. __________ is the connecting link between glycolysis & Krebs cycle. **(Acetyl Co-A)**

360. Acetyl Co-A is common intermediate between ________ & _______. **(Fat & Carbohydrate)**

361. Krebs cycle was discovered by ____________. **(H. A. Kreb)**

362. Optimum temperature for respiration is __________. **(20-35 °C)**

363. The minimum amount of oxygen at which aerobic respiration takes place and anaerobic respiration become extinct is called as ________. **(Extinction point)**

364. Oxygen concentration at which aerobic respiration & anaerobic respiration take place simultaneously is called as __________. **(transition point)**

365. The rapid increase in rate of respiration during ripening of fruits, senescence of leaves and plant organs is called as ___________. **(Climacteric respiration)**

366. Climacteric respiration is due to production of __________. **(Ethylene)**

367. Stomata are bound by a pair of guard cells and open guard cells are _________. **(Turgid)**

368. To enter or leave a cell, substances must pass through ___________. **(Plasma membrane)**

369. Alcohol is produced during the process of _________ respiration. **(Anaerobic)**

370. Precursor for ethylene biosynthesis is ____________. **(Methionine)**

371. The chemical nature of GA_3 is __________. **(Terpene)**

372. The net requirement for assimilation of CO_2 in C_4 plants is __________. **(Four)**

373. Formula for Leaf Area Index is __________. **(Leaf area/ ground area)**

374. Grana stacks of thylakoid membranes are high in ____________. **(PS-II)**

375. The first step of assimilation of sulphate is catalysed by ______________. **(ATP sulfurylase)**

376. Electron transport system takes place in __________ part of mitochondria. **(Cristae)**

377. The term 'biological clock' was coined by ___________. **(Bunning)**

378. The optimum pH of nutrient solution in nutrient solution culture is _________. **(Six)**

379. An enzyme associated with decarboxylation reaction in photosynthetic reaction of C_4 plant is ___________ enzyme. **(Malic enzyme)**

380. Enzyme used to cut double strand RNA Is ___________. **(Restriction endonuclease)**

14

Rangeland Management

1. 'Agrostoogical' refers to cultivation of____________. **(Grasses)**
2. Land on which the potential plant community is composed periodically of native grass, forbs and shrubs valuable for forage and in sufficient quantity to justifying grazing use is called as____________. **(Rangeland)**
3. The basic object of range management is to____________ and____________ the rangeland for higher production of fodder, rich in nutritive value and palatability. **(Improve & maintain)**
4. ICAR had carried a grassland survey of India in 1954 by Wnyte, Venkatramanan and Dabadghao had found____________ types of grass land in different environment. **(8)**
5. The main reason for the enormous increase in livestock population is____________. **(Social attitude of the people)**
6. All the vegetation consumed by grazing animals including fruits and twigs of trees and shrubs is called____________. **(Pasturage & forage)**
7. The practice by which the gap can be bridged between the requirement and availability of animal feeds and fodder is known as____________. **(Grass land techniques)**
8. The scientists____________ have recognised five type of grass covers in India. **(Dabadghao & Shankar Narayan, 1979)**
9. The maximum area of about 2.8 milloin square kilometre is covered under____________ type of grassland in India. **(Phragmites-Saccharum type)**
10. ____________refers to a comparison of current years forage production with that of the previous years. **(Range conditions)**
11. The most serious cause of the decline in the condition of the rangeland is____________. **(Over grazing)**

12. The scientists_____________who have grouped range conditions into five different classes in 1965. (**Stoddart *et al.***)
13. In range condition classification, where the climax vegetation is present but not dominant, palatable plants destroyed and replaced by less valuable plants is_____________. (**Class-3**)
14. The leaf and twig growth of shrubs, woody vines and trees available and acceptable for animal consumption refers to_____________. (**Browsing**)
15. Any non-injurious edible material having nutritive value for animals is called. (**Animal feed**)
16. _____________ denote to only that part of vegetation which is within reach of the livestock. (**Forage density**)
17. Tropical or subtropical grass land containing scattered trees or shrubs is. (**Savannah**)
18. The area covered with vegetation which is properly used by livestock is called. (**Forage acre**)
19. The manipulation or balancing the number of animals to be grazed with the carrying capacity of the area to accomplish desired results is known as_____________. (**Grazing management**)
20. _____________ is defined as preservation of green fodder in which the carbohydrates, proteins and salts have been converted into organic acids by bacterial action in the absence of oxygen. (**Silage**)
21. _____________is defined as conservation of green forage into dry form without deterioration of dry matter and nutrients having its natural green colour and palatability and can be stored over a long period of time. (**Hay**)
22. _____________ is defined as the preserved green fodder by packing fresh fodder in a suitable container and allowing it to ferment under anaerobic conditions without undergoing much loss of nutrients. (**Silage**)
23. The place or house where silage is stored called as_____________. (**Silo**)
24. The term applied to hay, silage, fodder, stover, pasture plants and roots used for animal feeding purpose referred as_____________. (**Roughages**)
25. _____________ refers to the summation of plant characteristics that determine the relish with which a particular plant species or plant part is consumed by an animal. (**Palatability**)

26. The proportion of the ground area covered by the canopy when viewed vertically is known as_____________. (**Ground cover**)
27. The process of making silage is called_____________. (**Ensilage**)
28. The most serious cause of decline in safe condition is grazing of cattle beyond _____________. (**Carrying capacity**)
29. _____________ are the best for erosion control in the grass land. (**Vegetative cover**)
30. A grassland is mainly covered by the plant species belonging to the family_____________. (**Poaceae**)
31. The climatic factors affecting grass cover distribution are_____________ and _____________. (**Rainfall and Temperature**)
32. The main factors responsible for depletions of range conditions are_____________. (**Over grazing**)
33. The best measures for the control of erosion in the safe lands are_____________. (**Vegetative measures**)
34. The optimum depth of sowing for small seeded grass should be_____________. (**0.5 to 1 cm**)
35. The deferred grazing management systems more suitable for_____________. (**Perennial grasses**)
36. Planting the rows of live plants on the boundaries of the area is called_____________. (**Life fencing**)
37. The conservation of green forage into dry form without affecting the quality is called _____________. (**Hay**)
38. The moisture content in hay is upto __________ percent. (**15%**)
39. Silage with higher content of_____________is considered to be of poor quality. (**Butyric acid**)
40. The forest floor consists of a large amount of_____________ which changes the physical and chemical properties of the soil. (**Organic matter**)
41. The main effect of repetition to improve the_____________of soil to control runoff during rainy season. (**Infiltration**)
42. The success of plantation programme depends upon and_____________ factors. (**Technical & Socio-economic factor**)

43. In case of artificial regeneration one of the most important decision is the_____________. **(Selection of species)**
44. An indigenous species is one that grows_____________ in a region. **(Naturally)**
45. _____________ species becomes susceptible to new insects pest and diseases. **(Exotic)**
46. For first growing species,_____________ soils should be chosen. **(Fertile)**
47. The effective bird repellent is_____________. **(Anthraquinone)**
48. Planting should preferably be done either_____________ and. **(Cloudy days, Evening)**
49. Soils which originates in humid region as a result of accumulation of organic matter in impeded drainage condition is called_____________ soils. **(Peat soils)**
50. When seed of some species are mixed with ash, earth and cow dung etc. is called _____________. **(Seed pelleting)**
51. To provide protection to the plants from the sun, frost, rain, hails etc. is called _____________. **(Shading)**
52. The pH of the forest soil is _____________in nature. **(Acidic)**
53. The tropical and sub-tropical grassland with sparsely scattered trees or bushes or none of them are called_____________. **(Savannas)**
54. Destruction of forest leads to the reduction in_____________. **(Precipitation)**
55. The repeated occurrences of drought are attributed to the loss of _____________. **(Forest)**
56. In tropical rain forests, the climate is_____________ and_____________. **(Warm and Humid)**
57. _____________ methods of sowing are most suitable when seeds are limited with the farmers. **(Vegetative propagation)**
58. _____________ weed is most common in paddy field in its appearance. (***Echinochloa crus galli***)
59. _____________ is a good fodder tree. **(Subabul)**

60. The full name of ICFRE is____________. (**Indian Council of Forestry Research & Education**)

61. When animals owner do not have a permanent place of residence and they keep on moving in search of forage is called____________. (**Nomadism**)

62. The practice of feeding fresh cut crops to households live stock is known as____________. (**Stall feeding**)

63. The economical way to remove species of thorny bushes and noxious plants from the grassland is____________. (**Burning of grassland**)

64. The various measures like pitting, basin listing, water spreading, tour trenching are used for____________. (**Soil &Water conservation**)

65. The most common practice of grazing the cattle in India is____________. (**Open grazing**)

66. Dredging and changing methods are used for the control of____________. (**Aquatic weeds**)

67. The uncultivated natural forest lands and grasslands which providing browsing and grazing grounds to live stocks is called____________. (**Rangeland**)

68. The systems of grazing developed in Germany to produce sufficient milk is called____________ of grazing. (**Hoheinhem method**)

69. ____________ of range is restored when the grass regeneration is inadequate and native hebetation has disappeared. (**Re-seeding**)

70. ____________ fencing is used to enclose the area from the animals who cannot take long leap. (**Trench**)

71. The main objectives behind soil and water conservation measures are to encourage the soil farming process to improve the____________ of soil. (**Nutrient status**)

72. Pitting the rangeland is quite an effective mechanical measure to reduce the runoff in ____________ areas. (**Semi-arid**)

73. In pasture establishment seeds mixture is preferred to____________ species. (**Single**)

74. In pellet sowing, the mixture of sand, clay, cow dung and fertilizer is prepared in the ratio of____________. (**3:1:1:1**)

75. The practice of feeding fresh cut crops to house hold livestock is known as_____________. (**Zero grazing**)

76. The coniferous trees intercept_____________ rainfall than deciduous trees in full leaf. (**More**)

77. The application of fertilizers has been found more effective on range lands situated in_____________areas. (**High rainfall**)

78. Broad leaved species require_____________ water than narrow leaved species. (**Less**)

79. Flooding method of irrigation is not good for_____________ seeds. (**Minute**)

80. In dense forest air circulation is_____________ and solar radiation is_____________which results in lower temperature. (**Decreased, Blocked**)

81. A miss-management of grassland result in_____________ of soil erosion. (**Increase**)

82. Permanent pasture consists chiefly of perennial grasses and legumes in regions. (**Humid and Sub-humid**)

83. Dicanthium-Cenchrus-Lasiurus type of grass land is normally found in regions. (**Sub-tropical & semi-arid**)

84. _____________type of grass cover is situated mostly in hilly region of the country. (**Themeda – Arundinella type**)

85. The first attempt at defining range condition classes was made in the year. (**1935**)

86. _____________ and_____________ in 1969 have done the range condition classification for the rangelands in Western Rajasthan. (**Bhimaya & Ahuja**)

87. The number of stock of a given classes expressed in livestock units which a rangeland unit will support for the period of grazing allowed is known as_____________. (**Carrying capacity**)

88. The number of animals that produces the greatest return without damage to the physical resources and in concert with other values received for the land is known as_____________. (**Grazing capacity**)

89. The actual number of animals on a specific area for a specific period of time, usually for a grazing season is called as_____________. (**Stocking rate**)

90. Grazing pressure is the ratio of____________ to____________. **(Feed demand to Feed supply)**

91. Forage arce =____________. **(Forage acre factor x Surface area)**

92. The scientific name of Paragrass/ Buffalo grass is____________. (***Brachiaria mutica***)

93. Thefoddercropsrichin____________andlowtomedium____________ content are ideal for silage making. (**Soluble carbohydrates, Protein**)

94. National forest policy ________ has emphasised the need to regulate the grazing in accordance with the carrying capacity of the forests. (**1988**)

95. FRI, Dehradun has indicated the general carrying capacity of forest as cow unit per 100 hactares. (**60**)

96. Under agroforestry,____________ system seems to be perhaps the best system for grassland management. (**Silvi-pastoral**)

97. ____________ is defined as the percentage of the present vegetation which is original vegetation of the site. (**Range condition**)

98. Carrying capacity =____________. (**Total annual forage production / Forage requirement animal-1 day-1 × 365 x area of the grazing land**)

99. The seed rate/ha of *Cenchrus ciliaris* is____________. (**4.5 kg/ha**)

100. *Dicanthium annulatum* is locally known as____________grass. (**Marvel/ Jarga/ Palwan**)

101. Indian Grass land and Fodder Research Institute (IGFRI) is located at _______. (**Jhansi**)

102. India supports about ________ % of the total livestock of the world. (**15%**)

103. The feed requirement used as the basis of comparison for different classes and species of stock is known as ____________. (**Livestock unit**)

104. 1 livestock unit requires approximately __________ quantity of good quality pasture dry matter per year. (**520 kg**)

105. 1 animal unit is equal to _________ steer. (**40 kg**)

106. Breeding cow and bull have __________ animal unit. (**2**)

107. Buffalo and camel have ________ and _________ animal unit respectively. (**2 & 8**)

108. The object of scientific management of grassland is to maintain the grassland in the ____________. **(Highest state of production)**

109. When the number of animals that are allowed to graze per unit area of rangeland is fixed accordance with rangeland carrying capacity and grazing is not allowed during plants are passing through the critical stage of growth it is called _________. **(Controlled grazing)**

110. When the animals are kept permanently on a given area of a rangeland and allowed to move freely over this area, due to this the crop composition is changed. This is called ___________. **(Continuous controlled grazing)**

111. When the grassland is divided into three compartments and the grazing being allowed in two compartments while third one is given rest it is called ____________. **(Deferred grazing)**

112. ___________ types of grazing system has been found more suitable for perennial grasses. **(Deferred grazing)**

113. The combination of deferred and rotational grazing is known as ____________. **(Deferred rotational grazing)**

114. Deferred rotational grazing system has increased the grazing capacity upto _________%. **(39%)**

115. Livestock population census is carried out in every ___________year. **(Five)**

15

Remote Sensing

1. The technique of acquiring information about some object by recording devices, without being in physical contact with the object under study is_____________. **(Remote sensing)**
2. The term remote sensing was coined by ____________. **(Ms. Evelyn Pruitt)**
3. Father of Indian remote sensing is _________________. **(Pisharoth Rama Pisharoty)**
4. In remote sensing_____________ type of radiation is utilized for photography.**(Infra red)**
5. First aerial photograph was taken in 1858 by Felix Tournachon, known as ________, from a tethered ballon over the Bievre Valley in France. **(Nadar)**
6. Remote sensing systems which provide their own source of energy for illumination are known as_____________. **(Active sensors)**
7. _____________ sensors are very useful in detecting and mapping forest fires. **(Thermal sensors)**
8. Land cover classification using three levels of the scale of 1 : 100000 was developed by _____________agency. **(NRSA)**
9. The forest survey of India (FSI) conducts forest cover survey once in every_____________. **(2 years)**
10. For preparation of inventories of large forest areas, of late_____________ is being increasingly used. **(Aerial photography)**
11. On the basis of the distance of the platform from the earth, remote sensing is broadly classified into two categories namely_____________ and_____________. **(Aerial and Space)**
12. Land surface feature has specific response to radiation which is termed as_____________. **(Reflection)**

13. Aerial photos are_____________ projection of earth surface. (**Central**)
14. IIRS is located at_____________. (**Dehradun**)
15. NNRMS head quarters in India is located at_____________. (**Hyderabad**)
16. Maps are_____________ projection of earth surface. (**Orthogonal**)
17. The oldest application of remote sensing technology is in_____________. (**Miltary**)
18. _____________is the agency for aerial photography in India. (**NRSA**)
19. Ability of sensor to differentiate different wavelengths is known as_____________ resolution. (**Spectral**)
20. Maps and aerial photo gives and dimensional view respectively. (**Two and three**)
21. _____________ photographs have tilt of more than 30. (**High oblique**)
22. _____________is a process of examining photographic images and judging their significance. (**Photo interpretation**)
23. The ideal time for taking aerial photograph is_____________. (**9 to 11 AM**)
24. Spectral signature of green vegetation image is. (**Red**)
25. Scanning system which collects the data over a range of wavelengths is called as. (**MSS**)
26. The distinguishable shade variation from white to black in crown of trees caused by differential reflection of light is called as_____________. (**Tone**)
27. _____________ is the first Indian satellite. (**IRS-IA**)
28. The frequency of change and arrangement of tones in an image is called as_____________. (**Texture**)
29. The spatial arrangement of object is called as_____________. (**Pattern**)
30. The situation of objects with references to site factors such as climate, topography and soil is called as. (**Location**)
31. The science which deals with achieving three dimensional effects with binocular vision is called as_____________. (**Stereoscopy**)
32. The art and science of obtaining reliable measurements and maps from aerial photographs is_____________. (**Photogrammetry**)

33. A point whose position is determined from photographs by photogrammetric methods and which is intended for use as a ground control point in the orientation of other photographs is____________. **(Pass point)**

34. The angle between the optical axis of the camera and the plumb line for a given photograph is called as____________. **(Tilt)**

35. The horizontal displacement of air craft due to wind from its originally determined course is____________. **(Drift)**

36. The drift and crab should not exceed____________ of the width of the photograph. **(10%)**

37. Contribution of picture element is____________. **(Pixel)**

38. The ability to distinguish closely spaced objects in an image or photograph is called as ____________. **(Resolution)**

39. The usual unit of measurement of wave length is____________. **(Micrometer)**

40. The maps show the objects by____________ symbols. **(Conventional)**

41. The scale of aerial photograph is the ratio of a distance on a photograph to its ____________ on the ground. **(Corresponding distance)**

42. The bands of the spectrum in which there is little or no absorption are known as. **(Atmospheric windows)**

43. An assemblage of overlapping aerial photographs whose edges have been cut and matched to form a continuous pictorial presentation of a portion of the earth surface is ____________. **(Mosaic)**

44. A set of photographs correctly oriented and mounted for stereoscopic viewing is ____________. **(Stereogram)**

45. The grid system generally adopted in India is____________. **(Lambert Grid System)**

46. For UTM (Universal Transverse Mercator) grid, the world is divided into north to south zones. **(60)**

47. Aerial or Satellite imagery are seen through____________ type of instruments to obtain three dimensional images **(Stereoscopic)**

48. Visible light is electromagnetic radiation with wavelengths between____________. **(0.4 μm to 0.7 μm)**

49. The____________ band is a valuable region for remote sensing. **(Microwave)**

50. Microwave band with wavelength ranges between____________ is used for active form of remote sensing. **(0.1 to 30 cm)**

51. ____________ wavelengths can penetrate through clouds, fog and rain, thus helps in active or passive mode of remote sensing. **(Microwave)**

52. Regions of the electromagnetic spectrum in which the atmosphere is transparent are called as____________. **(Atmospheric windows)**

53. The aerial photograph is____________presentation of the pattern of the landscape. **(True pictorial)**

54. The ALBEDO/ Whiteness coefficient =____________. **(All the light reflected by the surface/ All the light incident on the surface)**

55. HDDT stands for____________. **(High Density Digital Tape)**

56. ERTS stands for____________. **(Earth Resources Technology Satellite)**

57. Satellite image interpretation in forestry and other natural resources disciplines in India started in 1973 by____________. **(SAC, Space Application Centre)**

58. The National Remote Sensing Agency (NRSA) was established in 1975 at____________. **(Hyderabad)**

59. Visual image interpretation is greatly helped by image elements called____________. **(Qualitative image characteristics)**

60. The quickest possible way for the inventory and monitoring of the natural resources is through applications of____________ ,____________ and____________ technologies. **(Remote sensing, Geographic Information System & Global Positioning System)**

61. In remote sensing EMR stands for____________. **(Electro-magnetic Resolution)**

62. Ozone mainly confined to stratospheric heights with a maximum concentration near ____________ km. **(25)**

63. ____________is the only ozone band used to detect the oceanic constituents from space. **(Chappuis band)**

64. The spectral reflectance curves also known as____________. **(Spectral signatures)**

65. The spectral reflectance curves for vigorous vegetation manifests ______ configuration. (**Peak and Valley**)

66. Stefan-Boltzman law gives idea about____________ of a black body. (**Total spectral exitance**)

67. Most imaging sensor or imaging radars used for remote sensing is____________type. (**SLAR, Side looking Airborne radar**)

68. RADAR is an acronym derived from____________ and Ranging. (**Radio Detector**)

69. Surface roughness is a function of the incident angle and____________. (**Wavelength**)

70. The meaning of term 'nadir' used in remote sensing is____________. (**Look angle**)

71. A radar image displays the back scattering characteristics of the earth's surface in the form of a____________. (**Strip map**)

72. Indian Space Research Institute's Satellite Tracking and Ranging Station is located at ____________. (**Hassan**)

73. The greatest amount of distortion on an aerial photograph is at the____________. (**Edge of the photo**)

74. On a coloured topographic map, water is indicated by the____________ colour. (**Blue**)

75. The term biological control refers to pest control using____________. (**Natural agents**)

76. Pubscent and glaucous are terms used to describe leaf____________. (**Surface**)

77. Fields are depicted as____________ on a topographic map. (**White**)

78. The letters GPS stand for____________. (**Global Positioning System**)

79. To locate points and map points the best tool to use would be the____________. (**GPS receiver**)

80. To view an area in three dimensions there is use of____________ and ____________. (**Aerial photographs and a stereoscope**)

81. A GPS unit utilizes____________to locate points. (**Satellites**)

82. The oldest and most common method of remote sensing used in forestry is____________. (**Aerial photography**)

83. ____________ is the best season to differentiate dry deciduous and tropical evergreen forests through photography. (**Spring**)

84. Optimum season for aerial photography of sal forests in North India is____________. (**Mid January to 1st week of March**)

85. Optimum season for aerial photography for of teak in south India is____________. (**September to October**)

86. ____________is an autonomous regional satellite navigation system being developed by the Indian Space Research Organisation (ISRO) of the Indian Government. (**IRNSS**)

87. ISRO was formed in the year _________ and its headquarter at _________. (**1969 & Bengaluru, Karnataka**)

88. Aerial photographs are available with____________ and satellite data can be obtained from____________. (**Surveyor General of India, Dehradun & NRSA, Hyderabad**)

89. The surveyor General's office for aerial photography is situated in __________. (**Dehra Dun**)

90. Three energy interactions occurs when electromagnetic energy falls on any surface feature ie.,____________ ,____________ and____________. (**Reflection, Absorption & Transmission**)

91. The smallest dimension or minimum detectable area that can be resolved by the sensor is known as____________. (**Spatial resolution**)

92. Aerial photo interpretation is an act of examining____________and judging their ____________. (**Photographic image & Significance**)

93. Tone refers to the distinguishable shade variation from____________ to____________. (**White to Black**)

94. Rectangular and square objects normally designate the____________ structure where as conical shape confined to____________ mostly. (**Man-made, Trees**)

95. The digital image is composed of____________ of pixels. (**Two dimensional array**)

96. The LANDSAT and INSAT satellites provides data at an interval of____________ days. (**18–20**)

97. ____________ is a satellite based navigation system that provides continuous, real time, 3-dimensional positioning, navigation and timing worldwide. (**GPS, Global Positioning System**)

98. _____________is a computer based information system which is used for presenting and analyzing the geographic features present on the earth surface and the events that taking place on it. (**Geographic Information System, GIS**)

99. GIS experts and technical specialists who design and maintain the system for everyday work is referred as_____________. (**Liveware**)

100. _____________ data represents the boundaries or the course of the features as a series of points. (**Vector**)

101. _____________ data represents the features graphically. (**Raster**)

102. Space segment consists of_____________ satellites orbiting the earth at an altitude of about _____________ km. (**24, 36000**)

103. GPS satellites continuously broadcast satellite position and timing data by on two frequencies. (**Radio signals**)

104. _____________ and_____________ delays the satellite signals as it passes through the atmosphere. (**Ionosphere and Troposphere**)

105. GPS satellites can complete its orbit in_____________ hours. (**12**)

106. The oldest surviving aerial photograph is the one of Boston, taken by ___________ in 1860, using a tethered ballon. (**James Wallace Black**)

107. India launched its first satellite ______________ in 1975 with the help of 'Kapustin Yar' Rusian rocket launch site. (**ARYABHATA**)

108. In 1957, Russia became the first country to launch an artificial satellite _________ rocketed by __________. (**SPUTNIK 1 and SPUTNIK 8K71PS**)

109. The father of Indian space programme is ____________. (**Vikram Sarabhai**)

110. The most common and important source of energy used in remote sensing is ___________. (**Sun**)

111. The electromagnetic radiation occurring as a continuum of wavelength and frequency from short wavelength and high frequency cosmic waves to long wavelength and low frequency radio waves, is called ____________. (**Electromagnetic spectrum**)

112. The bands of spectrum in which there is little or no absorption found are called ___________. (**Atmospheric windows**)

113. In aerial remote sensing the platform height varies from ________________. **(500 m to 15000 m)**

114. In space remote sensing the platform height varies from ______________. **(300 km to 1000 km)**

115. In India, ______________, ______________ and ______________ agency are responsible for aerial photography. **(Indian Air Force, Messers Air Survey Company of India (Pvt.) Limited & National Remote Sensing Agency)**

116. Missers Airr Survey Company of India (Pvt.) Limited is located at _____________. **(Dumdum, Kolkata)**

117. The photography of sensitive or restricted area is carried out by ______________. **(Indian Air Force)**

118. Aerial photography lens are divided into _________ categories on their area coverage. **(Four)**

119. Most commonly used filter in forestry is ____________ while for water bodies ___________ is used. **(Yellow filter & Red Filter)**

120. _____________ film is used infra-red colour photograph. **(Kodak Aero-chrome Infrared)**

121. _____________ photograph is used for identification of species and diseases. **(Colour photograph)**

122. Infra red colour photograph is used for Green vegetation, which is seen _____________ colour in this film. **(Red to magenta)**

123. For multi-spectral imagery the device used is ______________. **(Optical mechanical scanner)**

124. ___________ scale of photograph is used for forest inventory preparation. **(1: 20,000 to 1: 40,000)**

125. Optimum season for aerial photography in coniferous forest in Himalayan is ______________. **(October-November)**

126. Optimum season for aerial photography in sal forest is ______________. **(Mid March-Mid April)**

127. Optimum season for aerial photography in Teak forest is ______________. **(September-December)**

128. Optimum season for aerial photography in tropical evergreen & moist deciduous forest is ______________. **(January-February)**

129. Aerial photographs show _____________ projection of earth surface. (**Perspcctive**)
130. Maps show _____________ projection of earth surface. (**Orthographical**)
131. Aerial photographs show ____________ view of earth. (**3-D**)
132. Maps show ___________ view. (**2-D**)
133. Oblique photograph is taken only in __________ area. (**Terrain**)
134. Area covered in vertical photograph is ___________ form. (**Rectangular**)
135. Area covered in oblique photograph is ____________ form. (**Trapezoidal**)

16

Forest Engineering

1. An imaginary line joining the points of equal elevation on the surface of the earth represents_____________. **(Contour lines)**
2. The vertical distance between consecutive contours lines is called_____________. **(Contour interval)**
3. When measured with a compass, the line of sight that run midway between due North and due West has a bearing of_____________. **(N 45 degree W)**
4. A survey of a large area which consider the earth's curvature is called a _____________ survey. **(Geodetic)**
5. The distinctive feature of topographic maps is the_____________ of the land. **(Shape or Contour)**
6. The colour used to depict fence lines and important roads on a topographic map is _____________ and_____________. **(Red and black)**
7. A surveyor's chain is divided into links and each link is_____________ long. **(0.66 feet)**
8. A body of water on a topographic map is presented by_____________ colour. **(Blue)**
9. On a topographic map a ploughed field would be sown as_____________ colour. **(White)**
10. The colour used to depict vegetation on a topographic map is_____________. **(Green)**
11. A hand-held instrument used for measuring angles of elevation or depression, exprcssed in degrees or per cent is_____________. **(Abney's level)**
12. A plot of land is 30 chains long and 16 chains wide which constitutes _____________ acre. **(48)**

13. The contour interval on a topographic map indicated the difference in _____________ represented by adjoining contour lines. (**Elevation**)

14. The most basic parts of an azimuth compass include a magnet, usually a needle, and a circle divided into_____________degrees. (**360**)

15. A hand held compass with numbers graduated up to 360 clockwise is called as_____________. (**Azimuth**)

16. If a topographic map has a scale of 1:24000 it means the scale is 1 inch equals to _____________ inches. (**24000**)

17. Contour lines on a topographic map provide information concerning_____________. (**Elevation**)

18. If you are travelling at an azimuth reading of 325 degree, you are heading_____________. (**Northwest**)

19. Contour trenches are usually made on slopes up to_____________ per cent. (**20**)

20. The contour interval is_____________ proportional to the scale of the map. (**Inversely**)

21. An imaginary line lying throughout the ground surface and having a constant inclination to the horizontal is called_____________. (**Contour line**)

22. The length of gunter's chain is_____________. (**66 ft**)

23. The horizontal distance between the vertical joints in two successive courses is called as _____________. (**Lap**)

24. For river crossing with span less than 3m and seasonal flow, the bridge recommended is _____________. (**Ford**)

25. _____________compass is used in plane table survey for orientation. (**Tough compass**)

26. Survey of India, Map Publication Directorate is located at_____________. (**Kolkata**)

27. In land capability map, the standard colour given for the land capability class VII is _____________. (**Brown**)

28. A line joining points of equal precipitation in a map is_____________. (**Isohyets**)

29. Engineer's scale is the relation between the distance on and the distance on is mentioned numerically in the style as 1 cm = 5 m, etc. (**Plan and Ground**)

30. Representative fraction (R.F.) is the ratio is worked out in such a way that numerator is____________ and the denominator is a____________ in the same unit of measurement as the numerator. (**Unity and fraction**)

31. ____________is a type of scale, it is possible to read only in two dimensions on paper such as meters, decameters, hundreds and tenths, units etc. (**Plain scale**)

32. In whole circle bearing system (W.C.B.), the angle is measured from the magnetic North in____________ direction and its value will therefore vary from 0° to 360°. (**Clockwise**)

33. Diagonal scale is a type of scale, it is possible to read in____________ dimensions on paper such as meters, decameters and centimeters; units, tenths and hundredths; etc. (**Three**)

34. ____________ is the fractional part of the smallest division of the main or primary scale. (**Vernier scale**)

35. Vernier scale was found out by____________ in 1631. (A.N. Vernier)

36. Vernier scale is of ____________ types. (**Five**)

37. The ratio of the value of the smallest division on main scale to the total number of divisions on vernier scale indicates the____________ of the vernier scale. (**Least count**)

38. Guner's chain is also called ____________ chain. (**Surveyor's**)

39. When the end points are visible from one another, intermediate ranging rods are placed in line by direct observation, the process is known as____________. (**Direct ranging**)

40. When the end stations of a line are not invisible due to high ground or intervening hill and also when the ends of a line are not distinctly visible from one another due to distance being too great, then____________ can be adopted. (**Indirect ranging**)

41. The Gunters chain is 66 ft long and is divided into____________ links. (**100**)

42. In____________ system, the angle is measured from North or South to East or West. (**Quadrantal bearing**)

43. Compass survey cannot be carried out with success in____________ areas of the town. (**Industrial**)

44. ____________is also used for accurately centring of compass or level over a station mark, and for testing the vertically of ranging poles. (**Plumb bob**)

45. Revenue chain is____________ long and divided into__________ links. (**33 ft and 16**)

46. Engineer's chain is____________long and is divided into 100 links and each link is equal to 1 ft. (**100 ft**)

47. ____________ is used for establishing the instrument stations only and thus helps in ascertaining the fact that the point plotted on plan is the station occupied by the plane table. (**Resection**)

48. ____________is defined as the angle made by the line with some reference direction/ meridian. (**Bearing of a line**)

49. In direct vernier, both the scale namely vernier and main move in the ____________direction and they are graduated in the same direction. (**Same**)

50. Prismatic compass is a circular in shape and its diameter varies from____________ mm to ____________mm. (**85 mm to 110 mm**)

51. Prismatic compass is made up of a/an____________metal. (**Non-magnetic**)

52. The instrument which is used for finding out the magnetic bearing of a line is known as ____________. (**Compass**)

53. ____________ resembles the work of a compass survey and it is useful for the survey work of roads, rivers etc. the traverse is run as usual and the details on the line are taken by radiation. (**Traversing**)

54. ____________ is the simplest method and it is useful only when the whole traverse can be commanded from a single station. (**Radiation**)

55. ____________ is useful where it is not possible to measure the distances on ground as in case of a mountainous. (**Intersection**)

56. The reading of line XY taken from point X is the Fore Bearing of line XY and the reading of the line XY taken from point Y is its____________. (**Back Bearing**)

57. In the quadrantal system, there will not be any changes in the value of Fore Bearing and Back Bearing, except that their respective quadrants will be____________. (**Interchanged**)
58. The angle formed by the intersection of two lines in a vertical plane is known as ____________. (**Vertical angle**)
59. ____________ in the field book shows roughly the area to be surveyed and important objects such as buildings, roads, streams etc. (**Sketch**)
60. The point which is to be set up at a given elevation or whose elevation is to be found out is known as____________. (**Station**)
61. The process of placing the contour lines proportionally between the plotted ground points is known as____________. (**Interpolation of contours**)
62. In____________ survey, the measurements of survey lines of the traverse and their plotting to a suitable scale are done simultaneously on the field. (**Plane table**)
63. ____________survey proves to be speedy, easy and accurate for preparing plans on a small scale. (**Plane table**)
64. Alidade and drawing board are the instruments required for____________ survey. (**Plane table**)
65. In retrograde vernier, its main scale and vernier scale moves in____________ direction and graduations are also marked in opposite directions. (**Opposite**)
66. In the process of____________ , the direction of the survey line is fixed by taking the angular measurements with suitable measurements. (**Traversing**)
67. The main principle of chain survey is to prepare a/an____________ of triangles because a triangle is a figure, which can be plotted on paper by measuring its sides only. (**Framework**)
68. Extended vernier is a type of vernier, which is similar to the direct vernier scale except that every____________ omitted. (**Second division**)
69. The process of____________ is not possible when the area is to be surveyed is large with irregular boundaries and many obstacles. (**Triangulation**)
70. Geodetic survey is also known as____________ surveying. (**Trigonometrical**)

71. A method of determining the relative elevations of points on the surface of earth for finding out the difference in heights is known as____________. **(Levelling)**

72. The average height of the sea for all states of the tides is known as____________. **(Mean Sea Level, MSL)**

73. The vertical distance of a point with respect to a given datum, either positive or negative is known as____________. **(Elevation)**

74. Backsight is the____________ reading from any set up of the instruments. **(First)**

75. Foresight is the____________ reading taken before disturbing the instruments from its set up. **(Last)**

76. The line joining the intersection of cross-hairs and optical centre of the object glass and its continuation is known as____________. **(Line of collimation)**

77. The elevation of the line of collimation, when the instrument is correctly levelled is known as the____________. **(Height of instrument)**

78. ____________is the point indicating the shifting of levels. **(Change point/C.P.)**

79. ____________is a last staff reading denoting the shifting of the level. **(Foresight)**

80. ____________ is a staff reading taken on a point of known elevation, as on a bench mark or a change point. **(Backsight)**

81. All sights taken between Backsight and Foresight are known as____________. **(Intermediate sight/I.S.)**

82. A fixed point of known elevation is known as____________. **(Bench mark/B.M.)**

83. The elevation of a point or its vertical distance above or below the datum is known as ____________. **(Reduced level)**

84. ____________ is used to determine the amount by which the station is above or below the line of sight. **(Levelling staff)**

85. The line normal to a level surface is known as a plumb line and the plane which contains a vertical line is called a____________. **(Vertical plan)**

86. Reconnaissance survey indicates____________ of the site of work. **(Preliminary inspection)**

87. ____________ is an instrument for finding gradients, setting out grade contours, preliminary survey of hill route for road alignment and also for contouring and levelling. (**Ghat tracer**)

88. The line drawn on a level surface is known as____________ line. (**Level**)

89. The process of plotting with scale is known as____________. (**Drawing**)

90. ____________ Is any arbitrarily assumed level surface or line from which the vertical distance is measured. (**Datum surface**)

91. Revenue chain is 33 ft long and divided into____________ links. (**16**)

92. In the quadrantal system, there will not be any changes in the value of fore bearing and back bearing, except that their respective quadrants will be____________. (**Interchanged**)

93. ____________ is a series of straight lines connecting a series of established points along a route of survey. (**Traverse**)

94. Classifications of surveying are of____________ type. (**Two**)

95. In plane surveying, the earth's surface is assumed as a____________. (**Plane**)

96. In geodetic surveying, the earth's surface is assumed as____________. (**Arc**)

97. Chain survey deals with____________ measurements. (**Linear**)

98. ____________is a system of chain surveying in which sides of various triangles are measured and no angular measurements are required. (**Triangulation chain survey**)

99. The actual north and south line of the earth is referred as____________. (**True meridian**)

100. Magnetic declination is the horizontal angle between the____________ and ____________. (**Magnetic meridian & true meridian**)

101. Plane table surveying (PTS) is a method of surveying in which the field observation, measurements and plotting are proceeding____________. (**Simultaneously**)

102. Topographical survey is used in preparing a____________. (**Contour map**)

103. Soil conservation research, demonstration and training centre for North Himalayan region is situated at ____________. **(Dehra Dun)**

104. Soil conservation research, demonstration and training centre for Southern hill high rainfall region is situated at ____________. **(Ooty, Tmilnadu)**

105. At any case one contour does not intersect the other lines except in the case of ______________in an area. **(Cliffs/ vertical rocks)**

106. _____________is the ratio of a unit distance on the map to the corresponding distance on the ground. **(Scale of the map)**

107. _____________ represents a line, which is subdivided into plain distance to the corresponding convenient units of length on the ground. **(Graphical Scale)**

108. ___________ vernier, main scale and vernier scale move in opposite directions and graduations are also marked in opposite directions. **(Retrograde)**

109. In ______________ survey indicates the preliminary inspection of the site of work. **(Rconnaissance)**

110. The institute of Forest Survey of India (FSI) is located at ___________. **(Dehra Dun)**

17

Forest Statistics

1. A measure of the items in a random sample is known as____________. **(Statistics)**
2. Most commonly used measure of central tendency is____________. **(Arithmetic Mean)**
3. Mean deviation is the____________, when calculated about median. **(Least)**
4. Arithmetic mean of central tendency is applied for calculation of____________ and ____________. **(Regression and Correlation coefficient)**
5. Arithmetic mean and variance are always equal in____________ distribution. **(Poisson)**
6. Geometric mean of central tendency is appropriate for____________. **(Index number)**
7. ____________ is the sum of all values/ number of values. **(Arithmetic mean)**
8. Middle most item of all values is____________. **(Median)**
9. ____________ measure of central tendency requires data arrangement in ascending or descending order for its estimation. **(Median)**
10. The most frequent occurred value of data when maximum is known as ____________. **(Mode)**
11. Measurement of central tendency used to study average rate of change in the population is called ____________. **(Geometric mean)**
12. The measure of central tendency to find out the average quantity of price is called____________. **(Harmonic mean)**
13. If the observations are____________ the relation between A.M., G.M. and H.M. are equal. **(Identical)**

14. The correlation coefficient is the_____________ between two regression coefficient. (**Geometric mean**)
15. In normal distribution mean, median and mode are_____________. (**Equal**)
16. If any items of the series is zero, then geometric mean of central tendency become _____________. (**Zero**)
17. Geometric mean can be easily calculated with help of_____________. (**logarithm**)
18. The square of the standard deviation is known as_____________. (**Variance**)
19. The relationship between independent and dependent variables is computed by _____________. (**Regression coefficient**)
20. If all the variate values are_____________ then the standard deviation will be positive. (**Negative**)
21. The difference between the highest and lowest values is known as_____________. (**Range**)
22. In a double sampling plan the decision to accept/ reject a lot is made on the basis of _____________ samples. (**Two**)
23. In RBD,_____________treatments can be adopted without any loss of efficiency. (**20**)
24. Indian Agricultural Statistical Research Institute is established at_____________. (**New Delhi**)
25. The number of independent constraints in the set of observation is called _____________. (**Degree of freedom**)
26. _____________ test is use for the testing the significance of correlation coefficient. (**t-test**)
27. For comparison of two means from independent samples, test is applicable. (**t-test**)
28. _____________ is used when sample size is small and population standard deviation is not known. (**Student t-test**)
29. For paired observation,_____________ test is used for testing the significance of a mean difference. (**Paired t-test**)

30. The lack of homogeneity in the size of distribution is called_____________. (**dispersion**)
31. _____________ is the simplest measure of dispersion. (**Range**)
32. The R.B.D. is available for a wide range of treatment is_____________ to_____________. (**2, 24**)
33. _____________ samples are required for getting higher precision. (**Larger**)
34. To test the agreement between observed frequencies and expected frequencies we use_____________. (**Chi square test**)
35. _____________ test is often used to refer specifically to the one sample location test. (**Z-test**)
36. _____________ test is used for test of significance of the difference between two means. (**Z-test**)
37. X^2 test is applicable for testing the_____________ of correlation coefficients. (**Homogeneity**)
38. The minimum sample size for using chi square test should be_____________. (**Fifty**)
39. When the fertility gradient of the field is in one direction, design is applicable. (**R.B.D.**)
40. When the fertility gradient of the field is in two directions,_____________ design is applicable. (**L.S.D.**)
41. Local control is not applicable for_____________ design. (**CRD**)
42. The regression coefficient is_____________of change of origin. (**Independent**)
43. The variation due to_____________ factors is known as experimental error. (**Uncontrolled**)
44. The analysis of covariance is a technique that combines the features of analysis of_____________ _____________ and_____________. (**Variance and Regression**)
45. The analysis of variance technique is used for comparing the means of_____________ population. (**More than two**)
46. During experimentation, some information lost and to get idea about these values, ________ techniques will be useful. (**Missing plot**)

47. The probability of committing_____________ error is known as level of significance. (**Type 1**)

48. S.D. was given by_____________. (**Karl Pearson**)

49. The square root of Arithmetic mean is _____________. (**Standard deviation**)

50. Coefficient of variation =_____________.
(Standard deviation/arithmetic mean X 100)

51. The principle of making use of greater homogeneity in groups of experimental units to reduce the experimental error is referred as_____________. (**Local control**)

52. Latin square design is suitable for comparing_____________treatments. (**5 to 12**)

53. In the field experiment,_____________is the commonly used design. (**R.B.D.**)

54. _____________design is most appropriate for the laboratory experiments. (**CRD**)

55. _____________ design is used when the number of replication is equal to number of treatments. (**Latin square**)

56. When there are several factors with different levels to be experimented simultaneously with the same level of precession,_____________ design is most appropriate. (**Factorial scheme**)

57. In an experiment, a treatment requires large area and another treatment requires smaller area, so_____________ design is most appropriate. (**Split plot design**)

58. When the values of correlation coefficient are_____________ then two variables are perfectly coefficient. (**1**)

59. When the distribution of data is_____________, normal distribution is applicable. (**Continuous**)

60. ________ is the difference between largest and smallest observation. (**Range of distribution**)

61. _____________ ,_____________ and_____________ are the basic principle of field experiment. (**Randomization, Replication and Local control**)

62. In variance ratio test, larger variance is kept as the____________. **(Numerator)**
63. When the observations are paired together____________ test is applicable. **(Paired t-test)**
64. ANOVA table stands for analysis of____________. **(Variance)**
65. Student t-test was given by____________. **(W. S. Gossett)**
66. The term variance was proposed by____________. **(R.A. Fisher)**
67. ____________ is consider as father of statistics. **(R.A. Fisher)**
68. The concept of ANOVA was given by____________. **(R.A. Fisher)**
69. In analysis of variance, all population from which samples have been drawn are ____________. **(Normal)**
70. The reciprocal of A.M. is____________. **(Harmonic mean)**
71. The repetition of the treatments under investigation is known as____________. **(Replication)**
72. Lack of symmetry of frequency distribution is known as____________. **(Skewness)**
73. Coefficient of skewness is calculated by____________. **(S_k = Mean-Mode/ S.D.)**
74. The minimum error degrees of freedom for RBD should be at least____________. **(12)**
75. ____________ helps in reducing experimental error and homogeneity of experimental units. **(Local control)**
76. The statistical procedure for estimating whether the difference under study is significant/ non-significant is known as____________. **(Test of significance)**
77. The range of correlation coefficient lies between____________. **(-1 to +1)**
78. When we want to conduct an experiment on a long strip of land,____________ design will be chosen. **(LSD)**
79. ____________ is a square of correlation coefficient. **(Coefficient of determination)**

80. LSD design gives precision high enough to reduce the standard error to ____________ per cent. (**Less than 1**)

81. Augmented design is most suited, when the experimental material is ______ and number of genotypes is____________. (**Very less and very large**)

82. The degree of flatness/ peakness of the frequency curve is known as. (**Kurtosis**)

83. The allocation of the treatments to the different experimental units in a random manner is known as____________. (**Randomization**)

84. ____________ design is appropriate, when the experimental material is limited and homogenous. (**CRD**)

85. ____________ is the most appropriate design when all factors are not of equal importance in experimentation. (**Split plot design**)

86. For the field experimentation the accurate idea of the fertility variation can be held by the help of____________. (**Uniformity trial**)

87. The regression coefficient lies between____________. ($-\infty$ **to** $+\infty$)

88. The number of independent constraints in the set of observation is called ____________. (**Degree of freedom**)

89. Binomial distribution was given by____________. (**James Bernoulli**)

90. Binomial distribution is otherwise called ____________. (**Bernoulli's distribution**)

91. The relation between two variables is studied by____________. (**Correlation**)

92. In case of absence of any relation between two variables, the value of correlation coefficient will be ____________. (**Zero**)

93. A statement about parameters which is to be rejected after testing is known as ____________. (**Null hypothesis**)

94. Term regression was first used by____________. (**Sir Francis Galton**)

95. ____________ test is used for enumeration or non-parametric/ qualitative data in Mendelian or population genetics. (**Chi-square test**)

96. Range of chi-square test varies between____________. (**0 to** ∞)

97. When the experimental material is limited and homogeneous, such as in pot analysis,____________ design is applicable. (**CRD**)

98. A statistical technique of partitioning the total variation into component variations and computing them by F-test is_____________. **(ANOVA)**

99. ANOVA stands for_____________. **(Analysis of Co- variance)**

100. _____________ is applicable, when the distribution of data is continuous. **(Normal distribution)**

101. _____________test is used to check the significance of the difference between two means. **(Z-test)**

102. The validity of coefficient correlation is tested by_____________. **(t-test)**

103. _____________design is preferred to conduct an experiment on a long strip of land. **(LSD)**

104. Rank correlation is tested by_____________. **(t-test)**

105. The relation between Mean, Median and Mode according to Karl Pearson is _____________. **(Mode = 3 Median – 2 Mean)**

106. The graphical method of studying dispersion is_____________. **(Lorenz Curve)**

107. The test used to check the randomness of a sample is _____________. **(Run test)**

108. In RBD, error or degree of freedom can be calculated by the formula. **((T-1) (r-1))**

109. If all variate value values are negative, the S.D. will be_____________. **(Positive)**

110. _____________ is used to study the association between two or more variables. **(Correlation)**

111. To reduce the experimental error, with heterogeneous material, we need more replication and use of_____________ technique. **(Local control)**

112. Under the assumption for analysis of variance, the parent population should be _____________. **(Normal)**

113. Experiments used to study the effects of two or more factors and their inter-relationship is _____________. **(Factorial experiments)**

114. Probability ranges from_____________. **(0 to 1)**

115. Allocation of treatments to the different experimental unit by a random process is called_____________. (**Replication**)

116. _____________design is used to study the effects of 2 or more factors requiring different plot sizes. (**Split plot Design**)

117. _____________ design is used to study two or more factors and their inter-relationship. (**Factorial design**)

118. The relation between independent and dependent variables is computed by _____________. (**Regression coefficient**)

119. Empirical relation between quartile deviation and standard deviation is. (**4 QD = 3 SD**)

120. Equity of two regression coefficient can be tested with_____________. (**ANOVA**)

121. Hypothesis that the population variance has a specified value can be tested by _____________. (**x^2 test**)

122. The value of x^2 always lies between _____________. (**0 to ∞**)

123. The t-test tends to be Z test when sample size becomes more than_____________. (**30**)

124. The probability of an impossible event is_____________. (**0**)

125. The limiting relative frequency approach of probability is known as. (**Statistical probability**)

126. _____________ is the basis of sampling theory. (**Randomization**)

127. Error degree of freedom in LSD is_____________. (**(n-a) × (n-2)**)

128. _____________ design is suitable to study the agronomic practices with fertilizer treatments. (**Split plot Design**)

129. When the fertility gradient of the field is in two directions, the most appropriate experimental design is_____________. (**LSD**)

130. _____________ is used to test the goodness of fit of a distribution. (**Chi-square test**)

131. Normal distribution is due to the work of _____________. (**Demoivre**)

132. Poisson distribution deals with ___________ variables. (**Discrete**)

133. National Institute of Agricultural Marketing (NIAM) is located in ____________. (**Jaipur**)

134. In hypothesis testing the most common level of significance that is used by statisticians is _______________. **(0.05)**

135. Principles of Experimental design was given by _____________. **(R. A. Fisher)**

136. When some of the plot data is missing the most suitable experimental design is ______________. **(Missing plot technique)**

137. The curve of normal distribution is ____________ shaped. **(Bell)**

138. Error degree of freedom for RBD is _____________. **{(n-1) (r-1)}**

139. Rejecting null hypothesis (H_0) when it is true or accepting alternative hypothesis (H_1) when it is false called is _______________. **(Type error (Alfa)-α)**

140. Accepting null hypothesis (H_0) when it is false or rejecting alternative hypothesis (H_1) when it is true called is _______________. **(Type error (Bita)-β)**

141. _____________ measure of dispersion is considered as best. **(Standard deviation)**

18

Wildlife Management

1. ____________ coined the term wildlife. (**William Hornday**)
2. National Biodiversity Board is situated at____________. (**New Delhi**)
3. The term biodiversity was coined by____________. (**Edward Wilson**)
4. The study of combination of animal behaviour and ecology is called as____________. (**Ethology**)
5. ____________ is the place where an organism or community lives. (**Habitat**)
6. ____________ is the living medium of the habitat in a wetland. (**Water**)
7. ____________can be defined as the functional status of an organism in an ecosystem or the role played by a species in its natural habitat. (**Niche**)
8. The food which is totally non-nutritive and is consumed by the animals only to fulfil its stomach is called ____________. (**Stuffing food**)
9. The period in which the food is not sufficiently available and causes trouble to the animal, is called ____________. (**Pinch period**)
10. Pinch period is the period during which the____________ factors are most effective. (**Limiting**)
11. A limiting factor is that factor which restrict the____________ of an animal population, when all other factors are favourable. (**Growth**)
12. The amount of an edge can be expressed as an index for a given area, indicated by ____________. (**Patton**)
13. ____________refers to a predictable change in the species composition of a plant community over a period of time. (**Biotic succession**)
14. ____________ is a measure of a system relation and its means the intermixing of units pertaining to different habitat type. (**Interspersion**)

15. The induced and inherent edges reflect the_____________ in an area. **(Total diversity)**
16. Study of snakes is known as_____________. **(Herpetology)**
17. Study of butterflies is known as_____________. **(Lepidopterology)**
18. Any event or activity which affects the quality of a habitat, by modifying it, is known as_____________. **(Qualifier)**
19. _____________ are consider as special places for wildlife and their value varies with season. **(Coverts)**
20. For practical purposes "Edge" can be used as a/ an_____________. **(Index of diversity)**
21. The succession which takes place after disturbances is known as_____________ succession. **(Secondary)**
22. Information about the number of plants and animal species, their frequency of occurrence is required for_____________. **(Standard diversity index)**
23. Similar niches may exist in different regions and these niche may be occupied by organisms which show similar, but not identical, habits, adjustments which called as _____________. **(Ecological equivalents)**
24. Animal adaption to climbing is known as_____________. **(Scansorial)**
25. Animal adaption to flight is known as_____________. **(Volant)**
26. Certain entities, created for management, such as water holes, salt licks, machans etc. can be seen as_____________on a map. **(Points)**
27. _____________is the wildlife symbol of WWF. **(Giant Panda)**
28. Indicator species gives us an indication of the_____________ with respect to the wild herbivores population present in the area. **(Habitat condition)**
29. The modification done in the habitat as per the requirement and benefit of the wildlife especially for food, water, shelter and area is called _____________. **(Habitat manipulation)**
30. The extent of an animal movement is consider outer boundary of a circle, half of this is treated as diameter to express the_____________. **(Cruising radius)**

31. The competitive exclusion principle in animal is called as_____________. **(Gause's principle)**

32. Forest fire have a tremendous influence on the _____________ and _____________ components of a habitat. **(Flora and fauna)**

33. A/an_____________ is the place of contact between plant communities or successional stages/ vegetative conditions. **(Edge)**

34. _____________ is crucial for assessing the natural richness, biological value and the uniqueness of a particular area. **(Species diversity)**

35. The area which is influenced by the transition between the communities is called as a/ an_____________. **(Ecotone)**

36. Many species of fungi, mosses and lichens are use_____________ as a substratum for growth. **(Decayed wood)**

37. The number of species present within a particular group, as found in a sample is known as_____________. **(Species richness)**

38. _____________ and_____________are comparatively richer in wild animals than the adjoining areas, and hence they are of much importance in wildlife management. **(Edges and Ecotone)**

39. The dropping of carnivores are called as _____________. **(Scat)**

40. The number of specific animals which a given area can support is known as_____________. **(Carrying capacity)**

41. In forest land management, a_____________ can be used to indicate the habitat diversity, for evaluating management alternatives with respect to the immediate and long term effects on diversity. **(Diversity index)**

42. The measurement of a particular attribute of a wild animal, to draw inference about the existing conditions of the habitat as well as the animal is known as_____________. **(Physiological indices)**

43. _____________ helps in studying social behaviour, breeding behaviour, predator-prey relationship and daily activity of the animal. **(Bio-telemetry)**

44. Habitat analysis means a report of the characteristics of a_____________ and it also includes an appraisal of the_____________ within the habitat. **(Given habitat, quality of the biomass)**

45. The first man made bird sanctuary is _____________. **(Keoladeo-Ghana)**

46. _____________ is called as a bird man of India. (**Dr. Salim Ali**)
47. The secretariat office of birdlife international is at_____________. (**Cambridge, UK**)
48. Objective of the modern zoo are_____________,_____________,_____________ and _____________. (**Conservation, education, research and recreation**)
49. A group of lions is called_____________. (**Pride**)
50. A group of wild dog is called_____________. (**Pack**)
51. The Periyar wildlife sanctuary is located at_____________. (**Kerala**)
52. Birds egg is measured by_____________. (**Oometer**)
53. Corbet national park is located at_____________. (**Uttrakhand**)
54. The World's only floating national park is _____________. (**Keibul Lamjao N. P.**)
55. The floating national park is situated at_____________state. (**Manipur**)
56. The keoladeo Ghana national park is situated in the state of_____________. (**Rajasthan**)
57. _____________ award is given for wildlife conservation in India. (**Amrita Devi Bishnoi Award**)
58. Barriers for the animals to feel_____________ in a well designed cage, especially when it finds the human beings at a close distance. (**Secure**)
59. _____________ and_____________ are the important tools for habitat evaluation. (**Remote sensing & High altitude photography**)
60. The staffing of wild animals is known as_____________. (**Taxidermy**)
61. Crocodile breeding project was undertaken from the_____________ for all the three crocodilian species found in India, based on the advice of an_____________. (**1st April 1975, FAO expert Bustard**)
62. The crocodile population in India declined considerably due to unabated persecution of these animals for their_____________. (**Skin**)
63. All the crocodile species comes under_____________ of the wildlife protection act. (**Schedule I**)
64. The one-horned rhinoceros is found in_____________ sanctuary. (**Kaziranga**)

65. The Asiatic lion is localized in the ____________ of Gujarat. (**Gir forest**)

66. ____________ census method is followed for Asiatic lion. (**Waterhole**)

67. The first marine national park is at____________ state. (**Gujarat**)

68. ____________ is a powerful tool for evaluating wildlife habitat, which combines several variables spatially on maps. (**Geographic Information System**)

69. ____________ and____________in a national park is a concept which was introduced for allowing tourism, so as to achieve the twin objectives of nature conservation and tourism development. (**Core zone and Buffering**)

70. ____________ is an extension of the nature reserve, where the spoiler of the total breeding populations of both plant animal species can establish, after exceeding the limits of the reserve boundaries. (**Extension buffering**)

71. The core areas were freed from all sorts of____________ and the buffer areas were subjected to____________ oriented land use. (**Human activities and Conservation oriented**)

72. Project tiger lunched in____________year. (**1973**)

73. India has____________ number of tiger reserve. (**39**)

74. Wildlife protection act was enacted in the year____________. (**1972**)

75. First Indian forest act was drafted in the year____________. (**1865**)

76. The world's largest private international nature conservation organisation is. (**WWF**)

77. Age and sexual maturation are important to known the____________ and of a population with respect to habitat, health and behaviour. (**Structure and status**)

78. The concept of____________ associates the occurrence of a certain disease with the existing landscape, which was enunciated by a Russian scientist____________. (**Landscape Epidemiology, Pablosky**)

79. In tiger protection____________ projects are engaged. (**47**)

80. Asiatic elephants have a life span of____________ years. (**60 – 70 years**)

81. ____________have sound legal status; which are created exclusively for the conservation of wildlife in their natural environment. (**National Parks**)

82. In National Parks,_____________ are not allowed and grazing and forestry operations are_____________. (**Human settlements, Prohibited**)

83. _____________ have sound legal status with a strong but not exclusive wildlife oriented management. (**Sanctuaries**)

84. The scientific name of king cobra is_____________. (***Ophiophagus hannah***)

85. The Olive Ridley turtle is scientifically called as_____________. (***Chelonia mydas***)

86. _____________ is the most serious threat to the wildlife. (**Habitat loss**)

87. _____________ is restricted and subordinated to the requirements of wildlife sanctuaries. (**Grazing**)

88. The status of a sanctuary can always be evaluated to that of a_____________. (**National Park**)

89. Special game is included in schedules _____________ of the wildlife protection act 1972. (**Schedule II**)

90. Cheeta from India become extinct due to_____________. (**Excess hunting**)

91. Scientific name of cheeta is_____________. (***Acinomyx jubatus***)

92. IUCN stands for_____________. (**International Union for Conservation of Nature and Natural resources**)

93. The headquarters of IUCN is located at _____________. (**Geneva**)

94. ZSI stands for_____________. (**Zoological Survey of India**)

95. IBWL stands for_____________. (**Indian Board of Wildlife**)

96. The IBWL was first constituted in the year_____________. (**1952**)

97. The wildlife week is celebrated on_____________ in India. (**1st week of October**)

98. _____________ zoo is famous for white tiger, which is situated at _____________. (**Nandankan Zoo, Odisha**)

99. _____________ sanctuary is called an oasis of water birds, which is located at. (**Chilika lake, Odisha**)

100. The Kanha National park is located at_____________ and established in the year __________. (**Madhya Pradesh, 1955**)

101. The Rajaji National Park is located at_____________ state and famous for________. **(Uttarakhand and Elephant)**

102. First biosphere reserve set up at_____________ in India during the year_____________. **(Nilgiri, 1986)**

103. _____________ numbers of biosphere reserve exist in India. **(18)**

104. A spider normally has_____________ eyes. **(8)**

105. _____________ is the largest bird in the World. **(Ostrich)**

106. _____________ is the largest bird in India. **(Sarus crane)**

107. _____________ is the longest reptile in the world. **(Reticulated Python, *Python reticulatus*)**

108. _____________ National Park in Kerala celebrates its 25th year of declaration as a national park in 2010. **(Silent Valley)**

109. _____________ is the only animal in the world to have four horns and its scientific name is_____________. **(Chausingha, *Tetracerus quadricornis*)**

110. Indian cobra is scientifically called as _____________. **(*Naja naja*)**

111. _____________ is the first wildlife sanctuary in India. **(Vedamthangal)**

112. _____________ is the first zoo in India. **(Alipur zoo, Calcutta)**

113. The emblem of WWF is_____________. **(Giant Panda)**

114. In our country the "Protected areas" for wildlife protection constitute_____________ per cent of geographical area. **(6%)**

115. "Rajaji National Park" is mainly associated for the protection of _____________. **(Elephants)**

116. ___________ is the first National park of India. **(Corbett National park)**

117. World's first national park is ______________, situated in ______________. **(Yellow stone N. P. & America)**

118. The concept of keeping forest reserve "Abhayaranya" was first given in India by _____________. **(Kautilya)**

119. An animal that feeds on fruit is called _____________. **(Frugivorous)**

120. _____________ continent is considered as heaven for wildlife. **(Asia)**

121. The largest and mostly widely spread Indian deer is_____________. **(Sambar)**

122. The tuskless elephant is known as_____________. (**Makna**)

123. Kedarnath Sanctury is well known for the protection of_____________. (**Musk deer**)

124. Sunderban Biosphere Reserve is a part of India's largest delta formed by the_____________ and_____________ river. (**Ganga & Brahmaputra**)

125. The national bird of the country is_____________. (**Peacock**)

126. Birds are migrate for_____________. (**Breeding and food**)

127. Wild buffalo belongs to schedule_____________ of wildlife (Protection) Act 1972. (**Schedule-1**)

128. Wild Asiatic water buffalo is distributed in the state of _____________. (**Assam**)

129. Manas wildlife sanctuary is located in_____________ state. (**Assam**)

130. World's largest bird, which can run very fast is_____________. (**Ostrich**)

131. Salim Ali was associated with_____________. (**Ornithology**)

132. _____________ is a mammal which can fly. (**Bat**)

133. The wildlife protection act was enacted in the year_____________. (**1972**)

134. Silent valley national park is situated in_____________ state. (**Kerala)**

135. The technical name of Earth summit 1992 was_____________. (**Convention on Biological Diversity**)

136. Wildlife protection act 1972 is amended in the year_____________. (**1991**)

137. The project tiger was initiated as per the advice of_____________ then PM of India. (**Indira Gandhi**)

138. National park for conservation of insectivorous plant Pitcher plant is. (**Bhagmara, Garo hills**)

139. Valley of flower –Nandadevi is a biosphere situated in_____________. (**Uttarakhand**)

140. Wetlands conservation is the significance of_____________. (**Ramsar convention**)

141. _____________ is the only Indian state where all three species of Indian crocodile are found wild. (**Odisha**)

142. ____________ species is protected in the Keibul- Lamjao national park. (**Dancing deer**)

143. ____________ is the smallest national park in India. (**South Button Island N.P.**)

144. ____________ is the biggest national park in India. (Hemis N.P.)

145. Jaldapara wildlife sanctuary is famous for ___________ wild animal. (**Rhinoceros**)

146. The zoological name of lion tailed macaque is____________. (***Macao silenus***)

147. ____________ is the smallest bird in India. (**Tickell's Flower pecher**)

148. The Himalayan mountain ranges is divided into____________divisions. (**5**)

149. Cheetah is found____________in India. (**Extinct**)

150. India signed in CITES during the year____________. (**1973**)

151. First bird sanctuary in India is____________. (**Vedanthangal**)

152. "Flamingo colony" is reputed as the largest flamingo colony in the world, which is located at____________. (**Khavda, Gujarat**)

153. The Similipal biosphere reserve is located at ____________. (**Odisha**)

154. Sariska national park is famous for ________. (**Tiger**)

155. The common methodology adopted for tiger census is _____. (**Pug mark**)

156. A book containing a record of animals & plants which are known to be in danger is called ________. (**Red Data Book**)

157. Red data book is maintained by ____________. (**IUCN**)

158. A book containing a list of rare plants in a protected area like as botanical garden is called ____________. (**Green data book**)

159. Ivory obtained from teeth of ___________. (**Elephant**)

160. Census of wild animal especially big cats based on _______. (**Pug mark**)

161. Tear mark or tear stripe is common in ________. (**Cheetah**)

162. Antler is common in ______. (**Deer**)

163. Largest living terrestrial animal in the world is _______. (**Elephant**)
164. Wild ass commonly present at ________. (**Rann of Kutch**)
165. Young one of carnivores is known as ______. (**Cub**)
166. Young one of monkey is known as _________. (**Infants**)
167. Dart gun is used for _______ purpose. (**Chemical immobilization**)
168. The chemical which causes behavioural change and brings emotional equity and reduces general activities is known as _________. (**Transquiliser**)
169. The drug used as a transquiliser is ________. (**Phenothiazines/ Acepromazine/ Valium**)
170. Droperidol is ______ type of drugs used for animal capturing. (**Transquiliser**)
171. Transquiliser is a ____ acting drug used for animal capturing. (**Central**)
172. Vulture population decreasing in an alarming rate due to the use of _______ drug. (**Diclofenac**)
173. The book 'Wildlife management techniques' was written by _____________. (**Robert H. Giles**)
174. Largest animal in the world of family Bovidae is ____________. (**Gaur-*Bos gaurus***)
175. First wildlife conservation movement in India was started by ________________. (**Bombay Natural History Society**)
176. Predators have an important role in regulating _________________. (**Balancing of nature**)
177. Study of analysis of the changes in the number of animals in wildlife population at a particular time is known as _____________. (**Population dynamics**)
178. Conservation of fauna and flora outside their native place is called _____________. (***In-situ* conservation**)
179. Conservation of fauna and flora outside their native place is called ______________. (***Ex-situ* conservation**)
180. Smallest tiger reserve in India is ___________, situated in ___________. (**Ranthambore N. P. & Rajasthan**)

181. Largest tiger reserve in India is ______________, situated in ______________. (**Nagarjuna Sagar sanctuary & Andhr Pradesh**)

182. Hot spot concept was given by ______________ in 1988. (**Norman Mayer**)

183. ____________ forest are called lungs of the planet. (**Amazon rain forest**)

184. India's first and only butterfly park was established in the year ___________ and is situated at ____________. (**1992 & Sikkim**)

185. Artificial salt licks are made by mixing the __________ with ___________. (**NaCl & Licking soil**)

186. Animals which can tolerate narrow fluctuation of the salt concentration is called ______________. (**Stenohaline**)

187. Animals which can tolerate wide range fluctuation of the salt concentration is called ______________. (**Euryhaline**)

188. The number of animals per unit area is called ____________. (**Density**)

189. The phenomena of regular movement of a species from one place to other and back is called ____________. (**Migration**)

190. The phenomena of movement of a species from one place to other permanently is called ____________. (**Dispersal**)

191. When the individual of a species **move into** an area leaving its original habitat is called ____________. (Immigration)

192. When the individuals of a population **move out** from its original habitat permanently is called ___________. (**Emigration**)

193. The rate of birth in a population is called ________ and rate of death of individuals in a population is called __________. (**Natality & Mortality**)

194. ___________ is used to indicate the capacity to bear young ones. (**Fecundity**)

195. Animals having singly is called as __________, it is found in Lion. (**Loner/ Solitary**)

196. The species which are in danger of extinction and whose survival is unlikely if the causal factors continue to be operating are called ____________. (**Endangered species**)

197. The species likely to move into the endangered categories in the near future if the causal factor continue to be operating are called ___________. **(Vulnerable species)**

198. The species with small population in the world at restricted area is called __________. **(Rare species)**

199. The distribution of the species and the differences in the diversity of the communities are called ______________. **(Species Evenness)**

200. The suitability of a vegetative cover type, for selected animal species, can be numerically characterized by determining the degree to which the life requisites are provided. The numerical value is termed as ____________. **(Habitat Suitability Index)**

201. Habitat suitability index varies from ____________. **(0 to1)**

202. ___________ and __________ are essential for population modelling and physiological assessment. **(Ageing & Sexing)**

203. One horned Rhinoceros is preserved in ______________ sanctuary, which is located at Assam. **(Kaziranga)**

204. Group of wild dogs are called __________. **(Pack)**

205. Humming bird can fast fly up to __________. **(60 miles per hour)**

206. Ictheologists is the study of______________. **(Fish)**

207. In wildlife protection act 1972, Nilgiri Tahr comes under __________. **(Schedule I Part I)**

208. The simplest behaviour of animal is ___________. **(Reflexes)**

209. Rhino usually produces one calf after a gestation period of __________ months. **(18)**

210. ___________ is the only state of India where all three species of Indian crocodiles are found wild. **(Odisha)**

19

Forestry Extension

1. The term extension is __________ word. (**Latin**)
2. _____________ is the father of extension. (**Leagnes**)
3. The term extension was first used in _____________. (**U.K.**)
4. The goal of the extension is to promote ______________. (**Scientific out-look**)
5. The success of rural projects depends upon ______________. (**Agricultural Extension**)
6. The National Extension Service was started in the year ___________. (**2nd Oct., 1953**)
7. Training of Rural Youth for self employment was started in the year _____________. (**1980**)
8. Agriculture extension was first started in ________________. (**Bihar Agricultural University, Sabour**)
9. T & V programme was introduced in ____________. (**1974**)
10. T and V system of extension was started by ____________. (**D. Benor**)
11. ______________ is the father of organic farming. (**Masanobu Fukuoka**)
12. The first Krishi Vigyan Kendra (KVK) was established at __________ in 1974. (**Ponduchery**)
13. The KVK was recommended by _______________ committee. (**Mohan Singh Mehta**)
14. Etawah pilot project was headed by _________ in 1948. (**Albert Mayer**)
15. Ntional Agricultural science Museum is located at _________ and was established in 2004. (**New Delhi**)
16. ICAR was established on the recommendation of ______________ commission. (**Lord Linlithgow**)

17. Lab to land programme was started by __________ in 1979. (**ICAR**)
18. Grow more food campaign was started in _________. (**1948**)
19. Television broadcast for rural development in India was started in _________. (**1977**)
20. Model village concept was given by _________________. (**Daniel Hamilton**)
21. CAMPA stands for ____________________. (**Compensatory Afforestation Fund Management and Planning Authority**)
22. ____________ is the system of farming in which farmers are managed by Govt. and the agricultural workers are paid wages generally on monthly basis. (**State farming**)
23. The political personality who got Norman Borlaug award was _____________. (**C. Subramanium**)
24. The cartagene Kyoto protocol for agriculture product came in the year ___________. (**2005**)
25. Directorate of plant protection, quarantine and storage is located at _____________. (**Faridabad**)
26. Nobel prize for Micro credit was given to _____________. (**Mohammad Yunus**)
27. NARP was launched by ICAR in ___________. (**1979**)
28. ____________ is the chairman for national development of council of India. (**Prime Minister**)
29. The journal 'Indian Farming' published by ___________. (**ICAR**)
30. _____________ is the first state that adopted two tier panchayat raj. (**Karnataka**)
31. Panchayati Raj was first started in ____________. (**Rajsthan**)
32. Food for work programme was renamed as _____________. (**NREP**)
33. NREP was started in the year _________. (**1982**)
34. ISARB stands for __________________. (**Institute for Studies on Agricultural and Rural Development**).
35. National policy for farmers was approved during __________. (**2007**)
36. The first fully Indian bank was ______________. (**Punjab National Bank**)

37. The community development programme was started in India on __________. (**2nd Oct., 1952**)

38. SIDBI stands for ________________. (**Small Industries Development Bank of India**)

39. Reserve Bank of India was established on ____________. (**1st April, 1935**)

40. Sevagram project was started by ____________. (**M. K. Gandhi**)

41. National Institute of Agricultural Marketing (NIAM) is located at _________. (**Jaipur**)

42. National Academy of Agriculture Research Management (NAARM) is located in _____________. (**Hydrabad**)

43. IVLP stands for ____________. (**Institutional Village Linkage Programme**)

44. NARS refers to ________________. (**National Agricultural Research System**)

45. ____________ is the functional unit of rural society. (**Village**)

46. NABARD is the Apex bank providing rural credit in ___________ field. (**Agricultural**)

47. Role of difference agencies for village development was included in _____________. (**Chapati diagram**)

48. Nationl food for work programme was launched on ________________. (**14th Nov., 2004**)

49. _____________ was the first chairman of National Commission on Farmers. (**Sompal**)

50. The govt. of India set up planning commission during ___________. (**1954**)

51. ATMA stands for _________________. (**Agriculture Technology Management Agency**)

52. Principles of Extension Education is _______________. (**Learning by doing**)

53. Goal of Extension Education is to promote ____________. (**Income of Farmers**)

54. ___________ rural development project was launched before independence. (**Co-operative movement**)

55. Cooperative society is basic institution for ___________ growth of the villages. (**Socio-economic**)

56. For a study of farming system, the best PRA exercise is ___________. (**Resource mapping**)

57. Community development project was started in India in ___________. (**1952**)

58. ___________ source of interview is the best in a rural society. (**Focussed group**)

59. High yielding variety programme was started in the year ___________. (**1966**)

60. 'Kisan Bharati' periodical is published from ___________. (**Pant Nagar**)

61. According to Balwantrai Mehta committee, Panchayati Raj should be a ___________ structure of local self government bodies from village to district. (**Three tier**)

62. To get the intelligence quotient (I.Q.) of the individual, the formula will be ___________. {**I.Q. = (Mental age/Chronological age) × 100**}

63. Communication is the process by which messages are transferred from source to ___________. (**Receiver**)

64. ___________ are the first farmers to adopt a new idea. (**Innovators**)

65. ___________ are the last to adopt new idea. (**Laggards**)

66. Normal rate of adoption requires ___________ years from the introduction of the innovation to its adoption throughout the community. (**6-10 years**)

67. ___________ methods of extension teaching is a way of showing people the value of an improved practice. (**Result demonstration**)

68. Leaflet is a single sheet of paper folded to make a ___________ page piece of printed matter. (**Four**)

69. For one talk, ___________ flash cards should be used. (**10-12**)

70. ___________ is the primary institution of society. (**Family**)

71. ___________ extension approaches has got the highest intensity of influence. (**Individual approach**)

72. The basic unit of development under IRD program is ___________. **(Family)**

73. First Agricultural University in India was established at ____________. **(Pant Nagar)**

74. Extension Education is _________ and ___________. **(Discipline & Profession)**

75. General meeting are broadly, the meeting of ____________ participation. **(Heterogeneous)**

76. Campaign is an ___________ teaching. **(Intensive)**

77. ___________ is a procedure of working with the people. **(Programme plnning)**

78. A series of still pictures on one roll is called __________. **(Film strip)**

79. A bulletin should contain _________ pages. **(24-28)**

80. ____________ is a unit of two or more people in reciprocal communication and interaction with each other. **(Group)**

81. The village panchayat act cme into existence in __________. **(1958)**

82. The rural System Research idea was motivated by M. S. Swaminathan in __________. **(1986)**

83. The Govt. of India set up planning commission in ___________. **(1950)**

84. The term community development appears to have originated from __________ in England in 1946. **(Cambridge)**

85. Among the community, voluntary agencies are characterized by _____________. **(Greater continuity)**

86. B. R. Mehta team stimulated an active consideration of __________ through democratic bodies in India. **(Decentralization)**

87. Most people retain _________ % of what they read. **(10-15)**

88. Most people kept in mind ____________% of what they seen. **(30-35)**

89. Symposium is a short series of lectures; usually by ___________ speakers. **(2-5)**

90. Small group interaction designed to encourage the free introduction of ideas on an unrestricted basis without any limitations to feasibility is known as ___________. **(Brain Storming)**

91. ____________ essentially follow the seminar method and the focus is on any particular subject or problem. (**Syndicate studies**)

92. News paper articles are included in ___________ method of extension. (**Mass**)

93. 'Stay at home' type of people can be contacted by ____________ method of extension. (**Individual**)

94. Less intensive and less effective method of extension is ___________. (**Mass**)

95. A working model is known as ___________. (**Mock-up**)

96. Real objects taken out of their nature settings is called as ___________. (**Specimens**)

97. The disc recording is made by ___________ process. (**Mechanical**)

98. Tape and wire recording are made by ____________ process. (**Magnetic**)

99. The movie film recording is made by _____________ process. (**Optical**)

100. Visual teaching aid flannel graph is also known as ___________. (**Khadder graph**)

101. A B C of posters means ______________. (**Attractive, Brief and Complete**)

102. ____________ consists of series of individual chart which are bound together and hung on a supporting stand. (**Flip charts**)

103. ____________ is the process by which two or more people exchange ideas, facts, feelings or impressions. (**Communication**)

104. "The communication process in Rural development" is written by ___________. (**Leagans**)

105. ______________ is a statement of situation, objectives, problems and solutions. (**Extension Programme**)

106. ___________ are expression of the ends towards which our efforts are directed. (**Objectives**)

107. ___________ is defined as the distance in any given direction one expects to go during a given period of time. (**Goal**)

108. The ultimate objective of extension work is the full development of ___________. (**Individual**)

109. The gap between the situation and objective is the area of ________. **(Needs)**

110. First step for making a programme planning includes analysis of the ________ and determining ___________. **(Situation & Problems)**

111. The main objective of evaluation in extension is to ________________. **(Facilitate effective decision making without jumping to conclusion)**

112. In a good administration the offices of a ___________ are arranges in a pyramid of authority and responsibility. **(Bureaucracy)**

113. Farm demonstration work began in ________ in USA by Dr. Seaman. **(1903)**

114. __________ is the dominant social institution permeating social and economic relations. **(Caste)**

115. ___________is one who has received specialized training in the field in which he work as full time and paid for his work. **(Professional leader)**

116. __________ is best method for selection of leader. **(Sociometry)**

117. __________ are those things which person ought or ought not to do in society. **(Mores)**

118. Adopter category which respect and regarded by many others in the social system as a role model are _____________. **(Early adopters)**

119. National Agricultural Science Museum was inaugurated by ______________. **(Dr. A. P. J. Abdul Kalam)**

120. NATP stands for _____________. **(National Agriculture Technology Project)**

121. NATP was started in ____________. **(1998-99)**

122. NATP was funded from ____________. **(World Bank)**

20

Notes

1. First in Indian Forestry

Events	Name	Year
First Inspector General of Forest	M. D. Chaturvedi	1949
First Chief Conservator of Forest	Ruther	1906
First Conservator of Forest	Watson	1806
First silviculturist in India	R S Troup	1909
First National Forest Act	Indian Forest Act	1865
First Forest Movement	Chipko Movement	1973
First Forest Policy	Indian Forest Policy	1894
First National Park	Jim Corbet National Park	1936
First Wildlife Sanctuary in India	Vedanthangal Wildlife Santuary	1898
First Working Plan in India	Nilambur Forest Working Plan	1895
First Zoo in India	Calcutta Zoo, West Bengal	1854
First Director of the forest school, Dehradun	F. Bailey	1878
First Forest Entomologist	Stebbing	1901
First issue of Indian Forester issued by	Powel and Schlich	1875
First President of Forest Research Institute and College, Dehradun	Eardley Wilmot	1906

2. Superlatives in Indian Forestry

a. Plants	
Biggest gymnosperm genus	Ephedra
Biggest leaf	*Tectona grandis*
Biggest plant family	Poaceac
Biggest tree (Girth)	*Cedrus deodara and Shorea robusta*
Largest flowering plants genus	Impatiens
Largest Fruit bearing tree	*Artocarpus heterophyllus*
Longest life span	*Cedrus deodara*

Most valuable root parasite	*Santalumalbum*
Shortest grass	*Mybra minima*
Tallest grass	*Bambusa arundinacea*
Tree without leaf for longest period	*Galeola spp.*
b. Wood	
Hardest wood	*Acacia chundra, Olea laurifolia*
Heaviest wood	*Soymida febrifuga*
Lightest wood	*Cryptomeria japonica*
Strongest wood	*Mesua ferrea*
Toughest wood	*Anogeissus pendula*
Weakest wood	*Cryptomeria japonica*
c. Bamboo	
Largest stem	*Dendrocalamus giganteus*
Smallest stem	*Arundinaria microphylla*
Strongest bamboo	*Dendrocalamus strictus*
Thickest bamboo	*Dendrocalamus brandissi*
Weakest bamboo	*Dendrocalamus mambraneus*
d. Forest cover (Area wise)	
State having largest Forest cover	Madhy Pradesh
U. T. having largest Forest cover	Andaman and Nicobar
Highest increase in Forest cover	Andhra Pradesh
Highest decrease in Forest cover	Mizoram
e. Tiger Reserves	
Biggest tiger reserve	Nagarjun Sagar, A. P.
Smallest tiger reserve	Pench, Maharashtra
f. Biosphere Reserves	
Biggest biosphere reserve	Gulf of Mannar, Tamil Nadu
Smallest biosphere reserve	Nokrek, Meghalaya
g. Animals	
Largest Antelope	Nilgiri
Fastest Antelope	Blackbuck
Largest crocodile	Estuarine crocodile
Largest Flying Squirrel	Large flying Squirrel
Largest Bat	Flying Fox
Largest Bovine	Buffalo
Largest Deer	Sambar Deer
Largest wild Sheep	Great Tibetan Sheep

Largest non-poisonous Snake	Reticulate Python
Smallest otter	Clawless Otter
h. Birds	
Largest resident Stork	Adjutant Stork
Commonest Birds	House crow, Sparrow
Fastest Fliers	White throated spinetail Swift
Heaviest Bird	Great Indian Bustard
Largest Bird	Sarus Crane
Largest Egg	Sarus Crane
Largest wingspan	Beared Vulture
Migratory bird with the longest distance covered	Wood cock
Rarest Birds	Mountain Quail, Pinkheaded duck
Smallest Bird	Scarlet backed Flower pecker
Smallest egg	Scarlet backed Flower pecker
Songster Bird	Malabar whistling thrush
Best talker Bird	Hill Myna
Tallest living Bard	Sarus Crane

3. List of India wetland included in the Ramsar Conservation

Ramsar site	Date of Inclusion	State	Are (ha)
Chilika Lake	01/10/81	Orissa	116.5
Harike Lake	23/03/90	Punjab	4.1
Keoladeo	01/10/81	Rajstan	2.873
Loktak	23/03/90	Manipur	26.6
Sambhar	23/03/90	Rajstan	24.0
Wular	23/03/90	Jammu & Kasmir	18.9

4. List of Endangered Flora in India

Endangered Trees		
Abies delavayi *Adinandra griffithii* *Anacotosa ilicoides* *Canarium mannii* *Commiphora wightii* *Cycas beddomei* *Dipterocarpus kerrii* *Elaeocarpus prunifolius*	*Ilex embeliodes* *Lagerstroemia hypoleuca* *Loropetalum chinense* *Magnolia griffithii* *Myristica andamanica* *Olax nan* *Picea brachytyla*	*Populus gamblei* *Potamela paradoxa* *Pterocarpus santalinus* *Pterospermum obtusifolium* *Rhododendron arizelum* *Santalum album* *Podocarpus wallichianus*

Endangered Shrubs		
Amblyanthus multiflorus *Apama barberi* *Camelia caduca* *Frerea indica* *Helwingia himalaicah*	*Hippocratea nocobarica* *Osmunda regalis* *Phyilostachys bambusoides* *Rouvolfia serpentine* *Rhododendron dalhousiae*	*Rhododendron nutallii* *Rhododendron santapaui* *Rosa involuctrata* *Zanthoxylum scandens*
Endangered Herbs		
Dianthus coschemiricus *Captis teeta* *Balanophora dioica* *Astemisia amygdalina* *Aconitum deinorrhzum* *Atropa acuminata*	*Nordostachys grandiflora* *Gentana kurroo* *Dipteris wallichii* *Drosera indica* *Lavatera kashmiriana* *Meconopsis betonifolia*	*Viola falconeri* *Psilotum nudun* *Rheum nobile* *Saussurea lappa* *Anoectolchilus sikkimensis*
Endangered Climbers		
Dioscorea deltoidea *Dischidia bengalensis*	*Dischidia rafflesiana* *Iodes hookeriana*	*Piper barberi* *Entada pursseatha*

5. List of Threatened Animals

Common Name	Scientific Name	Category
Asiatic Golden cat	*Felis temmincki*	I
Asiatic lion	*Panthera leo persica*	E
Asiatic wild Dog	*Cuon alpinus*	V
Asiatic Black Bear	*Selenarctos thibetanus*	E
Banteng	*Bos javanicus*	V
Bengal Fox	*Vulpes bengalensis*	I
Brow-antlered Deer	*Cervus eldi eldi*	V
Brown Bear	*Ursus arctos*	V
Brown Paim Civet	*Paradoxurus jerdoni*	V
Clawless Otter	*Anonyx cinerea*	K
Clouded Leopard	*Neofelis nebulosa*	V
Desert Cat	*Felis libyca ornate*	E
Sea cow	*Dugong dugon*	V
Ganges River Dolphin	*Platanista gangetica*	V
Gaur	*Bos gaurus*	V
Golden Langur	*Trachypithecus geei*	R
Great Indian Rhinoceros	*Rhinoceros unicornis*	E
Indian Wolf	*Canis lupus*	V
Himalayan Musk Deer	*Moschus chrysogaster*	V

Common Name	Scientific Name	Category
Himalayan Shrew	*Soriculus nigrescens*	V
Hoolock Gibbon	*Hylobates hoolock*	E
Indian Elephant	*Elephas maximus*	E
Indian Wild Ass	*Equus hemionus khur*	T
Jackel	*Canis aureus*	V
Hangul	*Cervus elaphus hanglu*	E
Leopard	*Panthera pardus*	T
Lesser Panda	*Ailurus fulgens*	V
Lion-tailed Macaque	*Macaca silenus*	E
Malabar Civet	*Viverra cievettina*	E
Marbled Cat	*Felis marmorata*	I
Markhor	*Capra falconeri*	E
Nayan	*Ovis ammon hodgsoni*	E
Nilgiri Leaf Monkey	*Presbytis johni*	E
Nilgiri Marten	*Martes gwatkinsi*	I
Nilgiri Thar	*Hemitragus hylocrius*	V
Phayre's Langur	*Trachypithecus phayrei*	I
Pygmy Hog	*Sus salvanius*	E
Red Fox	*Vulpes vulpes Montana*	T
Rusty-spotted Cat	*Prionailurus rubiginosa*	K
Serow	*Nemorhaedus sumatraensis*	E
Sloth Bear	*Melursus ursinus*	V
Smooth Indian Otter	*Lutra perspicillata*	K
Snow Leopard	*Uncia uncial*	E
Swamp Deer/ Barasingha	*Cervus duvauceli duvauceli*	E
Takin	*Budorcas taxicolor*	I
Tibetan Wild Ass	*Equus hemionus kiang*	V
Tiger	*Panthera tigris*	E

E- Endangered, V- Vulnerable, I- Indeterminate, K- Insufficiently Known, T- Threatened

6. Endemics Birds in India

Malabar Lark Intermediate parakeet Grey Jungle Fowl Green Avadavat Black-and- rufous Flycather Forest Owlet	Laughingthrush Rock Bush-Quail Nilgiri Pipit Rufous-tailed Lark Nilgiri Flycather Nilgiri Wood-pigeon	Spot-breasted Fantail White breasted Wren Babbler White cheeked Barbet White naped tit Yellow throated Bulbul

7. Endangered Birds in India

Indian Skimmer Indian Bustard Lesser Adjutant Long billed Bush Warbler Jerdon's Babbler Andaman Crake Bengal Florican Black necked Crane Black breasted Parrotbill Blyth's Kingfisher Forest Owlet Dark rumped Swift Kashmir Flycather	Wood Spine Marsh Babbler Nicobar Bulbul Pallas's Fish Eagle Rusty bellied Shortwing Spot billed Pelican Yellow Weaver White winged Duck White bellied Heron Tibetan Eared Pheasant Western Tragopan Pale capped Pigeon Masked Finfoot	White throated Bushcant Spoonbil Sandpiper Lesser Kestrel Dalmation Pelican Baikal Teal Grey sided Thrush Hooked Crane Oriental Stork Pale backed Pigeon Sociable Lapwing Dalmation Pelican

8. USDA classification of Soil Texture

Particles	Size (Diameter)
Very Coarse Sand	2.0 – 1.0 mm
Coarse Sand	1.0 – 0.5 mm
Medium Sand	0.5 – 0.25 mm
Fine Sand	0.25 – 0.10 mm
Very Fine Sand	0.10 – 0.05 mm
Silt	0.05 – 0.002 mm
Clay	< 0.002 mm

9. Types of Mycorrhiza

Mycorrhiza are of Three type i.e. Endotrophic Mycorrhiza, Ectotrophic Mycorrhiza and Ectendotrophic Mycorrhiza.

Differences between Ectotrophic, Endotrophic and Ectendotrophic Mycorrhiza

Particulars	Ectotrophic Mycorrhiza (Intercellular)	Endotrophic Mycorrhiza (Intracellular)	Ectendotrophic Mycorrhiza
Characters	• Heterozoic root system-long and short root • Short root become infected and turn mycorrhiza • Short root coloured, short lived • Root hair absent and replaced by a fungal covering • Hyphae growing in the inter cellular spaces forming Hartig net	• Homozoic root system • Roots having normal growth • Root hairs are present Fungal mantle is absent	Ectotrophic infection followed by intracellular penetration
Fungal Family	Basidiomycetes, Ascomycetes and fungi	Phycomycetes, Rhizoctonia	*E. stains*
Tree species	Betulaceae, Salicaceae, Abietinaceae, Pinaceae, Myrtaceae	Ericaceae, Orchidaceae, Araucaria, Taxus	Beech, Lodgepole, Pine
Occurance in Soil	Free mycelia, Hyphal Strands and Sclerotia	As Spores and Sporocarps	Non-sporing Mycorhiza, abundant in agricultural fields

10. Coppicing Power of Some Important Tree Species

Coppice strongly	Coppice fairly	Coppice weakly	Do not coppice
Tectona grandis *Shorea robusta* *Dalbergia sissoo* *Acacia catechu* *Anogeisus spp.* *Eucalyptus spp.*	*Terminalia tomentosa* *Hardwickia binate* *Quercus incana* *Pterocarpus marsupium*	*Acacia nilotica* *Madhuca indica* *Mombax ceiba* *Adina cordifolia*	*Pinus roxburghii* *Cedrus deodara* *Picea smithiana* *Abies pindrow*

11. Advantages of Micropropagation

- To multiply a tree which can not readily be multiply by seed or conventional vegetative methods
- To rapidly propagate large quantities of propagules of superior trees provenance
- To clean infected clonal plant material
- To rejuvenate older trees through repeated in-vitro micrografting
- To multiply pathogen free propagules for farmers

12. Harmful Insects in Plantation and Natural Forests

Borers		
Common Name	**Scientific Name**	**Family : Order**
Sal Heart wood borer	*Haplocerambyx spinicornis*	Coleoptera : Cerambycidae
Root Borer of Sal	*Lophosternus hugelii*	Coleoptera : Cerambycidae
Bamboo borer	*Dinoderus cervinus*	Coleoptera : Cerambycidae
Teak borer	*Dichomeris eridantis*	Coleoptera : Cerambycidae
Shisham borer	*Celosterns scrabator*	Lepidoptera : Gelechidae
Babul borer	*Hypsipyla robusta*	Coleoptera : Cerambycidae
Toon borer	*Plocaederus ferrugineous*	Lepidoptera : Pyralidae
Tendu stem borer	*Aprioma cinerea*	Coleoptera : Cerambycidae
Poplar stem borer	*Indarbela quadrinotata*	Coleoptera : Cerambycidae
Bark eating caterpillar	*Tonica niviferans*	Lepidoptera : Indarbelidae
Semul shoot borer		Lepidoptera : Oecophocidae
Defoliator		
Teak defoliater	*Hyblea puera*	Lepidoptera : Hyblaedae
Sal defoliator	*Lymantria Mathura*	Lepidoptera : Lymantriidae
Tendu defoliater	*Hypocola rostrata*	Lepidoptera : Noctuidae
Shisam defoliator	*Plecoptera dichomeris*	Lepidoptera : GEometridae
Deodar defoliator	*Ectropis deodarae*	Coleoptera : Chrysomelidae
Gamhar defo;iator	*Calopepla leayana*	Lepidoptera : Lisiocampidae
Chir defoliator	*Labeda nobilis*	Lepidoptera : Noctuidae
Ailantus defoliator	*Eligma narcissus*	Lepidoptera
Poplar defoliator	*Clostera cupreata*	Lepidoptera : Pyralidae
Kadam defoliator	*Arthroschista hilaralis*	
Others		
Sal semi-lopper	*Ahaea janata*	Lepidoptera : Noctuidae
Gamhar tingids	*Tingis beesoni*	Hemiptera : Tingidae

Teak skelitonizer	*Eutectona machaeralis*	Lepidoptera : Pyralidae
Ailantus web worm	*Atteva fabriciella*	Lepidoptera : Yponomeutidae
Casuarina bag worm	*Clania crameri*	Lepidoptera : Psychidae
Tendu gall former	*Trioza obsolete*	Homoptera : Psyllidae
Bark eating caterpillar	*Inderbela quadrinotata*	Lepidoptera : Indarbelidae
Phasus borer	*Phassus malabaricus*	Lepidoptera : Hepialidae
Red borer	*Zeuzera coffeae*	Lepidoptera : Cossidae
Chafer beetle	*Apogania coriacea*	Coleoptera : Scarabeidae
Canker grub	*Dihammus cervinus*	Coleoptera : Cerambycidae
Mosquito bug	Helopeltis antonii	Hemiptera : Miridae

13. Major Agroforestry Practices

Agroforestry Practices	Brief Description (Arrangement of Component)	Major group of component (W- woody, H-Herbaceous, F-Fodder and A-animal)	Agroecological adaptibility
Improved Fallow	Woody species planted and left to grow during fallow phase	W- fast growing preferably leguminous, H-common agricultural crop	In shifting cultivation areas
Taungya	Combined stand of woody and agricultural species during early stages of establishment of plantations	W- usually plantation forestry spp. H-common agricultural crop	All ecological region where taungya is practised
Alley Cropping (Hedge row intercropping)	Woody species in hedges, agricultural species in alleys	W- fast growing leguminous, coppicing vigorously, H-common agricultural crops	Sub humid to humid areas with high human populationpressure
Multilayer tree gardens	Multi species, Multilayer dense plant associations with no organised planting arrangements	W- different woody component of varying form and growth, H-Usually absent; shade tolerant ones sometime present	Areas with fertile soil, good availability of labour
Multi purpose trees on croplands	Trees scattered haphazardly or according to some systemetic pattern on bunds, terraces or field boundaries	W-multipurpose trees and other fruit trees, H- common agricultural crops	In all ecological region especially in subsistence farming
Plantation crops combinations	Intigrated multistory mixing of plantation crops/ mixture of plantation crops in alternate or regular arrangement/ intercropping with agricultural crop	W- plantation crops such as coffee, coconut etc, fuel wood or fodder tree species, H-Usually shade tolerant species	In humid low land or teopical humid/sub humid high lands

Agroforestry Practices	Brief Description (Arrangement of Component)	Major group of component (W- woody, H-Herbaceous, F-Fodder and A-animal)	Agroecological adaptibility
Home gardens	Intimate multistory combination of various trees and crops around homesteads	W- Fruit trees predominantalso other woody species	In all ecological regions, especially in areas of high population density
Trees in soil conservation	Trees on bunds, terraces, with or without grass strip	W- Multipurpose and/ or fruit trees H-common agricultural species	In sloping land especially in high land
Shelter belts and Wind breaks, live fences	Trees around farm lands	W- Combination of tall growing species, H- agricultural crop	In wind prone areas
Fuel wood production	Interplanting fire wood species on or around agricultural lands	W- firewood species, H-agricultural crops of the locality	In all ecological regions
Trees on Rangeland	Trees scattered irregularly or arranged according to some systematic pattern	W- multipurpose of fodder value, F- present	Extensive grazing areas
Protein Banks	Production of protein rich fodder on farm for cut and carry fodder production	W-leguminous fodder trees, H-present F-present	Usually in areas with high man : land ratio
Plantation crops with pastures and animals	Stylo grass under the mangium plantation	W-plantation crop, A-present, F-present	In areas with less pressure in plantation crop lands
Home gardens involving animals	Multi storey combination of various trees and crops and animals around homesteads	W-Fruit trees predominate/ other woody species A-present	In all ecological regions with high density of human population
Multipurpose woody hedges rows	Woody hedges for browse, mulch, green manure, soil conservation etc.	W-fast growing and coppicing fodder H- Soil conservation species	Humid to sub humid areas with hilly and sloping terrain

Agroforestry Practices	Brief Description (Arrangement of Component)	Major group of component (W- woody, H-Herbaceous, F-Fodder and A-animal)	Agroecological adaptibility
Apiculture with trees	Trees for honey production	W- Honey producing	Depending on feasibility of apiculture
Aqua forestry	Trees lining fish ponds, trees leaves being used a s forage for fish	W- trees and shrubs preferred by fish	Lowlands
Multipurpose wood lot	For wood fodder, soil protection etc	W-multipurpose species, special location specific species	Various

14. Casual Organisma of some Important tree Diseases In India

Disease	Causal Organism
Wilt disease	*Fusarium oxysporium, Verticillium*
Root rot	*Ganoderma spp*
Root and butt rot	*Fomes annosus, Armillaria mellea*
Polyporus root rot	*Polyporus shoreae*
Armillaria root rot	*Armillaria mellea*
White root rot	*Fomes lignosus*
Red root rot	*Ganoderma pseudoferreum*
Stump rot	*Fomes lignosus*
Heart rot	*Hymeno mychaetes*
Punk knots	*Fomes caryophylli*
Stem wilt	*Trichosporium vesiculosum*
Bamboo blight	*Sacrocladium oryzae*
Damping off	*Rhizoctonia, Pythium, Fusarium*
Heart rot	*Fomes caryophylli, Fomes badius*
Stem blister rot	*Cronartium abicola*
Pink disease	*Corticum salmonicolor*
Spike disease	*Mycoplasma like organism*
Powery mildew	*Uncinula tectonae*
Leaf spot	*Cercospora tectonae*

15. Nursery Disease management

Disease	Symptoms	Control
Damping off	General : Chlorosis, wilting, constriction of stem and root rot Pre emergence : Seeds or seedling are killed before they emerge, difficult to diagnose, low germination may be an indication Post emergence : Shortly after germination the young seedling are infected at the base of the stem. or just below Late damping off : can take place weeks or months after emergence. leaf chlorosis and wilting of the terminal bud result from root death	Cultural : create conditions that are not favourable to the development of the disease Chemical : a). Formalin fumigation @160 cc/10 ltr of water. b). Acidification with sulphuric acid @@ 130 cc/ 10 ltr of water. c). Application of fungicides such as Blitox 50 Dithane Z-78 @ 22 g/ m^2. d). Disinfect nursery soil with chemic als such as Methyl bromide.
Root rot	Causing after development of seedling. Death of secondary and tertiary roots. Yellowing, wilting and death of seedling	Same as damping off control
Powdery mildew	Airborne fungal disease causing leaves to be covered with a white powdery dust	Remove and burn affected seedling and leaves of older plants to avoid spreading of the disease Fungicides : Sulphur based
Leaf blisters	Airborne fungal disease that causes leave curling and blistering	Same as powdery mildew
Sooty mold	Fungi growing on aphid excrement causing a black mold on the leaves which curl up. Associated with aphids or other sucking insects.	Control aphids and other leaf-sucking insects
Fusarium	Causes rotting of seedling roots and cutting in nursery beds.	Sterilizing nursery soil Dip cutting in a fungicide solution

16. Classification of Silvicultural System

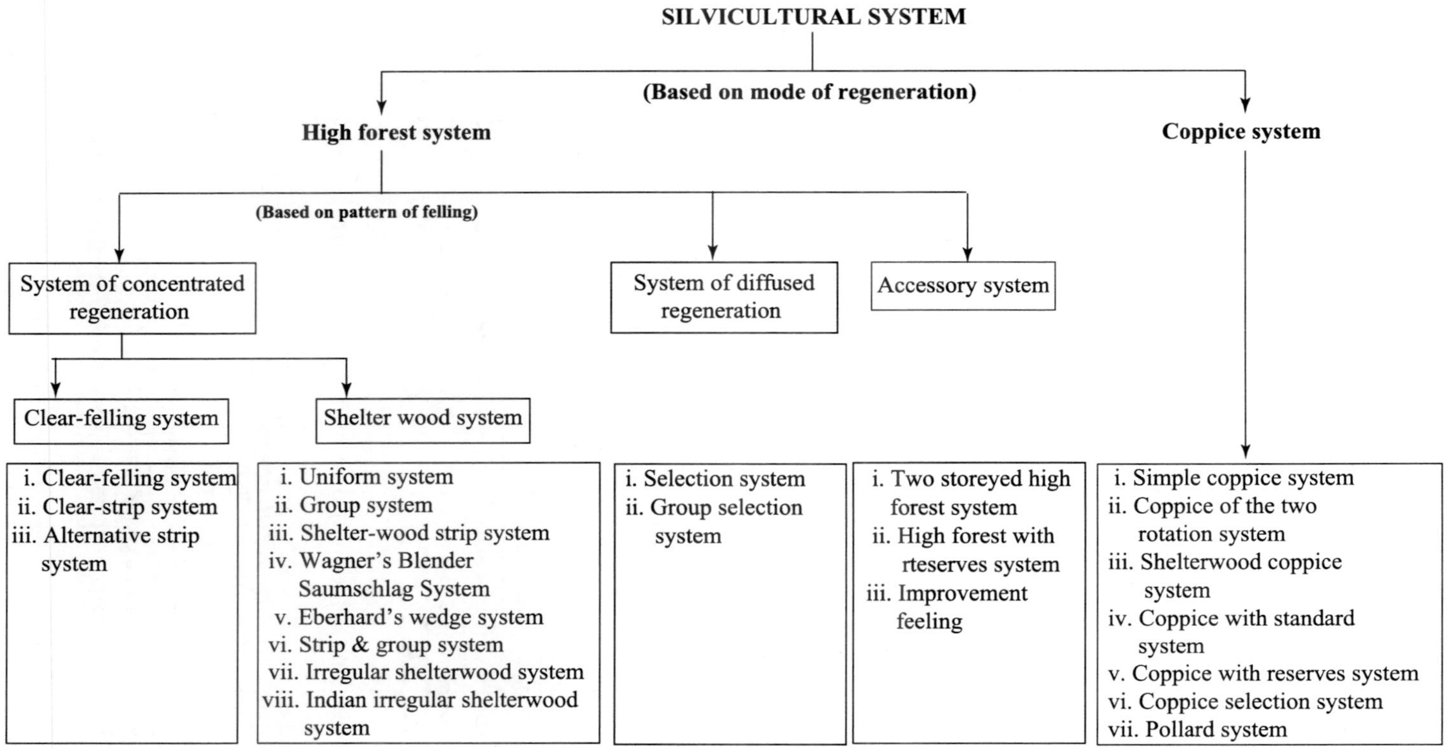

17.Choice of the suitable silvicultural system

Conditions	Suitable silvicultural system
Light demander species	Clear felling/ Clear strip
Shade bearer species	Shelterwood / Selection system
Species which are vigorous coppice	Coppice system
If natural regeneration is not possible and species produces seeds profusely every year	Clearfelling system followed by Artificial regeneration
Natural regeneration	Shelterwood system
If natural regeneration is not certain and artificial regeneration is not possible	Indian irregular shelterwood system
Forests growing on rocky area	Selection system
Forests in the catchment of rivers	Selection system/ Indian irregular shelterwood system
Snow damage areas	Indian irregular shelterwood system
Frost/Drought/ insect damage areas	Shelterwood system
Dense weed growth areas	Selection system
Production of fuel wood and small timber	Coppice system
Production of large sized timber	High forest system
For high concentration work	Clearfelling system
For early returns	Simple coppice/ Coppice with standard system
Forest situated in inaccessible areas	Selection system
Forest situated in accessible areas	Clear felling system
Less availability of skilled labour	Clear felling system
More availability of unskilled labour	Clearfelling system
More availability of skilled labour	Shelterwood system
For aesthetic purpose	Selection system

18. International forest and forest related organizations

Organization	Place
Food and Agriculture Organisation (FAO)	Rome
United Nations Development Program (UNDP)	Kenya
International Tropical Timber Organization (ITTO)	Yokohama
Centre for International Forestry Research (CIFOR)	Indonesia
International Union of Forestry Research Organisation (IUFRO)	Austria
World Wildlife Fund (WWF)	Switzerland
World Bank	Washington

Organization	Place
International Union for Conservation of Nature and Natural Resources (IUCN)	Switzerland
International Institute of Environment and Development (IIED)	Britain
International Institute of Sustainable Development (IISD)	Canada
Forest Research Supporting Programme for Asia the Pacific (FORSPA)	Indonesia
International Centre for Research in Agroforestry(ICRAF)	Nairobi, Kenya
World Trade Organization (WTO)	Geneva
World Tourism Organization	Spain
World Food Programme (WFP)	Rome, Italy
United Nations Environment Programme (UNEP)	Kenya
United Nations Framework Convention on Climate Change (UNFCCC)	Germany
International Strategy for Disaster Reduction 9ISDR)	Switzerland
International Labour Organisation (ILO)	Switzerland

19. Forest and forest related organizations in India

Organization	Place
A. FORESTRY	
National Research Centre for Agroforestry (NRCAF)	Jhansi
Tropical Forest Research Institute (TFRI)	Jabalpur
Temperate Forest Research Institute (TFRI)	Shimla
Institute of Forest Genetics and Tree Breeding (IFGTB)	Coimbatore
Institute of Forest Productivity (IFP)	Ranchi
Institute of Rain and Moist Deciduous Forest Research (IRMDFR)	Jorhat
Institute of Wood Science and Technology (IWST)	Bangalore
Arid Forest Research Institute (AFRI)	Jodhpur
CEntarl Arid Zone Research Institute (CAZRI)	Jodhpur
Forest Research Institute (FRI)	Dehra Dun
Forest Survey of India (FSI)	Dehra Dun
Indian Council of Forestry Research and Education (ICFRE)	Dehra Dun
Indian Institute of Forest Management (IIFM)	Bhopal
Indira Gandhi National Forest Academy (IGNFA)	Dehra Dun
Indian Plywood Industries Research and Training Institute (IPIRTI)	Bangalore
Indian Institute of Biodiversity (IIB)	Itanagar
Directorate of Forest Education (DFE)	Dehra Dun
Centre for Environment Management of Degraded Ecosystem (CEMDE)	New Delhi

Organization	Place
Advance Centre of Social Forestry and Eco-Rehabilitation (ACSFER)	Allahabad
National Environmental Engineering Research Institute (NEERI)	Nagpur
Centre for Ecological Research and Training (CERT)	Bangalore
Central Pollution Control Board (CPCB)	New Delhi
Biotechnological Centre for Tree Improvement (BIOTRIM)	Tirupathy
B. WILDLIFE	
Wild life Institute of India (WII)	Dehra Dun
Wildlife Protection Society (WPS)	New Delhi
Zoological Survey of India (ZSI)	Kolkata
Indian Board of Wildlife (IBWL)	New Delhi
Animal Welfare Board (AWB)	Chennai
Bombay Natural History Society (BNHS)	Mumbai
Central Zoo Authority (CZA)	Delhi
Salim Ali Centre for Ornithology and Natural History (SACON)	Coimbatore
World Wide Fund for Nature (WWF)	New Delhi
Wildlife Preservation Society of India (WPSI)	Dehra Dun

20. Instruments used in measurement of trees

Diameter measurement : Wooden scale, Caliper

Girth Measurement : Tape

Basal area measurement : Wedge prism, Spigel Relaskop

Height measurement :

(a) Instrument based on principle of similar triangle :- Christen's hypsometer, Smythie's hypsometer, Improvised caliper

(b) Instrument based on trigonometric principles :- Brandis hypsometer, Abney's level, Haga altimeter, Spiegel Relaskop, Ravi Altimeter, Ravi Multimeter, Clinometer

Age measurement : Pressler's increment borer

Bark thickness : Swedish bark gauge

21. Systems of Measurements

Metric system (C.G.S.)	British System (F.P.S.)
Units of Length	
10 millimetres = 1 centimetre	12 inches = 1 foot

10 centimetres = 1 decimetre	3 feet = 1 yard
10 decimetres = 1 metre	66 feet/ 22 yards = 1 chain
10 metres = 1 decametre	10 chains/ 220 yards = 1 furlong
10 decametres = 1 hectometre	8 furlong/ 1760 yards = 1 mile
10 hectometres = 1 kilometre	
Units of Area	
100 mm^2 = 1 cm^2	144 square inches = 1 square foot
100 cm^2 = 1 square decimetres	9 square feet = 1 square yard
100 square decimetres = 1 square metres	484 square yards = 1 square chain
100 square metres = 1 decametre square	10 square chains/ 4840 square yards = 1 acre
100 decametre square = 1 square hectometre	640 acres = 1 square mile
100 hectares = 1 square kilometre	
Units of volume	
1000 cubic millimetres = 1 cubic centimetre	1728 cubic inches = 1 cubic foot
1000 cubic centimetres = 1 cubic decimetre	27 cubic feet = 1 cubic yard
1000 cubic decimetres = 1 cubic metre	
Units of Weight	
10 grams = 1 decagram	16 ounces = 1 pound
10 decagrams = 1 hectogram	28 pounds = 1 quarter
10 hectograms = 1 kilogram	4 quarter 1 hundred weight
100 kilograms =1 quintal	20 hundred weight = 1 ton
1000 Kg / 10 quintals = 1 metric tonne	

22. Conversion of Measurements

C.G.S. to F.P.S.	F.P.S. to C.G.S.
1 centimetre = 0.393701 inch 1 metre = 1.09361 yards 1 kilometre = 0.62137 mile 1 square metre = 1.19599 square yards 1 hectare = 2.47105 acres 1 square kilometre =0.386101 square mile 1 square metre/hectare = 4.356 square feet/ acre 1 cubic centimetre = 0.0160 cubic inch 1 cubic metre =35.3147 cubic feet 1 cubic metre/ hectare = 14.291 cubic feet/ acre 1 kilogram = 2.20462 pounds 1 metric tonne = 0.98420 ton	1 inch = 25.4 mm 1 foot = 30.48 cm 1 yard = 0.9144 m 1 mile =1.609344 km 1 square inch = 6.4516 square cm 1 square feet =9.2903 square decimetre 1 square yards = 0.83613 square metre 1 acre = 0.40468 hectare 1 square mile = 2.58999 square km 1 cubic inches = 16.3872 cubic cm 1 cft = 0.0283 cubic metre 1 pound = 0.4535924 kg 1 ton = 1.01605 metric tonnes

23. Forestry Formulae

A. Forest Nursery:-

i. $\% \text{ of seed purity} = \frac{\text{Total weight of pure seed}}{\text{Total weight of sample}} \times 100$

ii. $\% \text{ of seed moisture} = \frac{\text{Weight of moisture}}{\text{Weight of moisture} - \text{Wt. of dry moisture}} \times 100$

iii. $\text{Quantity of seeds required for raising plantation} = W = \frac{A \times D}{S \times N \times G \times P \times M}$

W- Wt. of seed required, A- Area to be planted, D- Density of plantation, S-Survival percentage

N-Number of seeds per kilogram, G-Germination percentage, P-Purity of the seed, M-Mortality

B. Plantation Forestry

I. Method of calculating number of plants (N) per hectare

a. $\text{Line spacing} = \frac{100 \times 100}{\text{L to L distance} \times \text{ P to P distance}}$

b. $\text{Square planting} = \frac{100 \times 100}{\text{Square of the planting distance}}$

c. $\text{Triangular planting} = \frac{100 \times 100 \times 1.155}{\text{Square of the planting distance}}$

d. $\text{Quincunx planting} = \frac{100 \times 100 \times 2}{\text{Square of the side of the planting square}}$

II. Method of calculating spacing for given number of plants per hectare

a. $\text{Square planting} = \frac{100 \times 100}{\text{No. of plants per hectare}}$

b. $\text{Rectangular planting} = \frac{100 \times 100}{\text{No. of plants} \times \text{spacing of plants}}$

c. $\text{Triangular planting} = \frac{100 \times 100 \times 1.155}{\text{No. of plants/ha}}$

d. $\text{Quincunx planting} = \frac{100 \times 100 \times 2}{\text{No.of plants/ha}}$

III. Mechanical thinning formula

a. Glover's formula D = d (for Deodar)

b. Warren's formula D = $1^{1/2}$ d (for Deodar)

c. Sagreiya's formula D = 1.5 (d+4) (for Teak)

d. Sagreiya's formula D = 2(d+3) (for Teak)

e. Howard's formula D = 2d (for Sissoo)

C. Ecology

i. Net primary productivity (NPP) = Gross Primary Productivity (GPP) – Respiration (R)

ii. Importance value = Relative Density + Relative Frequency + Relative Basal Area

iii. Density = Number of trees per unit areas

iv. Relative density $= \dfrac{\text{Number of individual species}}{\text{Total individuals of all species}} \times 100$

v. Frequency = % of plots within a stand in which individuals of species are found

vi. Relative frequency $= \dfrac{\text{Frequency of species}}{\text{Sum of frequency value of all species}} \times 100$

vii. Relative basal area $= \dfrac{\text{Basal area of species}}{\text{Total basal area of all species}} \times 100$

viii. Growth efficiency index $= \dfrac{\text{Wood production in grams per year}}{\text{Foliage per unit area}}$

ix Phenotype (P) = Genotype (G) – Environment (E)

D. Soil Science

i. Universal soil loss equation = A = RKLSCP

Where, A – soil loss in ton per ha, R – Rainfall erosivity, K- soil erodibility factor, L- Length factor, S-Slope factor, C-Crop management factor

ii. Bulk density $= \dfrac{\text{Weight of dry soil (g/cc)}}{\text{Volume of dry soil}}$

iii. Particle density $= \dfrac{\text{Weight of dry soil (g/cc)}}{\text{Volume of soil solids}}$

iv. Per cent pore space =(100- B.D/ P.D) × 100

E. Forest management

a. Increment Per cent

i. Pressler's formula $= \text{p} = \dfrac{V-v}{V+v} \times \dfrac{200}{n}$

ii. Compound intrest formula for Volume increment per cent = p= 100 $[(V/v)^{1-n} - 1]$

iii. Schneider's formula $= \text{p} = \dfrac{400}{nD}$

Where p-Vol. of increment in per cent, V-Vol. of crop at present in m^3,

v- Vol.of crop n years ago in m^3, D-dbh at present in cm, n- No. of rings in the last cm

b. Estimation of growing stock (G.S.)

I. From M.A.I.

- Volume of G.S. at the beginning of the growing season = (1 × r/2) – ½
- Middle of the growing season = (1 × r/2)
- At the end of the growing season = (1 × R/2) ÷ 1/2

II. From yield table

- Vol. of G.S. at the beginning of the growing season = n(A+B+C+D/2) – D/2
- Middle of the growing season = n(A+B+C+D/2)
- At the end of the growing season = n(A+B+C+D/2) – D/2
- Where A,B, C & D are the vol. in ages n, 2n, 3n & 4n respectively n- Yield table interval

III. Fisher's formula for estimation of G.S. in regular shelter wood system

Vol. of N.G.S = $(V - V_1) \times p/2 \times D$

Where, V & V_1 are the G.S. at the initial and Final of regeneration period

p-Regeneration period, D-Crown density

F. Yield Regulation

1. Based on area only

a. Kotta's formula = $Ya = \frac{V}{P} - \frac{i}{2}$

Where, Ya- Annual yield, P-the period, V- Standing vol. in P.B.I, annual increment of the P.B.I

2. Based on Volume only

a. Von Mantle's formula = Annual yield = $\frac{2\ G.S}{r}$

b. Howard's formula $= Y = \frac{8V}{3r}$

c. Simmon's modification $= Y = \frac{2n^2 \times V}{r(n^2 - 1)}$

d. Smythei's modification $= Y = \frac{2V}{r - x}$

Where,

r-Rotation in years, V- Vol. of G.S. enumerated down to half rotation age in m^3, n-Rotation, Y- Annual yield, x-Age down to which enumeration & measurement were done

3. Based on Increment

a. Increment method $= Y = \frac{V \div a - Vn}{N}$

Where, Y- annual yield, V-Present yield, Vn – Vol. n years ago, a-Vol. removed during the previous n years

b. Swiss method $= Y = \frac{Y - Z}{c,c}$ and $c.\ c = \frac{Y - Z}{Xi - Yi - Zi}$

Where, X- Vol. of the dia. class below the fixed dia. limit, Y- Vol. of the dia. classes above the dia. limit and up to the limit which the c.a.i. is considered, Z- Vol. of the dia. classes above, i- the c.a.i of various dia. classes, c.c – the period during which the vol. of the oldest dia. classes should be removed

c. Biolley's 's' check method $= Y = \frac{(V_2 - N) - (V_1 - P)}{n}$

Where, V_2 – Present G.S. vol., V_1 – G.S. vol. n years ago, N-Vol. of trees cut, P– Vol. of recruitment

G. Mensuration

A. General

i. Growth index = $\frac{\text{Actual crop diameter}}{\text{Estimated crop diameter}} \times 100$

ii. Girth under bark (G.U.B) $g' = g - 2\pi t$

Where, G- girth over bark; G' – girth under bark; t- bark thickness

iii. Metzger's theory $S = \frac{W \times F \times L \times 32}{d^3} \times \frac{32}{\pi}$

Where, W-wind pressure per unit area; F- crown area, L- distance of a given cross section from the point of application of force; S- bending stress; d- diameter

B. Form Factor

i. Form factor = F = V/Sh

Form height = Fh= V/S

ii. Form quotient = F.Q. = Mid diameter/DBH

iii. Hojer's formula for tree form $= \frac{d}{d.b.h} = \frac{C \log c \div 1}{c}$

iv. Behre's formula for tree form $= \frac{d}{d.b.h} = \frac{1}{a-b}$

Where, V- Vol. of trees; S-basal area; h- height of the tree; d- distance at any point on the stem; l-distance from the top of the tree to the point at which d is measured; C,c,a& b constants for each form class

C. Measurement of Area

i. Area of circle = πr^2 or $\pi d^2/4$

ii. Area of rectangle = bl

iii. Are of square = b^2

iv. Area of a triangle = ah/2

Where, b-breadth; l-length; a-base of the triangle; h-height

v. Are of circular section = πr^2 x (θ/360)

vi. Area of Ellipse = πab

Where a & b are semi major and minor axis

vii. Area of parabola = 2/3 × ld

Where l-length of base, d- depth of parabola

D. Measurement of volume of logs

i. Samlian's formula $= V = \frac{S_1 + S_2 \times l}{2}$ (for paraboloid)

ii. Huber's formula = V= $S_m \times l$ (for paraboloid)

iii. Prismoidal/ Newton's formula = $V = \frac{(S_1 + 4S_m + S_2)}{6} \times l$ (for neiloid)

iv. Hoppus rule / Quarter girth formula = V= $(g/4)^2 \times 1$
(Gives 78.5 % of the true volume)

Where, S_1- sectional area at the thick end; S_2 – sectional area at the thin end; S_m – Sectional area at the middle; l-length of the log

H. Wild life

i. Estimation of weight of an animal = $W = \frac{\text{length of the animal} \times \text{girth}^2}{10.838}$

ii. Birth rate (b) $= \frac{\text{No. of individual born (B)}}{\text{Unit of time (t)}}$

iii. Death rate (d) = $\frac{\text{No. of deaths (D)}}{\text{Time unit (t)}}$

iv. Growth rate $= \frac{\text{No. of birth (b)} - \text{No. of deaths (d)}}{\text{Time interval}}$

I. Economics

i. B : C Ratio $= \frac{\text{Present value of benefits}}{\text{Present value of costs}}$

ii. Internal Rate of Return (IRR) = (Lower discount rate/ Difference between the discount rates) – (present worth of incremental net benefit at lower discount rate/ sum of the present worth of the incremental net benefit stream of the two discounted rates signs ignored)

iii. NPV / PNW = Present worth of benefits – Present worth of costs

J. Aerial Photography

i. Photo scale factor $= \dfrac{\text{Flying height}}{\text{Focal length of camera}}$

ii. Flying altitude = mean terrain elevation + Flying height

iii. Ground coverage = Negetive size × Scale factor

iv. Number of lines $= \dfrac{\text{Width of the area}}{\text{Line spacing}}$

v. Number of photos for given block = Minimum number of photos per run length × No. of lines

vi. Height of the tree = Z × Δp where, Z-flyinf height of the air craft;

$$\frac{bm - \Delta p}{bm - \text{Average photo base}}$$

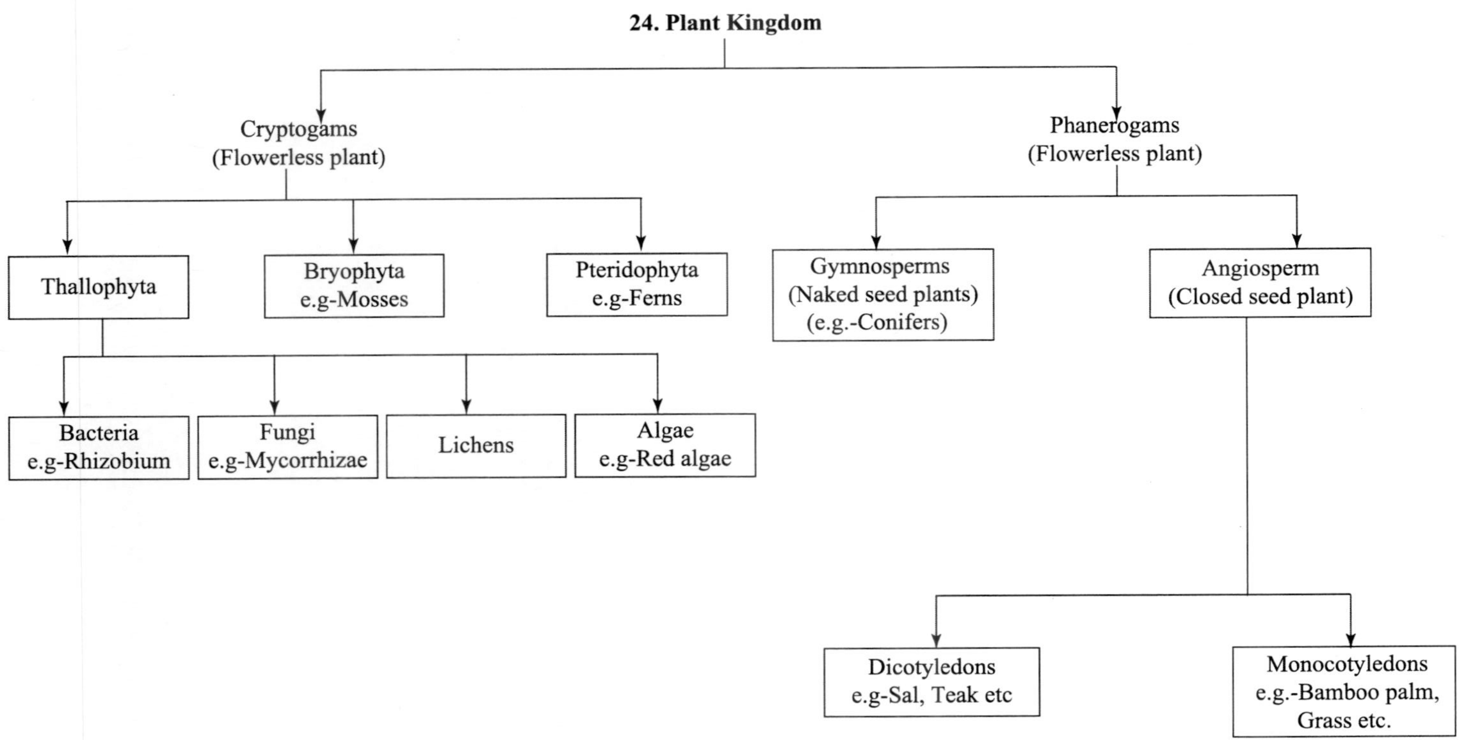
24. Plant Kingdom
Cryptogams
(Flowerless plant)
Phanerogams
(Flowerless plant)
Thallophyta
Bryophyta
e.g-Mosses
Pteridophyta
e.g-Ferns
Gymnosperms
(Naked seed plants)
(e.g.-Conifers)
Angiosperm
(Closed seed plant)
Bacteria
e.g-Rhizobium
Fungi
e.g-Mycorrhizae
Lichens
Algae
e.g-Red algae
Dicotyledons
e.g-Sal, Teak etc
Monocotyledons
e.g.-Bamboo palm,
Grass etc.

25. Major chemical pollutants : Sources and Environmental effects

Chemicals	Sources	Environmental Effects
Nitrogen Oxides (NO_2)	Combustion of oil, coal, gas in both automobiles and industry Bacterial action in soil Forest fire Volcanic action Lightning	Decreased visibility due to yellowish colour of NO_2 NO_2 contributes to heart and lung problems NO_2 can suppress plant growth Decr5ease resistance to infection
Ozone (O_3)	Formed from photolysis of NO_2 Sometimes result from stratospheric ozone instrusions	Bronchial constriction Coughing, wheezing Respiratory irritation Eye irritation Decreased crop yields Retards plant growth Harsh odor
Volatile organic compounds (VOCs)	Evaporation of solvents Evaporation of fuels incomplete combustion of fossil fuels Naturally occurring compounds like terpenes from trees	Eye irritation Some are carcinogenic Decreased visibility due to blue-brown haze
Peroxyacetyl Nitrate (PAN)	Formed by the reaction of NO_2 with VOCs	High toxicity to plant Damaging to proteins Respiratory irritation

26. Consrervation project

PROJECT	YEAR
Haungal project	1970
Lion projects	1972
Tiger project	1973
Crocodile projects	1975
Ibex and musk deer	1991
Projects elephant	1992
Great Indian bustard	1993
Siberian crane	1995
Snow lepord	2009

27. Classification of forest types (champion and seth 1968)

Major groups	Types
Tropical forest	Wet evergreen forest
	Semi evergreen forest
	Dry deciduous forest
	Moist deciduous forest
	Littoral and swamp forest
	Thorn forest
	Dry evergreen forest
Montane sub-tropical forest	Sub tropical broad leaved forest
	Sub tropical pine forest
	Sub tropical evergreen forest
Montane temperate forest	Montane wet temperate forest
	Himalayan moist forest
	Himalayan dry forest
Sub-alpine forest	Sub-alpine forest
	Moist alpine
	Dry alpine

28. Imporatant national forest institutes in India

Sl. No	Abbreviations	Full name	Location
1.	FRI	Forest Research institute	Dehradun
2.	AFRI	Arid Forest Research Institute	Jodhpur
3.	FRC	Forest Research Centre	Hyderabad
4.	RHRDC	Research & human research Development Centre	Chindwara
5.	HFRI	Himalayan forest Research Institute	Shimla
6.	IFGTB	Institute of Forest Genetics and Tree breeding	Coimbatore
7.	IFP	Institute of Forest productivity	Ranchi
8.	IWST	Institute of wood Science and technology	Bangalore
9.	RFRI	Rain Forest Research Institute	Johrat
10.	TFRI	Tropical Forest Research institute	Jabalpur
11.	CAZRI	Central Arid Zone Research Institute	Jodhpur
12.	CSWCRT	Central soil & water Conservation research and Training Centre	Dera Dun
13.	NRCAF	National Research Centre For Agroforestry	Jhansi
14.	SFRC	State Forest Service College	Dehradun
15.	IGNFA	Indian Gandhi National forest Academy	Dehradun
16.	DFE	Directorate Forest education	Dehradun
17.	IIFM	Indian Institute of forest management	Bhopal
18.	IPIRTI	Indian Plywood industries Research and Training institute	Bengaluru

Sl. No	Abbreviations	Full name	Location
19.	SACON	Salim Ali Centre for Orinthology and natural History	Coimbatore
20.	SFRC	State forest Service College	Coimbatore
21.	SFRC	State forest Service College	Burnihat
22.	SFRC	State forest Service College	Kurseong
23.	ZNP	National Zoological Park	New Delhi
24	CARTMAN	Centre for Animals and Environment	Bengaluru
25.	TBGRI	Tropical Botanic Garden and Research Institute	Thiruvanan
26	CEMDE	Centre for Environment Management of	Delhi
27.	CME	Degraded Ecosystem	Dhanbad
28.	CES	Centre of Mining Environment, Indian School of Mines	Bengaluru
29.	CEE	Centre for Ecological Sciences	Ahmedabad
30.	BIOTRIM	Centre for Environment Education	Trupathy
31.	CPCB	Biotechnological Centre for Tree Improvement	New Delhi
32.	CERT	Central Pollution Control Board	Bangalore
33.	CEMDE	Centre for Ecological Research and Training Centre for Environmental Management of Degraded Ecosystem	New Delhi
34.	FSI	Forest Survey of India	Dehra Dun
35.	ICFRE	Indian Council of Forestry Research & Education	Dehra Dun
36.	IIB	Indian Institute of Biodiversity	Itanagar
37.	NEERI	National Environmental Engineering Research Institute	Nagpur
38.	BNHS	Bombay Natural History Society	Mumbai
39.	CZA	Central Zoo Authority	Delhi
40.	WII	Wildlife Institute of India	Dehra Dun
41.	WPS	Wildlife Preservation Society of India	Dehra Dun
42.	WPS	Wildlife Protection Society	New Delhi
43.	WWF	World wide Fund for Nature	New Delhi
44.	ZSI	Zoological Society of India	Kolkata
45.	IBWL	Indian Board of Wildlife	New Delhi
46.	CSSRI	Central Soil Salinity Research Institute	Karnal
47.	IISS	Indian Institute of Soil Science	Bhopal
48.	NBSS& LUP	National Bureau of Soil Survey & Land Use Planning	Nagpur
49.	NRCMAP	National Research Centre for Medicinal & Aromatic Plants	Anand
50.	NRCO	National Research Centre for Oilpalm	Eluru
51.	NRCO	National Research Centre for Orchids	GAngtok
52.	GSI	Geological Survey of India	Kolkata
53.	NRSA	National Remote Sensing Agency	Hyderabad

29 Important days related to forestry

Date	Day
February 2	World Wetland Day
February 28	National Science Day
March 21	World Forestry Day
March 22	World Water Day
March 23	World Meteorological Day
April 7	World Health Day
April 18	World Heritage Day
April 22	Earth Day
April 24	World Lab Animals Day
May 3	International Sun Day
May 8	World Red Cross Day
May 20	Endangered Species Day
May 22	International Day for Biological Diversity
May 31	Anti-Tobacco Day
June 5	World Environment Day
July 11	World Population Day
September 16	World Ozone Day
September 18	World Bamboo Day
September 21	Biosphere Day
September 28	Green Consumer Day
October 3	World Animal Day
October 1 -7	World Wildlife Day
October 4	World Animal Welfare Day
October 13	International Day for Natural Disaster Reduction
November 10	Forest Martyrs Day
November 14 – 20	National Land Resources Conservation Week
November 19 - 25	World Heritage Week
November 21	World Fisheries Day
December 2	Bhopal Tragedy Day
December 3	World Conservation Day
December 14	National Energy Conservation Day
December 23	Farmer's Day

30. Important tree species in India, their Artificial regeneration and Uses

Sl No.	Species	Number of Seeds/kg	Sowing season	Germination %	Planting season	Uses
1	*Abies pindrow* (Silver fir)	27200	Oct.- Nov.	10-65	July	Timber
2	*Abies Spctabilis* (Himalayan fir)	16700	February	-	June-July	Timber
3	*Acacia auriculiformis* (Austrelian wattle)	40000	April	50-67	July-August	Timber, Fuel, Sand dune fixation, Ornamental
4	*Acacia catechu* (Khair)	40000	MarcApril	60-80	July	Fodder, Timber, Fuel, Gum, Katha, Soil conservation
5	*Acacia dealbata* (Silver wattle)	86500	October	68	July	Fuel, Gum, Tannin, Fodder
6	*Acacia decurrens* (Green wattle)	80375	October	50-90	June-July	Fuel, Tannin, Fodder
7	*Acacia farnesiana* (Pissi babul)	9700 - 12100	May	60-70	July	Fuel, Tannin, Fodder, Gum, Soil conservation
8	*Acacia mearnsii* (Black wattle)	95000	Sept.- Oct.	57	June-July	Tannin, Fuel, Soil conservation, Fodder
9	*Acacia melanoxylon* (Australian black wood)	66000-77000	Sept.- Oct.	60-70	-do-	Fuel, Timber, Tannin, Fodder
10	*Acacia modesta* (Phulai)	21179-35298	Feb.- March	60-90	-do-	Fuel,Tannin, Timber, Gum, Fodder, Soil conservation
11	*Acacia nilotica* (Babul)	7000-11000	March	88	-do-	Fuel, Timber, Tannin, Gum, Fodder
12	*Acacia Senegal* (Khor)	7165	June – July	60-70	July	Fuel, Fodder, Tannin, Soil conservation, Gum

Sl No.	Species	Number of Seeds/kg	Sowing season	Germination %	Planting season	Uses
13	*Acacia tortilis* (Israeli kikkar)	12000	Feb. – March	40-60	July	Fodder, Fuel
14	*Acer caesium* (Mapple)	12142	Feb. – March	-	Winter	Fuel, Timber
15	*Acer campbellii* (Mapple)	1800000	March-April	75	June-July	Fuel, Timber
16	*Acer oblongum* (Mapple)	20120-40480	March	50	July	Agricultural implements, Timber
17	*Acrocarpus fraxinifolius* (Mundani)	32000	May-June	75	June-July	Fuel, Box wood, Planks, Boards
18	*Adina cordifolia* (Haldu)	11000000	May	50	August	Bobbins, Timber
19	*Aegle marmelos* (Bel)	5300	Soon after collection	56	July	Fuel, Fruit, Gum, Bark and fruit
20	*Aesculus indica* (Horse chestnut)	36	-do-	80-90	July	Timber, Fodder, Fruit, Ornamental
21	*Ailantus altissima* (Tree of haven)	2100	March-April	80-90	Monsoon	Ornamental
22	*Ailantus excels* (Maharukh)	9500	Soon after collection	70-90	July-August	Timber, Packing cases, Boards, Bark, Fishing floats
23	*Ailanthus integrifolia* (Gogul)	1235-1750	May-June	60	July	Plywood, Ornamental
24	*Albizia amara* (Lallei)	14472	-	-	July	Fodder, Timber, Tool handles, Agricultural implements
25	*Albizia chinensis* (Sirin)	32143	Soon after collection	50	-	Timber, Fodder, Fuel
26	*Albizia falcataria* (Sengon)	42240	-	70	-	Fuel, Match wood, pulp
27	*Albizia lebbeck* (Siris)	8000-13000	Feb.- July	60-94	July	Timber, fodder, fuel, Medicinal
28	*Albizia odoratissima* (Kala siris)	15600	April	30-45	August	Timber, Fodder

Sl No.	Species	Number of Seeds/kg	Sowing season	Germination %	Planting season	Uses
29	*Albizia procera* (Safed siris)	21786	May	80-90	July	Fuel, Fodder, timber
30	*Alnus nepalensis* (Alder)	570000	March	70	July	Timber, Fuel, Soil conservation
31	*Alnus nitida* (Sarol)	885987	February	-	Jan.-Feb.	Timber, Fuel, Soil conservation
32	*Amora wallichii* (Amari)	132 - 155	June –July	90	June-July	Plywood, Fuel
33	*Anacardium occidentale* (Cashewnut)	145 – 165	May-June	90	June	Fuel, Fruits
34	*Anogeissus latifolia* (Axelwood)	135800	April- May	-	July	Fuel, Fodder
35	*Anogeisus pendula* (Kardhai)	96400	June	-	August	Fuel, Fodder
36	*Anthocephalus chinensis* (Kadam)	932142	February	-	June-July	Timber, Plywood, Ornamentals
37	*Artocarpus chaplasha* (Chaplash)	2000	July- August	80	July-August	Timber, Fruit, Fodder
38	*Artocarpus hirsutus* (Aini)	1412	-	60	-	Timber, Fruit, Fodder
39	*Artocarpus heterophyllus* (Kathal)	43 – 50	June –July	75	July	Timber, Fruit, Fodder
40	*Azadirachta indica* (Neem)	3330	July	100	July	Medicinal, Insecticidal, fertilizer, Fodder, timber
41	*Bambusa arundinacea* (Bans)	75000-105000	April	50	July	House construction, Boats and rafts
42	*Barringtonia acutangula* (Hijal)	1412	-	90-100	July-Oct.	Fuel
43	*Bauhinia purpurea* (Khairwal)	4000-5000	March-April	80-100	-	Gum, Fodder, Fuel

Sl No.	Species	Number of Seeds/kg	Sowing season	Germination %	Planting season	Uses
44	*Bauhinia racemosa* (Jhinjhora)	7000-9000	April – May	60-95	-	Tannin, Fibre, Gum
45	*Bauhinia variegate* (Kachnar)	2800-3520	May	95	June-July	Gum, Dye, Medicinal, fodder
46	*Betula alnoides* (Bhojpatra)	13875000	September	10	June-July	Plywood Furniture, Tool handles
47	*Betula cylindrostachys* (Birch)	14000000	Soon after collection	10	July	Essential oils, Fuel, Timber, Charcoal
48	*Bischofia javanica* (Uriam)	58400-104000	-do-	70-90	July	Timber, Fuel
49	*Bombax ceiba* (Semul)	21430-38500	May	14-75	Jan.-Feb.	Match, plywood
50	*Butea monosperma* (Dhak)	9850-14790	May	75-100	July-August	Reclamation of saline soil, Host tree for lac insect, Gum, Board
51	*Cassia fistula* (Amaltas)	6000-7090	March-April-	22-60	May-June	Medicinal, Timber, Tannin
52	*Cassia javanica* (Mazeli)	5460-6400	-	60	-	Fuel, Ornamental
53	*Cassia siamea* (Kassod)	37040	June –August	98	July	Fuel, Timber
54	*Casuarina equisetifolia* (Beef wood)	760000	Jan. – Feb.	70	June	Timber, Fuel, Ornamental
55	*Cedrus deodara i(Deodar)*	7000-9000	Nov.-Dec.	80-90	October	Timber
56	*Celtis australis* (Kharik)	-	Oct.-Nov.	60-70	Sept.-Oct.	Timber, Fodder, Fuel, Sports goods, Edible fruits, Medicinals
57	*Chloroxylon swietenia* (Satinwood)	35000-60000	Soon after collection	70	July	Furniture, Building timber, Agricultural implements, Sleepers
58	*Chukrasia velutina* (Chickrassi)	100000	Jan. –Feb.	70-90	June-July	Timber, Furniture, Decorative plywood

Sl No.	Species	Number of Seeds/kg	Sowing season	Germination %	Planting season	Uses
59	*Cinnamommum camphora* (Camphor)	5000-11000	March	47	June-July	Camphor oil, Medicinal
60	*Cinnamomum cecidodaphna* (Rohu)	2000-3000	Soon after collection	5-25	July-August	Cabinet making, Planks, Construction work, Essential oil
61	*Cryptomeria japonica* (Suji)	338000	February	80	Dec.-Jan	Timber, Pulpwood
62	*Cupressus torulosa* (Himalayan cypress)	201786	June	50-60	July	Timbers, Furniture, Oil
63	*Dalbergia latifolia* (Rosewood)	18500-40000	March-April	45-80	Mar-April	Timber, Furniture, Cabinet
64	*Dalbergia sissoo* (Shisham)	53000	March	90-100	July-August	Timber, Furniture, Sleepers, Plywood, Fuel, Fodder
65	*Dendrocalamus strictus* (Narbans)	32000	May-June	25-80	-	Paper, Cottage industry
66	Diospyrus melanoxylon *(Tendu)*	880-1410	March- April	30-60	-	Timber, Fuel, bidi industry
67	*Diospyrus peregrine* (Kala tendu)	786-895	July	12	July	Gum, Tannin, Dyeing, Medicine, Construction wood
68	*Dipterocarpus macrocarpus* (Hollong)	120	Soon after collection	25-80	May-June	Timber, Plywood, Sleepers
69	*Dipterocarpus turbinatus* (Common gurjan)	154	-do-	82	-	Timber, Railway sleepers
70	*Emblica officinalis* (Amla)	68000-89000	March	70	July	Fruit, Fodder, Tannin, Timber, Fuel

Sl No.	Species	Number of Seeds/kg	Sowing season	Germination %	Planting season	Uses
71	*Eucalyptus citriodora* (Lemmon sented spotted gum)	155000	Sept.- oct.	100	July	Fuel, Oil-medicinal
72	*Eucalyptus globulus* (Blue gum)	230000	March- April	30-80	Aug.-Oct.	Timber, Paper, essential oil
73	*Eucalyptus tereticornis* (Mysore gum)	367400	Sept.- oct.	90	July-August	Timber, Paper, Essential oil, Fuel, Plywood
74	*Exbucklandia populnea* (Pipli)	270000	March	75	July-August	Soil conservation, Timber, Furniture, Plywood, Ornamental
75	*Fraxinus excelsior*(European ash)	-	Nov.-Jan.	-	Feb.-March	Timber, Furniture, Gum, Carriages, Sports good
76	*Fraxinus floribunda* (Angu)	9000	Sept.- Oct.	75	-	-
77	*Garuga pinnata* (Kharpat)	4000-5000	July	44	Aug.-Sept.	Timber, Furniture, Bark tannin, Fodder
78	*Gmelina arborea* (Gamhar)	2500-2600	March-April	13-85	June	Timber, printing blocks, Musical instruments, Fuel, Medicinal
79	*Grevillea robusta* (Silver oak)	100000	Nov.-Dec.	60-80	July	Gum, Resin, Timber, Fuel, Ornamental
80	*Hardwickia binate*(Anjan)	3900	April-May	79	July	Timber, Posts, Beams, Fibres, Oleoresins
81	*Heavea brasiliensis* (Pararubber tree)	-	Soon after collection	-	July	Rubber
82	*Holoptelia integrifolia* (Kanju)	27000	-do-	60	July	Fuel, Charcoal, Light Furniture, Cabinets, Plywood, fodder

Sl No.	Species	Number of Seeds/kg	Sowing season	Germination %	Planting season	Uses
83	*Hymenodictyon excelsum* (Bartu)	142000-172000	April-May	30-35	April-March	Match wood
84	*Juglans regia* (Walnut)	50-100	November	70-80	Nov.-Dec.	Timber, Gunsticks, Fruits
85	*Kydia calycina* (Pula)	32000	Soon after collection	13-16	July	Timber, match box, Plywood, Fibre, light packing cases
86	*Lagerstroemia parviflora* (Lendia)	28000	Feb.-March	-	June	House posts, Beams, Frames for door and windows, carts and boats, Sleepers
87	*Lagerstroemia speciosa* (Jarul)	120000	Feb. –March	90	June	Timber, Medicinal, Fodder, furniture, Poles, Sleepers
88	*Lophopetalum fimbriatum* (Rumu)	8000	Soon after collection	40	May	Tools, Boxes, Aeroplane frames
89	*Machilus edulis* (Dudri)	29	February	50	June	Timber
90	*Machilus gammeiana* (Lali)	282	February	90	July	Timber
91	*Madhuca longifolia* (Mahua)	450	Julu-August	-	July	Timber, Fuel, Fruits and flower edible
92	*Mangifera indica* (Mango)	10-237	June –July	-	July	Edible fruits, Ply wood and Shoe heels
93	*Melia azedarach* (Bakain)	750-917	Feb.-March	70-80	July	Box planks, fuel, Paper pulp, Medicinal
94	*Mesua ferrea* (Nahor)	260	Soon after collection	90	July	Timber, Fuel, Fatty oil, Resin
95	*Michelia champaca* (Champa)	15000	-do-	70-80	June-July	Decorative timber, Linear, fuel, Ornamental planks, Furniture

Sl No.	Species	Number of Seeds/kg	Sowing season	Germination %	Planting season	Uses
96	*Michelia doltsopa* (Safed champa)	8470-10580	-do-	25-40	April-May	House construction, Decorative timber, Furniture, Fuel, Planks
97	*Michelia lanuginose* (Gogay champa)	10000	-do-	60	June	Timber, fuel, planks, Charcoal
98	*Moringa oleifera* (Sanjna)	8000-9000	-	66	July-August	Edible fruits, Medicinal, fodder
99	*Morus alba* (Mulbery)	428000-465000	June	70	July	Edible fruits, Timber, Sports goods, Fodder
100	*Morus indica* (Tutri)	512500	May	40	August	Timber, cultivation of silk worms
101	*Morus laevigata* (Bola)	443000	Soon after collection	50-80	July-August	Timber, furniture, planks, Sports goods, Fodder
102	*Olea ferruginea* (Indian olive)	9000	Sept.-oct.	45-65	July-August	Fodder, fatty oil, Articles, walking sticks
103	*Ougeinia oogeinensis* (Sandan)	28000-33000	May-June	75	July-August	Cart carriage, Building construction, Fodder, Agricultural implements
104	*Phoebe attenuate* (Bansum)	3000	Soon after collection	65	June	Timber, furniture, Plywood, Fuel
105	*Phoebe hainesiana* (Angani)	9000	-	80	June-July	Plywood, Fuel, Planks, Celling and partition boards
106	*Picea smithiana* (Himalayan spruce)	63928	June-July	22-65	June	Planks, Packing cases, joinery
107	*Pinus elliottii* (Slash pine)	28204-30000	Sept. – Oct.	70-92	July-August	Timber, Resin, pulpwood
108	Pinus kesiya *(Khasi pine)*	49000-58000	February	87-100	July	Timber, resin, Pulpwood

Sl No.	Species	Number of Seeds/kg	Sowing season	Germination %	Planting season	Uses
109	*Pinus patula* (Patula pine)	88123-132275	Feb.-March	30-60	August-Sept.	Timber, Resin, Pulpwood
110	*Pinus ruxburghii* (Chir)	8800-12300	Soon after collection	80-90	August	Timber, Resin, pulpwood
111	*Pinus wallichiana* (Kail)	19240	March-April	-	July-August	Timber, resin, pulpwood
112	*Pithecellobium dulce* (Jangli jalebi)	6700	Feb.-March	45	-do-	Packing cases, Fuel, Fodder, edible fruits, Fatty oil
113	*Pongamia pinnata* (Karanj)	800-1500	July	80	June-July	Furniture, Fuel, Fodder, Medicinal, Turnery articles
114	*Prosopis chilensis* (Kikkar)	12500	March	85-95	July	Fuel, Fodder, Small timber, Soil conservation
115	*Prosopis cineraria* (Jand)	25000	June	65	July	Fodder, house construction, Fuel and charcoal, Soil conservation
116	*Prunus nepalensis* (Sailong)	1350	March-April	85	June	Planks, Fuel
117	*Pterocarpus dalbergioides* (Padauk)	1450	May	37-40	June-July	Billiard tables, Furniture, Musical instruments, Decorative timber, Ornamental
118	*Pterocarpus marsupium* (Bijasal)	1590-1940	June	40-90	June-July	Gum, Kino-medicinal, railway carriages, Fibre wood, Fodder and constructional purpose
119	*Pterocarpus santalinus* (Red sanders)	1000	May-June	10-80	July	Carving, Dye, Medicinal, Fuel, Fodder, Musical instruments
120	*Pterocymbium tinctorium* (Papita)	4940	Soon after collection	90	July	Match wood, Packing cases, Plywood, Fuel

Sl No.	Species	Number of Seeds/kg	Sowing season	Germination %	Planting season	Uses
121	*Pterygota alata* (Buddha narikal)	1060-16660	-	80-90	July	Ornamental, Timber
122	*Quercus lineata* (Indian oak)	170	Feb.-March	40	June	Fuel, Bark and leaf tannin
123	*Robinia pseudoacacia* (Black locust)	3000-50000	Pril-May	80-90	July	Agricultural implements, Tools handle, Sports goods, Sleepers, Light construction furniture
124	*Santalum album* (Chandan)	7000-10000	Feb.- March	75	July-August	Sandal oil, Carving work, Medicinal, Fodder, ornamental
125	*Schima wallichii* (Needle wood)	300000	Soon after collection	-	June-July	Plywood
126	*Schleichera oleosa* (Kusum)	1410-2190	-do-	58	July	Timber, Turnery articles, Constructional purposes, Sleepers
127	*Shorea assamica* (Makai)	2000	-do-	83	July-August	Timber, Constructional puposes, Plywood, Paper pulp
128	*Shorea robusta i(Sal)*	575-1000	-do-	75-90	July-August	Beams, planks, Bridges, Carts, raiway sleepers
129	*Sterculia villosa* (Udal)	5600-6000	May-June	50-70	July-August	Fibres for ropes and coarsed bags
130	*Sterculia urens i(Karaya)*	4410-6360	Soon after collection	88	July	Timber, Toys, match boxes and splints, Fuel, Seed edible
131	*Streospermum suaveolens i(Paral)*	27356	March-May	45	-	Timber, Carriages, Wagons, furniture, Fuel, Charcoal medicinal

Sl No.	Species	Number of Seeds/kg	Sowing season	Germination %	Planting season	Uses
132	*Swietenia macrophylla* (Mahogany)	2150	Soon after collection	20-40	July	Jetty piles, Furniture, Plywood
133	*Syzygium cumini* (Jaman)	1200	-do-	90	July	Fruit edible and medicinal, Timber, plywood, Fuel, Fodder, Match wood
134	*Tamarindus indica* (Imli)	1800	April	66	July	Fruit pulp- Sauces and chutney, medicinal, Fodder, Timber, Fuel charcoal
135	*Tectona grandis* (Teak)	1850-3100	April	10-60	July-August	Timber, Railway carriages, wagons, Cabinets, furniture
136	*Terminali alata* (Laurel)	13000	April	35-70	July	Telegraphic and electric poles, Constructional purpose, Agricultural implements, Pulp, Fodder
137	*Terminalia arjuna* (Arjun)	775	April-May	50-60	June-July	Timber, plywood, Bark tannin, Fodder, Leaves for tasar silk worms
138	*Terminalia bellirica (Bahera)*	423	March-April	86-100	June-July	Timber, pulp, Plywood, Fruit, Medicinal, Fodder
139	*Terminalia chebula* (Harra)	141-220	June-July	60	July	Fruit, Medicinal, Timber
140	*Terminalia crenulata* (Laurel)	500	Soon after collection	70	July	Timber, fodder, Bark-oxalic acid
141	*Terminalia myriocarpa*(Hollock)	500000	-do-	20-30	July	Timber, plywood, Jute mill rollers, Fuel, Fodder

Sl No.	Species	Number of Seeds/kg	Sowing season	Germination %	Planting season	Uses
142	*Toona ciliata* (Toon)	550000	-do-	60-80	July	Timber, plywood, fodder, ornamental
143	*Trewia nudiflora* (Gutel)	5360-8200	Feb.- March	90-100	June-July	Timber, plywood, Fuel, Medicinal
144	*Tsuga dumosa* (Hemlock)	400000	February	80-90	July	Ornamental, timber bobbins and reels
145	*Xylia dolabriformis* (Jambu)	4300	February	80	July-August	Railway sleepers, telegraph posts, Agricultural implements, Carts Tool handle
146	*Xylia xylocarpa* (Irul)	3300-4000	June-July	80	-	Sleepers, construction of boats, ship building, Beams, Posts, Agricultural implements
147	*Ziziphus mauritiana* (Ber)	1240-1760	April-May	31-95	July	Fruits, Fuel, Tannin, Medicinal, Fodder

31. Some Important Tree species their scientific name and synonyms name and family.

Sl. No.	Common Name/ English Name	Scientific Name	Synonyms name	Family
1	Silver fir	*Abies pindrow*	*Pinus pindrow*	Pinaceae
2	Himalayan fir	*Abies Spctabilis*	*Pinus spectabilis*	Pinaceae
3	Austrelian wattle/ Sonajhuri	*Acacia auriculiformis*	-	Leguminosae
4	Khair/ The cutch tree	*Acacia catechu*	*Mimosa catechu, Acacia wallichiana*	Leguminosae
5	Silver wattle	*Acacia dealbata*	*Acacia decurrens, Racosperma dealbatum*	Mimosaceae
6	Green wattle	*Acacia decurrens*	*Mimosa decurrens*	Mimosaceae
7	Pissi babul	*Acacia farnesiana*	-	Leguminosae
8	Black wattle/ Australia Acacia	*Acacia mearnsii*	*Acacia mollissima, Acacia decurrens*	Leguminosae
9	Australian black wood/ Blackwood Acacia	*Acacia melanoxylon*	*Racosperma melanoxylon*	Leguminosae
10	Phulai	*Acacia modesta*	*Mimosa dumosa*	Leguminosae
11	Babul/ Acacia Gum-Arabic tree	*Acacia nilotica*	*Acacia Arabica*	Leguminosae
12	Khor/ Gum-Arabic tree	*Acacia Senegal*	*Acacia rupestris*	Leguminosae
13	Israeli kikkar/ Sayal	*Acacia tortilis*	*Acacia raddiana*	Leguminosae
14	Mapple	*Acer caesium*	*Acer giraldii*	Aceraceae
15	Himalayan Mapple	*Acer campbellii*	-	Aceraceae
16	Mapple	*Acer oblongum*	-	Aceraceae
17	Mundani/ red ceder/pink ceder	*Acrocarpus fraxinifolius*	-	Leguminosae
18	Haldu	*Adina cordifolia*	*Nauclea cordifolia*	Rubiaceae

Sl. No.	Common Name/ English Name	Scientific Name	Synonyms name	Family
19	Bel	*Aegle marmelos*	*Belou marmelos, Crateva marmelos*	Rutaceae
20	Horse chestnut	*Aesculus indica*	*Pavia indica*	Hippocastanocae
21	Tree of haven/ China sumac	*Ailantus altissima*	*Ailantus glandulosa*	Simaroubaceae
22	Maharukh	*Ailantus excels*	*Ailanthus altissima*	Simaroubaceae
23	Gogul	*Ailanthus integrifolia*		Simaroubaceae
24	Lallei	*Albizia amara*	*Acacia amara/ Mimosa amara/ Acacia wightii*	Leguminosae
25	Siris/ Sau tree	*Albizia chinensis*	*Albizia stipulate*	Leguminosae
26	Sengon/ Batai	*Albizia falcataria*	*Albizia falcate*	Leguminosae
27	Siris/ women's tongue tree/ East Indian walnut	*Albizia lebbeck*	Ailanthus blancoi, *Ailanthus moluccana*	Leguminosae
28	Kala siris/ Ceylon Rosewood	*Albizia odoratissima*	*Acacia odoratissima, Mimosa odoratissima*	Leguminosae
29	Safed siris/ White siris	*Albizia procera*	*Mimosa elata*	Leguminosae
30	IndianAlder/ Nepalese Alder	*Alnus nepalensis*	*Alnus mairei, Clethropsis Nepalensis*	Betulaceae
31	Sarol	*Alnus nitida*	*Clethropsis nitida*	Betulaceae
32	Amari	*Amora wallichii*	*Amoora Spectabilis*	Meliaceae
33	Cashewnut/ Cashew	*Anacardium occidentale*	-	Anacardiaceae
34	Axelwood/ Bakali/ Button tree	*Anogeissus latifolia*	*Anogeissus latifolia*	Combretaceae
35	Kardhai	*Anogeisus pendula*	*Anogeissus acuminata*	Combretaceae
36	Kadam/ Miracle tree	*Anthocephalus chinensis*	*Anthocephalus cadamba*	Rubiaceae

Sl. No.	Common Name/ English Name	Scientific Name	Synonyms name	Family
37	Chaplash	*Artocarpus chaplasha*	*Artocarpus chama*	Moraceae
38	Aini	*Artocarpus hirsutus*	*Artocarpus hirsute*	Moraceae
39	Jack Fruit tree/ Jacquir	*Artocarpus heterophyllus*	*Artocarpus integrifolia*	Moraceae
40	Neem/ Margosa tree	*Azadirachta indica*	*Melia azadirachta*	Meliaceae
41	Bans	*Bambusa arundinacea*	-	Bambusaceae
42	Hijal	*Barringtonia acutangula*	*Eugenia acutangula*	Lecythidaceae
43	Khairwal/ Butterfly tree/ Geranium tree	*Bauhinia purpurea*	*Bauhinia platyphylla, Bauhinia triandra, Bauhinia castrata*	Leguminosae
44	Jhinjhora/ Kanchan	*Bauhinia racemosa*	*Bauhinia parviflora*	Leguminosae
45	Kachnar/ Mountain Ebony	*Bauhinia variegata*	*Bauhinia chinensis, Bauhinia decora, Phanera variegate*	Leguminosae
46	Bhojpatra/ Alder-birch/ Indian bearch	*Betula alnoides*	*Betula acuminata*	Betulaceae
47	Birch/ Low level Darjeeling Birch	*Betula cylindrostachys*	*Betula cylindrostachys*	Betulaceae
48	Uriam/ Bishopwood	*Bischofia javanica*	*Bischofia trifoliate*	Biischofiaceae
49	Semul/ Cotton wood	*Bombax ceiba*	*Bombax malabaricum*	Bombacaceae
50	Dhak/Palas/ Flame of the forest	*Butea monosperma*	*Butea frondosa*	Leguminosae
51	Amaltas/ Indian Laburnum/ Rajbriksh	*Cassia fistula*	*Cassia excels, Cassia fistuloides, Cassia bonplandiana*	Leguminosae
52	Mazelia/ Javamese Cassia	*Cassia javanica*	*Cassia bacillus*	Leguminosae
53	Kassod/ Iron wood tree	*Cassia siamea*	*Cassia florida*	Leguminosae

Sl. No.	Common Name/ English Name	Scientific Name	Synonyms name	Family
54	Beef wood/ Casuarina	*Casuarina equisetifolia*	*Casuarina muricata*	Casuarinaceae
55	Deodar	*Cedrus deodara*	-	
56	Kharik/ Nettle tree	*Celtis australis*	*Celtis caucasia*	Ulmaceaea
57	East-IndianSatinwood	*Chloroxylon swietenia*	*Swietenia chloroxylon*	Meliaceae
58	Chickrasy/ Chittagong Redwood	*Chukrasia velutina*	*Chukrasia tabularis*	Meliaceae
59	Camphor/ Kapur	*Cinnamommum camphora*	*Laurus camphora*	Lauraceae
60	Rohu	*Cinnamomum cecidodaphna*	*Cinnamomum glaucescens*	Lauraceae
61	Suji	*Cryptomeria japonica*	*Cryptomeria fortune, Cupressus japonica*	Taxodiaceae
62	Himalayan cypress	*Cupressus torulosa*	*Cupressus torulosa*	Cupressaceae
63	Rosewood/ Bombay blackwood	*Dalbergia latifolia*	*Amerimnon latifolium*	Leguminosae
64	Shisham/ Sissoo	*Dalbergia sissoo*	*Amerimnon sissoo* *Dalbergia pseudo-sissoo*	Leguminosae
65	Narbans	*Dendrocalamus strictus*	*Dendrocalamus strictus* Nees	Poaceae
66	Tendu/ Ebony	Diospyrus melanoxylon	*Diospyros dubia*	Ebenaceae
67	Kala tendu	*Diospyrus peregrine*	*Diospyros malabarica*	Ebenaceae
68	Hollong	*Dipterocarpus macrocarpus*	*Dipterocarpus retusus*	Dipterocarpaceae
69	Common gurjan/ Wood oil tree of Assam	*Dipterocarpus turbinatus*	*Dipterocarpus laevis*	Dipterocarpaceae
70	Amla/ Indian goosberry	*Emblica officinalis*	*Phyllantus emblica*	Euphorbiaceae
71	Lemmon sented spotted gum	*Eucalyptus citriodora*	*Eucalyptus maculata citriodora*	Myrtaceae

Sl. No.	Common Name/ English Name	Scientific Name	Synonyms name	Family
72	Blue gum/ Southern blue gum	*Eucalyptus globulus*	*Eucalyptus cordata*	Myrtaceae
73	Mysore gum	*Eucalyptus tereticornis*	*Eucalyptus umbellatus* Gaertn	Myrtaceae
74	Pipli	*Exbucklandia populnea*	*Bucklandia populnea*	Hamamelidaceae
75	European ash	*Fraxinus excelsior*	-	Oleaceae
76	Angu/ East Indian Ash	*Fraxinus floribunda*	*Fraxinus floribunda* Wall	Oleaceae
77	Kharpat	*Garuga pinnata*	-	Burseraceae
78	Gamhar/White teak/ Candahar tree	*Gmelina arborea*	-	Verbenaceae
79	Silver oak/ Silky oak	*Grevillea robusta*	-	Proteaceae
80	Anjan	*Hardwickia binata*	*Hardwickia trapeziformis* R.Grah	Leguminosae
81	Pararubber tree/ Hevea	*Heavea brasiliensis*	*Siphonia brasiliensis*	Euphorbiaceae
82	Kanju/ Indian Elm	*Holoptelia integrifolia*	*Ulmus integrifolia*	Ulmaceae
83	Bartu/ Kuthan	*Hymenodictyon excelsum*	*Hymenodictyon orixense*	Rubiaceae
84	Walnut	*Juglans regia*	*Juglans duclouxiana, Juglans Orientis, Juglans kamaonia*	Juglandaceae
85	Pula	*Kydia calycina*	*Kydia fraternal*	Malvaceae
86	Lendia/ Benteak	*Lagerstroemia parviflora*	-	Lythraceae
87	Jarul/ Pyinma	*Lagerstroemia speciosa*	*Logerstroemia flos-reginae*	Lythraceae
88	Rumu	*Lophopetalum fimbriatum*	Euonymus fimbriatus, Euonymus ovatus	Celastraceae
89	Dudri	*Machilus edulis*	*Persea fructifera*	Lauraceae

Sl. No.	Common Name/ English Name	Scientific Name	Synonyms name	Family
90	Lali	*Machilus gammeiana*		
91	Mahua/ Butter tree	*Madhuca longifolia*	*Madhuca indica*	Sapotaceae
92	Mango	*Mangifera indica*	*Mangifera domestica, Mangifera sativa*	Anacardiaceae
93	Bakain/ Persian lilic	*Melia azedarach*	*Melia sempervirens/ Melia japonica*	Meliaceae
94	Nahor/ Iron wood tree	*Mesua ferrea*	*Mesua penduncalata/ M. Scleroptylla*	Guttiferae
95	Champ/ Champak	*Michelia champaca*	*Michelia aurantiaca*	Magnoliaceae
96	Safed champa	*Michelia doltsopa*	*Michelia excels*	Magnoliaceae
97	Gogay champa	*Michelia lanuginose*		Magnoliaceae
98	Sanjna/ Drum stick tree/Horse redish tree	*Moringa oleifera*	*Moringa pterygosperma/ Moringa moringa*	Moringaceae
99	Mulbery	*Morus alba*	*Morus tatarica*	Moraceae
100	Himalayan Mulbery	*Morus serrata*	*Morus alba*	Moraceae
101	Bola	*Morus laevigata*	-	Moraceae
102	Indian olive	*Olea ferruginea*	*Olea cuspidate*	Oleaceae
103	Sandan	*Ougeinia oojeinensis*	*Ougeinia dalbergioides*	Fabaceae
104	Bansum	*Phoebe attenuate*	-	Lauraceae
105	Angani	*Phoebe hainesiana*	-	Lauraceae
106	Himalayan spruce	*Picea smithiana*	*Picea morinda*	Pinaceae
107	Slash pine	*Pinus elliottii*	-	Pinaceae
108	Khasi pine	Pinus kesiya	*Pinus insularis* Endl	Pinaceae

Sl. No.	Common Name/ English Name	Scientific Name	Synonyms name	Family
109	Patula pine	*Pinus patula*	-	Pinaceae
110	Chir	*Pinus ruxburghii*	*Pinus longifolia*	Pinaceae
111	Kail	*Pinus wallichiana*	*P. chylla. P. excelsa. non Lamb. P. griffithii.*	Pinaceae
112	Jangli jalebi	*Pithecellobium dulce*	*Inga dulcis*	Fabaceae
113	Karanj	*Pongamia pinnata*	*Cytisus pinnatus , Dalbergia arborea, Galedupa indica, Galedupa pinnata*	Leguminosae
114	Kikkar	*Prosopis chilensis*	*Ceratonia chilensis*	Mimosaceae
115	Jand/ Khejri	*Prosopis cineraria*	*Prosopis spicigera*	Leguminosae
116	Sailong	*Prunus nepalensis*	-	Rosaceae
117	Padauk/Andaman red wood	*Pterocarpus dalbergioides*	*Lingoum dalbergioides, Pterocarpus advenus*	Leguminosae
118	Bijasal/ Kino tree	*Pterocarpus marsupium*	-	Leguminosae
119	Red sanders/ Saunderswood	*Pterocarpus santalinus*	-	Leguminosae
120	Papita	*Pterocymbium tinctorium*	*Sterculia campanulata*	Sterculiaceae
121	Buddha narikal	*Pterygota alata*	*Sterculia alata*	Sterculiaceae
122	Indian oak	*Quercus lineata*	*Quercus hilldebrandii*	Cupulifarae
123	Black locust/ False Acacia	*Robinia pseudoacacia*	*Robinia pseudoacacia*	Leguminosae
124	Chandan	*Santalum album*	*Breynia album*	Santalacea
125	Needle wood	*Schima wallichii*	*Schima bancana, Schima crenata*	Theaceae

Sl. No.	Common Name/ English Name	Scientific Name	Synonyms name	Family
126	Kusum	*Schleichera oleosa*	*Schleichera trijuga, Pistacia oleosa*	Sapindaceae
127	Makai	*Shorea assamica*	*Shorea siamensis*	Dipterocarpaceae
128	Sal	*Shorea robusta*	*Vatica robusta*	Dipterocarpaceae
129	Udal	*Sterculia villosa*	Sterculia ornata , Sterculia armata	Sterculiaceae
130	Karaya	*Sterculia urens*	*Sterculia urens Roxb.*	Sterculiaceae
131	Padal	*Streospermum suaveolens*	*Bignonia suaveolens*	Bignoniaceae
132	Mahogany	*Swietenia macrophylla*	*Swietenia humilis, Swietenia mahagoni*	Meliaceae
133	Jamun	*Syzygium cumini*	*Eugenia cumini/ Syzygium jambolanum*	Myrtaceae
134	Imli/ Tamarind tree	*Tamarindus indica*	*Tamarindus officinalis, Tamarindus occidentalis*	Leguminoseae
135	Teak	*Tectona grandis*	*Jatus grandis,Tectona theca*	Verbenaceae
136	Laurel	*Terminali alata*	*Terminalia tomentosa*	Combretaceae
137	Arjun	*Terminalia arjuna*	*Pentaptera arjuna/ Terminalia glabra*	Combretaceae
138	Bahera/ Belleric myrobalan	*Terminalia bellirica*	*Myrobalanus bellirica* Gaertn	Combretaceae
139	Harad/ Gall nut/ Chebulic myrobalan	*Terminalia chebula*	*Terminalia reticulate, Myrobalanus chebula*	Combretaceae

Sl. No.	Common Name/ English Name	Scientific Name	Synonyms name	Family
140	Laurel	*Terminalia crenulata*	*Terminalia tomentosa, Terminalia elliptica*	Combretaceae
141	Hollock	*Terminalia myriocarpa*	-	Combretaceae
142	Toon/ Red Cedar	*Toona ciliata*	*Cedrela toona*	Meliaceae
143	Gutel/ False white teak	*Trewia nudiflora*	*Mallotus nudiflorus*	Euphorbiaceae
144	Hemlock	*Tsuga dumosa*	*Pinus dumosa* D.Don	Pinaceae
145	Jambu	*Xylia dolabriformis*	*Xylia dolabriformis* Benth.	Fabaceae
146	Irul	*Xylia xylocarpa*	Xylia dolabriformis/ Mimosa xylocarpa	Leguminosae
147	Ber/ Jujube	*Ziziphus mauritiana*	*Ziziphus jujube*	Rhamnaceae

32. List of Medicinal Plants in India and their uses

Name	Uses
Abrus precatorius	Seed paste is applied externally in various skin diseases. Seed paste when placed in cotton and inserted into vagina causes abortion.
Achyranthes aspera	Fresh leaf juice is poured into both the ears and gives relief to inflamed eyes. The decoction of the root is given to women after menstrual period as an antifertility agent. The root when tied on the waist is said to cure dysentery.
Acorus calamus	The leaf paste is applied externally on the wounds of animals to kill the worms. The rhizome is chewed to cure bronchitis, cough and cold.
Adhatoda vasica	Used in treatment of dry cough. Put 2-3 tender leaves of it in cup of water. Boil and reduce to half. Drink ½ cup thrice daily for 7 days.
Aegele marmelos	The ripe fruit pulp is used as cooling drink. The pulp of green fruit is used in stomach disorders. The ash of the leaves is used to kill worms of animal wounds.
Azeratum conyzoides	The powdered leaf of the white flowered form when mixed with black pepper is made into snuff which cures headache.
Aloe barbadensis	Used in treatment of wet cough. Fry 2 tsp of leaf pulp in half tsp of ghee. Eat with a little sugar in the morning, noon and night for 3 days. Treatment of constipation- Dry the leaf pulp and make powder. Mix two pinches of powder with equal quantity of turmeric powder. Dissolve in ¼ cup of water and drink for 3 to 4 days. Treatment of cuts and wounds- Take pulp by splitting open the leaf and apply over the wound. Repeat twice daily till wound heals.
Alstonia scholaris	The decoction of stem bark is used in the treatment of fever.
Anisomeles indica	The decoction of the plant is used in animal dysentery.
Annona squamosa	The leaf juice is used in the treatment of animal wounds as an antiseptic.
Argemone Mexicana	The latex is used in eye disease.
Asparagus racemosus	Used in the treatment of acidity. 6 tsp of juice prepared from the tubers taken in the morning and evening for 7 days with sugar.

Name	Uses
Azadiracta indica	The leaves and bark is used is applied externally on boils as an antiseptic. The young leaves are eaten in the treatment of various skin diseases. Treatment of stomach- Crush (7.5X 7.5 cm) piece bark and keep in 1 cup of hot water for 15 minutes and strain. Take 1 cup of the infusion in the morning and evening for 2 days. Removal of worms- Make decoction of bark. Take 6 tsp of decoction twice daily for 7 days.
Butea monosperma	The seed paste is used for cooling effect. The dried flowers soaked in water are used for taking bath for preventing sunstroke.
Calotropis gigantean	The leaves warmed in lahi oil are applied in inflamed parts of the body.
Capparis zeylanica	The paste of root bark is used in the treatment of boils and in the swelling of testicles.
Cassia fistula	The fruit pulp is used in liver disorders. The seeds are used in constipation and treatment of jaundice.
Cassia tora	The plant paste is applied externally in skin diseases. The seeds are given to animals for inducing fatness.
Celastrus panicularus	The seed oil is used as hair tonic.
Centella asiatica	Treatment of scanty urination with burning sensation- Crush whole plant to extract juice. Take 3 tsp of juice with honey thrice daily for 7 days.
Centella asiatica (Hydrocotyle asiatica)	The leaf is used in the treatment of dysentery and is believed to improve memory.
Cissampelos pareira	The decoction of leaves, lemon and garlic juice mixed with salt is used in severe stomach pains.
Clausena pentaphylla	The decoction of leaves is used in cough and cold. Root paste is used in treatment of ulcers and fever.
Cleome viscose	The leaf paste is used externally in the treatment of boils.
Cordia dichotoma	The decoction of leaves is used in the treatment of cough and cold.
Crisium arvense	The root paste is used for diuretic purposes.
Cuscuta reflexa	The paste of the whole plant is applied to cure swellings of the testicles and also headache.
Cynodon dactylon	Crush whole plant to extract juice and apply over the wound twice daily till wound heals.
Eclipta prostrate	The aqueous extract of the mixed with black pepper and sugar taken twice a day for 15 days in body inflammation.

Name	Uses
Embilica officinalis	The aqueous extract of fruit mixed with fruits of Harr and Bahera is used in constipation.
Flemingia strobilifera	The decoction of leaves is used as tonic for sick animals.
Helicteres isora	The aqueous extract of the seeds is used in dysentery and stomach pains.
Hemidsmus indicus	Treatment of scanty urination with burning sensation- Crush roots, boil and make decoction. Take ½ cup of decoction with sugar twice daily on empty stomach for 7 days.
Hibiscus rosa-sinensis	The stamina column is used as a diuretic in kidney trouble.
Holarrhena antidysentrica	Treatment of diarrhea- wash and boil half cup of root bark in 4 cups of water and reduce to 1 cup. Drink half cup in the morning and evening until cured.
Lawsonia inermis	Treatment of fungal infection- Make paste from the leaves and apply it over the affected areas in the morning and evening for 5 to 6 days.
Mallotus philippensis	The red powder obtained from the fruit and mixed with coconut oil is applied externally in bone fractures, inflammations and sprains.
Martynia annua	The leaf paste is applied externally on the wounds of animals to kill worms.
Mimosa pudica	Treatment of cuts and wounds- Crush whole plant to extract juice and apply over the wound twice daily till the wound heals.
Moringa oleifera	Treatment of fungal infections- Take gum by cutting stem and dissolve it in little water and apply over affected areas.
Musa paradisica	The fibre extracted from the inner part of pseudostem is made into a ball. The latter is half burnt and put on the swollen umbilicus of a newly born baby for subsiding. The root is used in severe stomach pains.
Ocimum sanctum	The decoction of leaves is used in fever. The leaves are taken with tea and taken internally for cough and cold.
Oxalis corniculata	The leaf juice is applied externally on the backbone of infants in case of rickets.
Piper longum	The powdered fruit is mixed with honey and used in cough and cold.
Pogostemon benghalensis	The ash of the whole plant mixed with mustard oil is used to kill worms in animal wounds.
Pterocarpus marsupium	The water kept overnight in the wooden tumbler made from this species is taken internally in chest pains.

Name	Uses
Punica granatum	Used in treatment of acidity problems. Make decoction of fruit covers. Take ½ cup with honey 3 times a day for 7 days. Treatment of diarrhea- put 3long pieces of long barks in 4 cups of water. Boil and reduce to 1 cup and strain. Talk half cup in the morning and evening until cleared.
Rauwolfia serpentina	The roots are used in fever and as an antidote in snake bite.
Ricinus communis	The leaves coated with mustard oil are warmed and applied externally on the chest and stomach during pneumonial fever. Treatment of constipation- mix 1 to 2 tsp of castor oil with milk and take at bedtime for 2 or 3 days.
Shorea robusta	The gum mixed with curd is given in dysentery.
Sphaeranthus indicus	The leaves are eaten as a precautionary measure and believe to cure eye diseases and specially inflamed eyes.
Vanda tasellata	The plant paste is applied in the treatment of bone fractures.
Vetivera zizanoides	Paste prepared from the roots is applied externally to relieve the pain at the temples.
Withiana somnifera	Powder prepared from the dried roots(5g) with ½ cup of warm milk for 7 to 45 days in the treatment of chronic headache.
Zinziber officinale	Paste of dry rhizome applied externally to relieve headache. It is also used in the treatment of wet cough. 1 tsp of juice prepared from fresh rhizome taken with honey for 7 days.

33. Green house gases and their contribution to warming

Green House Gasses (GHG)	Contribution to warming (%)
CO_2	55 %
CH_4	15%
CFC	24%
Nitrous oxide	6%

34. NON-WOOD FOREST PRODUCTS

Essential Oils			
Wood Oils		D. Root oils	
Agar oil	*Aquillaria agallocha*	Costus oil	*Saussurea lappa*
Linoloe oil	*Bursera penicillata*	Indian valarian oil	*Valarian jatamansi*
Sandal oil	*Santalum album*	Curcuma oil	*Curcuma aromatica*

Deodar oil	*Cedrus deodara*		
Turpentine oil	*Pinus roxburghii*		
Leaf oils		**E.Flower oils**	
Camphor oil	*Cinnamomum camphora*	Rose oil	*Rosa dauriascena*
Citriodora oil	*Eucalyptus citriodara*	Jasmine oil	*Jasminum officinale*
Patchouli oil	*Pogostemon cablin*	Keora oil	*Pandanus tectorius*
Pine needle oil	*Pinus roxburghii*	Ylang-ylang oil	*Cananga odorata*
Cinnamomum oil	*Cinnamomum zeylanicum*	Others	*Michelia champaca Juniper macropoda Mimosops elangii*
Eucalyptus oil	*Eucalyptus globulus*		
Mint oil	*Mentha arvensis*		
Grass oils			
Citronella oil	*Cymbopogon nardus var. winterianus*	Palmarosa oil	*Cymbopogon martini var. Motia*
Ginger oil	*Cymbopogon martini var. sofia*	Vetiver oil	*Vetivera zizanoides*
Lemon grass oil	*Cymbopogon flexuosus*		
Fibre			
Sterculiaceae	*Sterculia urens , Sterculia villosa , Sterculia foetida*		
Asclepidaceae	*Calotropis gigantea ,Calotropis procera*		
Tiliaceae	*Grewia teliafolia , Grewia oppositifolia*		
Ulmaceae	*Trema orientalis*		
Moraceae	*Morus alba , Ficus religiosa , Ficus bengalensis*		
Leguminaceae	*Bauhinia vahilii , Bauhinia racemosa , Hardwickia binnata*		
Flosses			
Ceiba pentandra ,Bombax ceiba , Cochlospermum religiosum , Calotropis procera , Calotropis gigantean			
Tans			
Bark Tans			

Acacia mearnsii , Acacia nilotica , Cassia fistula , Cassia auriculiformis , Rhizophora mucronoides,Ceriops roxburghiana ,Terminalia arjuna			
Fruit Tans			
Terminalia chebula , Terminalia bellerica , Acacia nilotica , Emblica officinalis , Zyziphus xylocorpus , Caeselpinea coriarea			
Leaf Tans			
Anogeissus latifolia , Emblica officinalis , Carissa spinatum			
Gall Tans			
Tamarix spp			
Dyes			
Wood dyes			
Cassia sappan, Acacia catechu , Pterocarpus santalinus , Artocarpus heterophyllus			
Bark dyes			
Acacia leucophloea , Acacia coincinna , Acacia farnesiana ,Terminalia alata , Mannilkara littoralis , Rubia cordifolia			
Flower & Fruit dyes			
Annota dye-Bixa orellana , Kamala dye-Mallotus philippensis , Toona ciliata ,Butea monosperma ,Wrightia tinctoria , Nyctanthus arbortristis			
Root dyes			
Punica granatum , Rubia cordifolia			
Leaf dyes			
Lawsonia inermis , Indigofera tinctoria			
Gums			
Bengal gum	*Butea monosperma*	Gum Arabic	*Acacia senegal*
Salai gum	*Boswellia serrata*	Kino gum	*Pterocarpus santalinus*
Indian gum Arabic	*Acacia nilotica , Acacia modesta , Acacia catechu*	Gum Traganth	*Sterculia urens*
Waxes			
Sopia sebiferum , Rhus succedarea			
Resins			
True resins			
Black dammar	*Canarium strictum*	Rock dammar	*Hopea odorata*
Green dammer	*Shorea tumbuggaia*	White dammar	*Vateria indica*
Amber	*Pinus succinifera*	Mastic	*Pistacia leutiscus*
Lacquer	*Rhus vernicifera*	Sand arac	*Callitiris quadrivalvis*

Shellack	*Laccifera lacca*		
Oleoresins			
Salai gum	*Boswellia serrata*		
Gum resins			
Asafoetida	*Ferula assofoetida*	Galbanum	*Ferula galbaiflua*
Gambage	*Garicinia spp.*	Myrrah	*Commiphera spp.*

35. Protected Areas of India

	Number	Total Area	Coverage % of Country
National parks (NP)	102	40074.36	1.22
Wildlife Sanctuaries (WLSs)	528	125295.3	3.81
Conservation Reserves (Con. Res.)	57	2017.94	0.06
Community Reserves (Com. Res.)	4	20.69	0
Total Protected Areas (PAs)	**691**	**167408.3**	**5.09**

36. International and National organizations concerning wildlife Conservation

Government Organizations	Non-Government Organization/ Voluntary
CITIES : Convention on International Trade in Endangered Species of Wild Fauna and Flora **GTF:** Global Tiger Forum	**WWF:** Worlwide Find for Nature **BNHS :** Bombay Natural History Society **WPSI:** Wildlife Preservation Society of India **IBWL:** Indian Board for Wildlife **IUCN:** International Union for Conservation of Nature and Natural Resources **CBSG:** Conservation Breeding Specialist Group **SSC:** Species Survival Commission **PHVA:** Population & Habitat Viability Assessments **CAMP:** Conservation Assessment & Management Plan

37. Comparison of different protected areas

National Park	Sanctuary	Biosphere Reserves
National park is hitched to the habitat of the particular animal like tiger, lion, Rhino etc. The size ranges is 0.04 to 3162 km^2. Most common size is 100 to 500 km^2. Boundaries are circumscribed by legislation Except buffer zone, no biotic interference Tourism is permissible Research and Scientific management is lacking No attention is given for gene pool conservation	This generally species oriented. The size range is 0.61 to 7818 km^2. Most common size is 100 to 500 km^2. Boundaries are not sacrosanct. Limited biotic interference. Tourism is permissible Research and scientific management is lacking No attention is given for gene pool conservation.	Biosphere reserves consider the whole ecosystem. Size range over 5670 km^2. ----- Boundaries are circumscribed by legislation Except the buffer zone , no biotic interference Tourism is normally not permissible. Research and scientific management is available Attention is given for gene pool conservation.

38. Trees suitable for different wastelands

Waste land type	Species suitable
Marshy areas	*Acacia nilotica, Terminalia arjuna, Derris indica,* Bamboo, *Eucalyptus grandis,*
Sandy areas	*Dalbergia sisoo, Casuarina equisetifolia, Melia azadiracta, Thespesia populnea, Acacia auriculiformis, Anacardium occidentale*
Acid soils	*Albizia procera, Gmelina arborea, Gliricedia sepium, Ailantus ultisima*
Alkaline soils	*Leuceana leucocephala, Terminalia arjuna, Prosopis juliflora, Acacia nilotica, Acacia excels, Tamarindus indica*
Saline soils	*Acacia catechu, Ailantus exelsa, Prosopis juliflora, Salvadora persica, Acacia nilotica*
Dry clay soils	*Acacia nilotica, Acacia tortolis, Albizia lebbeck, Inca dulse*
High altitude areas	*Ailantus altisima, Grevillea robusta*
Dry areas	*Acacia spp. Prosopis juliflors, Eucalyptus cammoldulensis*

39. Benefits of agroforestry

(i). Environmental benefits

- Agroforestry reduce the pressure on forest and enhance better protection of ecological systems.
- It auguments biological diversity by favouring environmental conditions.

- It improves microclimate of the area by lowering of soil temperature, reduction of evaporation and maintenance of soil moisture.
- It reduces runoff, soil erosion.
- It improves the soil fertility through addition of litter fall.
- It enhances efficient recycling of nutrients with in soil at different depth.
- It improves soil structure.

(ii). Economic benefits

- Agroforestry yields increased outputs per unit area.
- Agroforestry system reduces the total crop failure.
- All these benefits resulted in increase of farm income.

(iii).Social benefits

- Agroforestry improves the rural living standards by increasing their income in a sustained manner.
- Multiple outputs without the risk of total failure stabilize the farm communities.

40. Crop composition, Colour, Density and age class distribution of the stock map

Crop Composition	
Crop composition	**Classification**
More than 75 % by one species	Pure crop
50 – 75 % by one species	Main crop
25 – 50 % by one species	Mixed crop
Less than 25 % by one species	Miscellaneous crop
Colour	
Forest type	**Colour**
Sal	Blue
Teak	Burnt sirnna
Barren land	Yellow
Mixed forest	Green
Cultivation area	Colourless
Crop Density	
Density	**Canopy class**
1.0	Closed
0.8 – 0.99	Dense

Crop Composition	
0.6 – 0.79	Fairly dense
0.4 – 0.59	Thin
0.2 – 0.39	Open
< 0.2	Sparse
Age class distribution	
Age class Density	**Symbol**
Seedling crop	0
Thickest of Seedling	00
Mature high forest	-
Over mature forest	=

41. Different rays and their wavelength

Rays	Wavelength (µm)
Cosmic ray	$< 10^{-7}$
Gamma ray	10^{-7} to 10^{-4}
X ray	10^{-4} to 10^{-2}
UV ray	10^{-2} to 0.4
Visible ray	0.4 to 0.76
Infrared ray	0.76 to 10^{2}
Microwave	10^{2} to 10^{6}
Radio wave	$>10^{6}$

42. Difference between heartwood and sapwood

Heartwood	Sapwood
Cells are dead due to deposition of polysaccharides and process of embolism Dark colour woods Polyphenol present in heartwood Resistant to insects, pests and degradation Less hygroscopic nature Specific gravity is higher Wood is strong, so used for timber purpose	Most of the cells are alive 9Normally 50 %) Light coloured wood Polyphenol absent in sapwood Less resistant to insects, pests and degradation Highly hygroscopic in nature Specific gravity is less Wood is less strong

43. Difference between seedling seed orchards (SSO) and Clonal seed orchards (CSO)

Seedling seed orchards	Clonal seed orchards
It is established from seedlings It is propagated by seeds It has broader genetic base but less selection differential An outstanding genotype will appear only once Flowering and fruiting takes longer time Applicable for species with early physiological flowering It gains larger genetic gain	It is established from clones It is a kind of vegetative propagation It has narrow genetic base but higher selection differential An outstanding genotupe will appears many times continually Flowering and fruiting occurs earlier Applicable for species with longer physiological flowering It gains narrow genetic gain

44. Difference between agroforestry and social forestry

Object	Agroforestry	Social forestry
Defination	It is an land use management system by where woody perenials are grown along with the agricultural crops and/ or animals in the same land management unit with some spatially or temporally	Social forestry is the practice of forestry of the people, forestry by the people and forestry for the people
Crop composition	Trees, Agricultural crops, Animals	Only trees
Economic consideration	Combined yield of trees and agricultural crops	Yield from trees
Management	Difficult to manage	Quite simpler
Need for research	Intensive research needed to combine all three components	Quite less

45.Important characteristics of salt affected soil

Characteristics	Saline soil	Sodic soil (Alkaline)	Saline-sodic soil
Content in soil	Excess soluble salts of sodium	Presence of excess exchangeable sodium	Soil contains Na-clay as well as soluble salts
Exchangeable Na/Ca	Exchangeable Calcium	Exchangeble sodium	
Colour	Wite	Black	---------------
Dominant salts	SO_4^{2-}, Cl^-, NO_3^-	NA_2CO_3	------------
Sodium Absorption Ration(SAR)	Less than 13	More than 13	More than 13

Characteristics	Saline soil	Sodic soil (Alkaline)	Saline-sodic soil
Exchangeable sodium percentage (ESP)	Less than 15	More than 15	More than 15
Soil p^H	Less than 8.5	5.5 – 10.0	More than 8.5
Physical condition of the soil	Flocculated condition, permeable to water and air. Soil structure optimum	Deflocculated condition, permeability to waterand air is poor. Very poor soil structure	Flocculated or Deflocculated depending upon presence of sodium salts and Na-clay
Morphological character	White crust on the soil surface	Due to presence of high amount of exchangeable sodium and high p^H clay colloids get dispersed and migrated down ward and block the soil pore spaces with clay deposition	-------
Organic matter content	Slightly less than normal soils	Very low due to the presence of sodium carbonate	Variable
Total soluble salt content	More than 0.1 per cent	Less than 0.1 per cent	More than 0.1 per cent
Electrical conductivity	$> 4\ dsm^{-1}$	$< 4\ dsm^{-1}$	$> 4dsm^{-1}$

46. Effect of different pollutants on vegetation

Dust: Decrease photosynthesis and respiration.

Noise: Causes physiological and biochemical influences on plant metabolism.

Light: It alters the photoperiodism and affects growth, flowering and normal functioning of plants.

Ozone: Chlorosis, Abscission, Suppression of growth with out injury.

Oxides of nitrogen: Leaf injury and Reduction in growth.

Fluorides: Blocking of stomatal pores, reduction of photosynthesis, weak respiration and poor transpiration.

SO_2: Pollen injury, chlorosis, necrosis, stomatal injury, reduction of chlorophyll content, adverse effect of photosynthesis.

47. Trees for different pollution

Pollution type	Trees suitable
Dust pollution	All broad leaved species (e.g.- Teak, Mango, Terminalia etc.)
Noise pollution	*Acer negundo, Alstonia scholaris, Butea monosperma, Alnus indica, Terminalia arjuna, Grevellia robusta etc.*
Lead pollution	*Cassia siamea, Zizyphus mauritiana*
Ozone	*Acer campestral, Quercus rubra, Acer platanoides*
Oxides of nitrogen	*Robinia pseudocacia, Taxus buccata, Ulnus Montana, Pinus spp., Quercus rubra*
So_2	*Ailantus excelsa, Ficus spp., Terminalia arjuna, Albizzia lebbeck, Quercus palustrie*

48. Essential plant nutrient element require, their function and Deficiency symptoms

Nutrients	Function	Deficiency symptoms	Available forms of Nutrients in the soil
Nitrogen (N)	Component of chlorophyll Chemical constituents in proteins, amino scids	Yellowing of leaves Stunted growth	NH_4^- NH_3^-
Phosphorous (P)	Components of phospholipids, ATP. Enhance seed germination Root growth improvement	Stunted growth	$H_2PO_4^-$ HP_4^{--}
Pottassium (K)	Regulation of membrane permeability Resistance to disease Activetes enzyme and photosynthesis	Dieback Leaf scorching Marginal chlorosis	K^+
Calcium (Ca)	Stability of cell membrane Component of cell wall Neutralizes organic acids Prevent lodging of plants	Die back Tip hooking Blossom end rot	Ca^{++}
Magnesium (Mg)	Constituent of chlorophyll and chromosome Essential for carbohydrate metabolism	Interveinal chlorosis Sand drown disease	Mg^{++}

Nutrients	Function	Deficiency symptoms	Available forms of Nutrients in the soil
Sulphur (S)	Increase oil content Promote root growth and nodulation	Upward curling	SO_4^{-}
Boron (B)	Cell development Translocation of sugars Polysaccharide synthesis	Sterility and malformation Decrease water absorption Die back	BO_3^{--}
Iron (Fe)	Constituent of cytochrome Catalytic role in enzymatic activity	Interveinal chlorosis Lime induced chlorosis	Fe^{++} Fe^{+++}
Manganese (Mn)	Constituent of chlorophyll Oxidation and reduction function Helps in ion movement	Interveinal chlorosis Gum packets in citrus	Mn^{++}
Zinc (Zn)	Essential for synthesis of chlorophyll, IAA and Cysteine Oxidation and reduction function	Rosette of fruit trees Little leaf	Zn^{++}
Copper (Cu)	Necessary for synthesis of Vitamin 'A' Catalytic role in respiration	Multiple bud formation Fruit crack Exanthema of fruit trees Reclamation disease	Cu^{++}
Molybdenum (Mo)	Constituent of nitrate reductase and nitrogenase enzyme Enhance nitrogen fixation in plants	Affect nodulation Scald of legumes	MoO_4^{--}

49. Characteristic of tree for agroforestry system

- Tree species should not interfere with agricultural crops with respect to soil moisture, nutrients, solar energy and agricultural operations.
- Tree species should have fast growing habit and easy management.
- Tree species should have high survival percentage and easy establishment.
- Tree species should have easily decomposable leaves
- Tree species should have ability to fix atmospheric nitrogen

- Tree species should not have any toxic effects on soil and on associated crops.
- Tree species should have adequate shade regulation, upright growth, respond to pruning, pollarding and coppicing.
- Tree species should have their multiple uses
- tree species should have high yield potential and higher profitability
- Tree species should have wide acceptability by local people

50. Rotation period of some important tree species

Botanical name	Rotation period (Years)	Botanical name	Rotation period (Years)
Abies pindrow	150	*Eugenia grandis*	100
Acacia catechu	10	*Eugenia zeylanica*	100
Acacia auriculiformis	8	*Ficus bengalensis*	50
Acer acuminatum	80	*Ficus racemosa*	50
Albizzia amara	80	*Hardwickia binata*	100
Albizzia lebbek	50	*Hopea parviflora*	80
Albizzia species	50	*Lagerstroemia parviflora*	40
Alnus nepalensis	50	*Lannea coromandelica*	40
Anogeissus latifolia	15	*Pinus excelsa/wallichiana*	80
Anthocephalus cadamba	15	*Pinus roxburghii/longifolia*	80
Butea monosperma	60	*Prosopis ceneraria*	50
Bombax ceiba	15	*Quercus species*	80
Buchanania latifolia/lanzan	20	*Rhododendron species*	50
Cedrus deodara	120	*Schima wallichii*	50
Cinnamomum species	60	*Shorea assamica*	80
Citrus medica	30	*Shorea robusta*	100
Cupressus species	50	*Taxus baccata*	60
Dalbergia paniculata	60	*Tectona grandis*	80
Diospyros microphylla	60	*Terminalia tomentosa*	80
Diospyros melanoxylon	30	*Terminalia paniculata*	80
Dipterocarpus macrocarpus	7	*Terminalia travancorensis*	80
Eucalyptus spp.	7	*Wrightia tinctoria*	50

51. Suitable tree species for top feed in silvipastural system

Tree species	Crude protein (%)	Tree species	Crude protein (%)
Acacia nilotica	13.9	*Calligonum polygoniodes*	7.4
Prosopis cineraria	13.9	*Azadirachta indica*	12.4
Salvadora oleoides	9.6	*Grewia tenex*	10.3
Zizyphus nummularia	11.7	*Gymnosporia spinosa*	9.3
Acacia Senegal	10.3	*Prosopis juliflora*	21.4
Albizia lebbeck	20.2	*Tecomella undulate*	12.2
Anogeissus latifolia	7.5	*Leucaena leucocephala*	27.6
Anogeissus pendula	7.6	*Holoptelia integrifolia*	13.7

52. Phenology of some Important tree species

Species	Leaf fall	Leaf r enewal	Flowering	Fruiting	Seed collection
Tectona grandis	Nov. – Jan.	Apr. – June	June – Sept.	Nov. – Jan.	Jan. – Feb.
Dalbergia sisoo	Feb. – Mar.	Apr. - May	April	Dec. – Apr.	Feb. – March
Acacia catechu	Feb. – Mar.	Apr. – May	July – Aug.	Sept. – Oct.	Nov. – Dec.
Acacia auriculiformis	Evergreen	After old leaf shed	July – Sept.	Oct. – Dec.	Jan. – March
Casuarina equsetifolia	Periodically	Periodically	Feb. - Apr. / Sep.- Oct.	June/ Dec.	Oct. – Nov.
Eucalyptus spp.	Periodically	Periodically	May – June / Oct. – Nov.	July – Aug. / Dec. – Jan.	Feb. March / Oct. – Nov.
Terminalia arjuna	Evergreen	Evergreen	Apr. – June	Feb. – May	Apr. – May
Terminalia bellirica	Nov. – Jan.	Apr. – May	Apr. – June	Nov. – Feb.	Nov. - Feb.
Terminaliachebula	Feb. – Mar.	March – May	Apr. – June	March	Oct. – March
Terminalia cranulata	Jan. – Feb.	June – July	May- July	October	Apr. – May
Ceiba pentandra	Nov. – Dec.	December	Dec. Jan.	Feb. – March	March – May
Gmelina arborea	Feb. – Apr.	April – May	Feb. – Apr.	Apr. – June	June – July
Shorea robusta	Jan. – Mar.	Feb. – May	Feb. – Apr.	June	June – Aug.
Adina cordifolia	February	May – June	June – Aug.	October	March – Apr.

Species	Leaf fall	Leaf r enewal	Flowering	Fruiting	Seed collection
Albizia lebbeck	Oct. – Nov.	March – Apr.	March – May	October	Jan. – March
Albiziz procera	Apr. – June	Apr. – May	June – Aug.	Oct. – Jan.	Jan. – Apr.
Santalum album	Evergreen	Evergreen	May	October	November
Azadirachta indica	Feb. – March	March – Apr.	April	June – Aug.	June – July
Ailantus excels	Cold season	March – Apr.	Feb. – March	May – June	June
Anogeissus latifolia	Feb. – Apr.	April – May	June – Sept.	Dec. – March	Dec. – March
Artocarpus hirsutus	Evergreen	Evergreen	Dec. – March	March - Apr.	Apr. – may
Bamboo spp.	Dec. March	April	Gregareous & sporadic	March	Apr. – June
Acacia nilotica	March – june	After old leaf shed	Apr. – June	Feb. – March	May – June
Bombax ceiba	December	March – Apr.	Jan. – Feb.	March – May	Mid March – Mid May
Derris indica	March – Apr.	April	Apr. – June	March – May	Apr. – June
Lagerstroemia parviflora	Feb. – March	Apr. – May	Apr. – June	Dec. – Jan.	Feb. – March
Lagerstroemia speciosa	Feb. – March	Apr. - May	Apr. – June	Nov. – June	Dec. – Feb.
Prosopis juliflora	Evergreen	Evergreen	Sept. – Oct.	Nov. – Jan. / Apr. – June	Apr. – June
Syzygium cumini	Evergreen	After old leaf shed	March – May	June – Aug.	Aug. – Sept.
Swietenia macrophylla	February	March – Apr.	Apr. - May	Oct. – Dec.	Dec. Jan.
Cassia siamea	Evergreen	Evergreen	Apr. - May	April	May - June
Cedrus deodara	April	March – Apr.	June – Aug.	Aug. – Sep.	Sep. – Nov.
Emblica officinalis	Nov. – Dec.	March – Apr.	March - May	October	Nov. – Feb.
Populus spp.	Oct. - March	March – Apr.	April	May-June	June

Species	Leaf fall	Leaf r enewal	Flowering	Fruiting	Seed collection
Salmalia malabaricum	December	March – Apr.	January	Feb. - March	March - May
Terminalia tomentosa	Jan. – Feb.	June - July	June – Aug.	Feb. – Apr.	March - May
Butea monosperma	Nov. – Dec.	April	Jan. – Apr.	May - June	May-June
Chukrasia tabularis	Evergreen	Evergreen	Apr. - May	Nov. – Dec.	Dec. - March
Dipterocarpus indicus	Feb. - March	March – Apr.	Dec. – Jan.	Feb. – Apr.	Apr. - June
Hardwickia binata	December	April	July – Sept.	Dec. – Feb.	Apr. - May
Pinus ruxburghi	January	May - June	Jan. – Feb.	Apr. - June	June - July
Pterocarpus marsupium	Apr. - May	May - June	Oct. - Nov	Nov. – Dec.	Jan. - March
Tamarindus indica	Apr. - May	March – Apr.	Apr. - June	Jan. – Feb.	Feb. - March

53. Silvicultural character of some important tree species

Species	Light	Drought	Frost	Fire	Coppicing	Root sucker	Temperature (°C)
Tectona grandis	Very strong	Sensitive	Sensitive	High resistant	Great power	-	25 – 35
Dalbergia sisoo	Moderate	Killed	Prone to damage	Less resistant	Well	Produced	Max.-25–47.8 Min.-0-15.6
Acacia catechu	Strong	High mortality	More resistant	Moderately resistant	Well	Produced	Max.-40 - 50 Min.-2.5-7.5
Acacia auriculiformis	Light demander	Resistant	Tender	-	-	-	26 - 30
Casuarina equsetifolia	Strong	Less resistant	Very sensitive	Very sensitive	Poor	Sometime produced	18 - 26
Eucalyptus spp.	Strong	Sensitive	Moderately resistant	Damage occure	Vigorous	Ligotuber	Max.-22-32 Min.-2-17
Terminalia arjuna	Moderate shade bearer	Sensitive	Sensitive	-	Satisfactory	Produced	Max.-38-48 Min.-1-15
Terminalia bellirica	Light demander	Hardy	Sensitive	-	Fairly well	-	Max.-35-47 Min.-0-17.5
Terminalia chebula	Strong	Fairly resistant	Fairly resistant	Withstand fire	Very well	-	Max.-35-47.5 Min.-0-17.5
Terminalia cranulata	Light demander	Killed	Sensitive	Fairly tolerant	Well	Produced	Max.-35-47.5 Min.-1-16
Ceiba pentandra	Strong	Hardy	Fairly resistant	-	-	-	18-38
Gmelina arborea	Strong	Moderately hardy	Moderately resistant	-	Vigorous	Produced	Max.-38-48 Min.-1-16

Species	Light	Drought	Frost	Fire	Coppicing	Root sucker	Temperature (°C)
Shorea robusta	Light demander	Not resistant	More resistant	Hardy	Good	-	Max.-25-40 Min.-5-10
Adina cordifolia	Strong	Moderately resistant	Moderadely resistand	Tender	Good	-	Max.-35.5-50 Min.-2.5-15
Albizia lebbeck	Light demander	Sensitive	Not very tender	-	Well	Produced	-5-45.6
Albiziz procera	Light demander	Fairly resistant	Tender	-	Well	Produced	Max.-37-46 Min.-10-18
Santalum album	Light demander	Sensitive	-	Extremly sensitive	Fairly well	Produced	17.8-33.8
Azadirachta indica	Strong	Can withstand	Very sensitive	Succeptible	Well	Produced	Max.-40-47.5 Min.-0-15
Ailantus excels	Strong	Resistant	Suceptible	-	Well	Produced	-
Anogeissus latifolia	Light demander	Fairly resistant	Hardy	Succeptible	Well	Produced	Max.-38-45 Min.- -1-16
Artocarpus hirsutus	Light demander	Sensitive	Sensitive	Sensitive	-	-	-
Bamboo spp.	Light demander	Extremly resistant	Hardy	-	-	-	-
Acacia nilotica	Strong	Fairly resistant	Tender	Not resistant	Poorly	-	Max.-40-48 Min.- -1-15
Bombax ceiba	Intense light demander	Fairly resistant	Sensitive	Resistant	Good	Produced	Max.-37.5-50 Min.- -2.5-17.5
Derris indica	Shade bearer	Resistant	Fairly resistant	Moderately resistant	Excelent	Produced	Max.-27-30 Min.- -1-16

Species	Light	Drought	Frost	Fire	Coppicing	Root sucker	Temperature (°C)
Lagerstroemia parviflora	Light demander	Hardy	Resistant	Extremly resistant	Vigorously	-	Max.-38-48 Min.-1-7
Lagerstroemia speciosa	*Moderate light demander*	Hardy	Sensitive	-	Well	-	Max.-34-47.5 Min.-2.5-19
Prosopis juliflora	Strong	Resistant	Hardy	Tender	Fairly well	Produced	4.-45
Syzygium cumini	Shade bearer	Hardy	Hardy	Sensitive	Well	-	Max.-30-43.5 Min.-5-23.9
Swietenia macrophylla	Light demander	Sensitive	Sensitive	Sensitive	-	-	16-37
Cassia siamea	Strong light	Hardy	Tender	Sceptible	Well	-	20-49
Cedrus deodara	Light demander	Sensitive	Moderately frost hardy	-	No	-	-12 - 38
Emblica officinalis	Light demander	Sensitive	Sensitive	-	Well		5 - 49
Populus spp.	Light demander	Hardy	Hardy	Tender	Well	Produced	20 - 45
Salmalia malabaricum	Light demander	Resistant	sensitive	Moderate fire hardy	Well	Produced	10 - 45
Terminalia tomentosa	Light demander	sensitive	sensitive	sensitive	Well	-	10 - 40
Butea monosperma	Light demander	Resistant	Hardy	-	Well	Produced	17 – 42
Chukrasia tabularis	Light demander	Sensitive	Moderately resistant	-	Well	-	18 - 42

Species	Light	Drought	Frost	Fire	Coppicing	Root sucker	Temperature (°C)
Dipterocarpus indicus	Light demander	-	Sensitive	Tender	Poor	-	15 - 38
Hardwickia binata	Moderate light demander	-	Hardy	Hardy	Well	Produced	10 - 45
Pinus ruxburghi	Light demander	Resistant	Hardy	Resistant	-	-	15 - 38
Pterocarpus marsupium	Light demander	Hardy	Hardy	Hardy	Well	Produced	20 - 42
Tamarindus indica	Strong light demander	Hardy	Suceptible	Suceptible	Well	Produced	17 - 42

54. Countries with highest acreage under forest.

Sl. No.	Country	Forest Area (Million Ha.)
1	Russia	809
2	Brazil	519
3	Canada	310
4	USA	304
5	China	206
6	Kango	154
7	Australia	149
8	Indonesia	94
9	Sudan	69
10	India	69

55. Forest and tree cover of India in 2021

Class	Area (km^2)	% of Geographical Area
Forest Cover		
Very Dense Forest	99,779	3.04
Moderate Dense Forest	3,06,890	9.33
Open Forest	3,07,120	9.34
Total Forest Cover	**7,13,789**	**21.71**
Scrub	46,539	1.42
Non Forest	25,27,141	76.87
Total Geographical Area	**32,87,469**	**100.00**

Source : ISFR, 2021

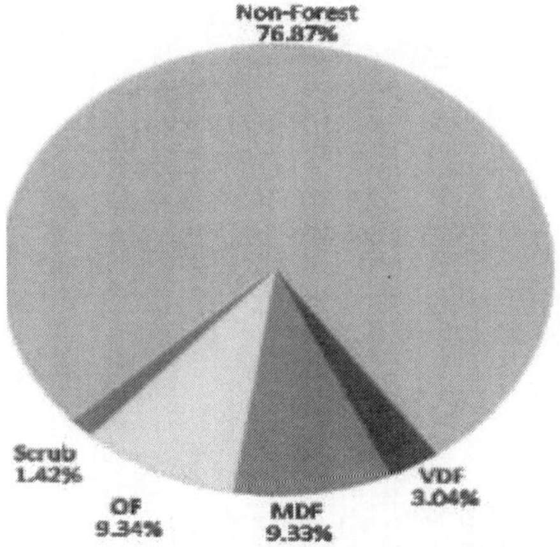

56. Forest cover in States/ UTs in India (Area in km^2)

States/ UTs	Geographical Area	2019 Assessment				% of Geographical Area	Change in Forest Cover wrt ISFR 2019	Change % w.r.t ISFR 2019 assessment	Scrub
		VDF	MDF	OF	TFC				
Andhra Pradesh	162968	1994	13929	13861	29784	18.28	647	2.22	8276
Arunachal Pradesh	83743	21058	30176	15197	66431	79.33	-257	-0.39	797
Assam	78438	3017	9991	15304	28312	36.09	-15	-0.05	228
Bihar	94163	333	3286	3762	7381	7.84	75	1.03	236
Chhattisgarh	135192	7068	32279	16370	55717	41.21	106	0.19	615
Delhi	1483	6.72	56.60	131.68	195	13.15	-0.44	-0.23	0.38
Goa	3702	538	576	1130	2244	60.62	7	0.31	0
Gujarat	196244	378	5092	9516	14926	7.61	69	0.46	2828
Haryana	44212	28	445	1130	1603	3.63	1	0.06	159
Himachal Pradesh	55673	3163	7100	5180	15443	27.73	9	0.06	322
Jharkhand	79716	2601	9689	11431	23721	29.76	110	0.47	584
Karnataka	191791	4533	20985	13212	38730	20.19	155	0.40	4611
Kerala	38852	1944	9472	9837	21253	54.70	109	0.52	30
Madhya Pradesh	308252	6665	34209	36619	77493	25.14	11	0.01	5457
Maharashtra	307713	8734	20589	21475	50798	16.51	20	0.04	4247
Manipur	22327	905	6228	9465	16598	74.34	-249	-1.48	1215
Meghalaya	22429	560	9160	7326	17046	76	-73	-043	663
Mizoram	21081	157	5715	11948	17820	84.53	-186	-1.03	1
Nagaland	16579	1272	4449	6530	12251	73.90	-235	-1.88	824

States/ UTs	Geographical Area	2019 Assessment				% of Geographical Area	Change in Forest Cover wrt ISFR 2019	Change % w.r.t ISFR 2019 assessment	Scrub
		VDF	MDF	OF	TFC				
Odisha	155707	7213	20995	23948	52156	33.50	537	1.04	4924
Punjab	50362	11	793	1043	1847	3.67	-2	-0.11	34
Rajasthan	342239	78	4369	12208	16665	4.87	25	0.15	4809
Sikkim	7096	1102	1551	688	3341	47.08	-1	-0.03	296
Tamil Nadu	130060	3593	11034	11792	26419	20.31	55	0.21	758
Telangana	112077	1624	9119	10471	21214	18.93	632	3.07	2911
Tripura	10486	647	5212	1863	7722	73.64	-4	-0.05	33
Uttar Pradesh	240928	2627	4029	8162	14818	6.15	12	0.08	563
Uttarakhand	53483	5055	12768	6482	24305	45.44	2	0.01	392
West Bengal	88752	3037	4208	9587	16832	18.96	-70	-0.41	156
Andaman & Nicobar	8249	5678	683	383	6744	81.75	1	0.01	1
Chandigarh	114	1.36	13.51	8.01	22.88	20.07	0.85	3.86	0.38
Dadra & Nagar Haveli and Daman & Diu	602	1.40	85.56	140.79	227.75	37.83	0.10	0.04	4.85
J & K Shapefile Area* (54624)	222236	4155	8117	9115	21387	39.15	29	0.14	284
Ladakh Shapefile Area* (168055)		2	512	1758	2272	1.35	18	0.80	279

States/ UTs	Geographical Area	2019 Assessment				% of Geographical Area	Change in Forest Cover wrt ISFR 2019	Change % w.r.t ISFR 2019 assessment	Scrub
		VDF	MDF	OF	TFC				
Lakshadweep	30	0	16.09	11.01	27.10	90.33	0	0	0
Puducherry	490	0	17.53	35.77	53.30	10.88	0.89	1.70	0
G. Total	**3287469**	**99779**	**306890**	**307120**	**713789**	**21.71**	**1540**	**0.22**	**46539**

VDF- Very Dense Forest, MDF- Doderate Dense Forest, OF- Open Forest, TF- Total Forest. Source: FSI (2021)

*Area of shapefile provided by Survey of India (August, 2021). Notified geographical areas for individual UTs from SoI are awaited.

57. Interpretation key for forest type delination using Landsat FCC

Forest/ Land Cover Classes	Tone/ Colour	Size	Shape	Texture	Pattern	Location	Association
Evergreen	Bright red	Large	Irregular	Fine	Irregular	High hill elevation	Sorrounded by semievergreen forest
Semi-evergreen	Medium red	Small	-do-	Mottled	-do-	High to medium elevation	Intermediate between evergreen & moist deciduous forest
Moist deciduous	Brownish red	Large	-do-	Coarse	-do-	Medium elevation	Within agricultural land and open area
Dry deciduous	Brown	Small	-do-	-do-	-do-	Low elevation	Mainly along agricultural land
Degraded forest	Reddish brown	-do-	-do-	Very coarse	-do-	On hills & plains	Plantation and agriculture etc.
Scrub/ Shrub	Light red	-do-	-do-	Coarse	Patchy	Along sea shore, low hills, plain	Plantations, agriculture areas and barren land
Non-forest							
Crop land	Medium red	Very small	-do-	Fine	Linear	Along water bodies	Within low land
Fallow/ Barren	White to grey	Large	-do-	Smooth	Irregular	Plains to hills	Associated with almost all types of land
Waterbody	Dark blue	Small	Regular	Fine	-do-	Hilly and plain areas	With any of the above types

58. List of Important Instruments

Device	Quantity Mesured
Actinometer	Heating power of sun light
Altimeter	Altitudes
Anemometer	Wind speed
Audiometer	Hearing
Barometer	Air pressure
Bomb calorimeter	Heat of combustion
Calorimeter	Heat of chemical reactions
Chronometer	Time
Compas	Direction of North
Crescograph	Plant growth
Densimeter	Specific gravity of Liquid
Dumpy level	Horizontal levels
Evaporimeter	Rate of evaporation
Fathometer	Ocean depth
Glucometer	Blood glucose
Hydrometer	Specific gravity of liquids
Hygrometer	Humidity
Clinometer	Angle of a slope
Lactometer	Specific gravity of milk
Manometer	Pressure of gas
Mercury Barometer	Atmospheric pressure
Odometer	Distance travelled
Osmometer	Osmotic strength of solution
Pedometer	Steps
pH meter	pH
Photometer	Light intensity
Potometer	Transpirtion
Planometer	Area
Psychrometer	Humidity
Pycnometer	Fluid density
Pyranometer	Solar radiation
Rain Gauge	Measuring of rain
Saccharometer	Amount of sugar in a solution
Seismometer	Seismic waves

Device	Quantity Mesured
Spectrophotometer	Intensity of light as a function of wavelength
Sphygmomanometer	Blood pressure
Tensiometer	Surface tension of liquid
Theodolite	Measuring angles in horizontal and vertical planes
Thermometer	Temperature
Tintometer	Colour
Wind vane	Wind direction
Zymometer	Fermentation

59. List of Important Revolutions

Name	Related With
Green revolution	Rice and wheat production
White revolution	Milk production
Blue revolution	Fish production
Yellow revolution	Oilseeds production
Red revolution	Tomato/ Meat production
Golden revolution	Fruit production
Round revolution	Potato production
Black revolution	Biofuel / Jatropha
Pink revolution	Onion production
Rainbow revolution	All sector of agriculture

60. Agroclimatic zones of India

India has fifteen agroclimatic zones.

Sl No	Agro-climatic zone	Annual rain-fall (cm)	Occupied region	Agricultural crops
1.	Western Himalayan region	75-10	J & K, Himachal Pradesh and the hill region of Uttarakhand	Crops-Rice, maize Fruit crops- Apple, Peaches, apricot, pears, cherry, almond, litchis, walnut, saffron
2.	Estern Himalayan region	200-400	Arunachal Pradesh, Assam, Sikkim, Meghalaya, Nagaland, Manipur, Mizoram, Tripura andDarjeeling district of West Bengal	Crops- Rice, maize, potato, tea Fruit crops- Pineaapple, Litchi, orange and lime Soil-Red-brown Jhuming (Shifting cultivation) prevails in the hilly areas.

Sl No	Agro-climatic zone	Annual rain-fall (cm)	Occupied region	Agricultural crops
3.	Lower Gangetic plain region	100-200	West Bengal (Except the hilly area) eastern Bihar and the Brahmaputra valley	Crops- Rice, Jute, maize, potato and pulses. Fruit crops- Banana, mango and citrus fruits
4.	Middle Gangetic plain region	100-200	Large parts of Utter Pradesh and Bihar	Soil-Fertile alluvial Crops- Rice, Maize, millets, wheat, gram, barley, peas, mustard and potato
5.	Upper Gangetic plain region	75-150	Central and Western parts of Utter Pradesh and Haridwar and Udham nagar district of Uttarakhand.	Soil-Sandy loam Crops- Wheat, rice, sugarcane, millets, maize, gram, barley, oilseeds, pulses and cotton
6.	Trans-Ganga plains region	65-125	Punjab, Haryana, Chandigarh, Delhi, Ganganagar district of Rajasthan	Soil-Alluvial Crops- Wheat, sugarcane, cotton, rice, gram, maize, millets, pulses and oilseeds
7.	Eastern Plateau and Hills	80-150	Chhotanagpur Plateau, extending over Jharkhand, Odisha, Chattisgarh and Dandakaranya	Soils- Red and Yellow soils Crops- Rice, millets, maize, oilseeds, ragi, gram, potato, tur groundnut, soyabean and castor
8.	Central Plateau and Hills	50-100	Bundelkhand, Baghelkhand, Bhander Plateau, Malwa Plateau and Vindhyachal Hills.	Soils- Red, yellow and black Crops- Millets, wheat, gram, oilseeds, cotton and sunflower
9.	Western Plateau and Hills	25-75	Southern part of Malwa plateau and Decan plateau (Maharastra)	Soil- Regul (black) soil Crops- Wheat, gram, millets, cotton, pulses, groundnut, oilseeds, sugarcane, rice, jower and bajra Fruit crops- Oranges, grape, bananas, ber, pomegranate, mango and guava (Sprinklers and drip system)

Sl No	Agro-climatic zone	Annual rain-fall (cm)	Occupied region	Agricultural crops
10.	Southern Plateau and Hills	50-100	Interior Decan and includes parts of southern Maharastra, the greater parts of Karnataka, Andhra Pradesh and Tamil Nadu	Crops- Millets, oilseeds and pulses Coffee, tea cardamom and spices
11.	Eastern Costal Plains and Hills	75-150	Coromandal and northern Circar coasts of Andhra Pradesh and Odisha	Soils- Alluvial, loam and clay Crops- Rice, jute, tobacco, sugarcane, maize, millets, groundnut and oilseeds, pepper and cardamom.
12.	Western Costal Plains and Ghats	> 200	Malabar and Konkan costal plains and the Sahyadris	Soils- Laterite and alluvial Crops- Rice, coconut, oil seeds, sugarcane, millets, pulses and cotton
13.	Gujarat Plains and Hills	50-100	The hills and plains of Kathiawar and the fertile vallys of Mahi and Sabarmati rivers.	Soils- Regur & lluvium Crops- Groundnut, cotton, rice, millets, oilseeds, wheat and tobacco.
14.	Western dry region	< 25	Extending over Rajasthan, west of Aravallis	Crops- Bajra, jower, moth, wheat and gram Fruit crops- Water melon, guava and date palm
15.	Island region	< 300	Andaman-Nicobar and Lkshadweep	Soil- Sandy loam Crops- Rice, maize, millets, pulses, arecanut, turmeric, cassava

(According to Planning Commission of India)

61. Location of National Parks in India

State	National Park
Assam	Dibru-Saikhowa N. P. Rajiv Gandhi Orang N. P. Kaziranga N. P. Manas N. P. Nameri N. P
Andhra Pradesh	Sri Venkateswara N. P.
Arunachal Pradesh	Namdapha N. P. Mouling N. P.

State	National Park
Bihar	Valmiki N. P.
Chattisgarh	Indravati N. P Guru Ghasi Das N. P. Kanger Valley N. P
Gujarat	Blackbuck N. P., Velvadar Gir N. P Marine N. P., Gulf of Kutch Bansda N. P.
Goa	Bhagwan Magavir N. P. (Mollem)
Haryana	Kalesar N. P. Sultanpur N. P.
Himachal Pradesh	Pin Valley N. P. Inderkilla N. P. Great Himalayan N. P. Khirganga N. P. Simbalbara N. P.
Jammu & Kasmir	Dchigam N. P. Salim Ali N. P. Kishtwar N. P. Hemis N. P.
Jharkhand	Belta N. P.
Karnataka	Bandipur N. P. Anshi N. P. Kudremukh Natiokeynal N. P. Bannerghatta N. P. Nagarhole N. P
Kerala	Pamapadum Shola N. P. Anamudi Shola N. P. Eravikulam N. P. Mathikettan Shola N. P. Silent Vally N. P.
Maharashtra	Chandoli N.P. Gugamal N. P. Tadoba N. P. Navegaon N. P.
Madhya Pradesh	Kanha N. P. Madhav N. P. Van Vihar N. P. Mandla Plant Fossils N. P. Panna N. P. Pench N. P. Sanjay N. P. Satpura N.P.

State	National Park
Manipur	Keibul Lamjao N. P
Meghalaya	Balphakram N. P. Nokrek N. P.
Mizoram	Murlen N. P. Mountain N. P. Phawnguni Blue N. P
Ngaland	Intanki N. P.
Odisha	Bhitarkanika N. P. Similipal N. P
Rajasthan	Sariska N. P. Keoladeo N. P. Ranthambore N. P. Darrah N. P. Desert N. P.
Sikkim	Khangchendzonga N. P.
Tamil Nadu	Mudumalai N. P. Guindy N. P. Mukurhti N. P. Indira Gandhi N. P. (Annamalai) Gulf of Mannar Marine N. P.
Telengana	Mrugavani N. P.
Tripura	Clouded Leopard N. P. Bison N. P. (Rajbari)
Uttarakhand	Gangotri N. P. Jim Corbett N. P. Valley of Flowers N. P. Rajaji N. P.
Utter Prdesh	Dudhwa N. P.
West Bengal	Buxa N. P. Sundarbans N. P. Neora Valley N. P. Singalila N. P.

62. List of Biosphere Reserves in India

Sl No.	Name of the Biosphere Reserves	Location
1.	Nilgiri	Tamil Nadu
2.	Nanda Devi	Uttarakhand
3.	Manas	Assam
4.	Achanakmar-Amarkantak	Madhya Pradesh
5.	Agasthyamalai	Kerala

Sl No.	Name of the Biosphere Reserves	Location
6.	Cold desert	Laddakh
7.	Dehang-Dibang	Arunachal Pradesh
8.	Great Nicbar	Andaman & Nicobar
9.	Gulf of Mannar	Tamil Nadu
10.	Kachchh	Gujarat
11.	Khangchendzonga	Sikkim
12.	Nokrek	Meghalaya
13.	Panchmarhi	Madhya Pradesh
14.	Seeshachalam Hills	Andhra Pradesh
15.	Similipal	Odisha
16.	Synderbans	West Bengal
17.	Panna	Madhya Pradesh
18.	Dibru-Saikhowa	Assam

63. Succession in different communities

Succession in frest water	Hydrosere
Succession in salty water	Halosere
Succession in acidic water	Oxalosere
Succession in dry region	Xerosere
Succession on rock	Lithosere
Successionon sand	Psamosere
Succession at moistened region	Mesosere
Succession of microbes on decomposed matter	Serula

64. Important association between the organisms

Interaction Types	Beneficial	Harmful	Relationship
Mutualism (+, +)	Both	-	Symbiosis/Obligatory
Commensalism (+, 0)	One	Unaffected	Non-obligatory
Protoco-operation (+, +)	Both	-	Non-obligatory
Parasitism (+, -)	One	One (Larger)	Non-obligatory
Predation (+, -)	One	One (Smaller)	Non-obligatory
Amensalism (0, -)	Unaffected	One	Non-obligatory
Allelopathy (0, -)	Unaffected	One	Non-obligatory

65. Wood wastage during various logging Operations.

Sl No.	Wood wastage form	Wastage percentage
1.	Felling: Stumps Damaged to surrounding tree by felled tree Stem splitting Felling of leaning trees without any control	4-6 2-3 1-2 2-3
	Total wastage due to all felling	9-14%
2.	Sawing: Hand sawing Pulp Wood	27-32 4-5
	Total wastage due to sawing	31-37 %
3.	Minor Transportation: Chutes and dry slides ____________ Wet slides ____________ Telescopic floating	1 2-3 2-3
	Total wastage due to minor transportation	5-6 %
4.	Major Transportation – Normal river loss	3-5
	Total wastage due to transportation	8-11%
	Grand total of wastage	48-63 %
	Total average wastage	56 %

66. Important features of paper making process

Sl No.	Process		Process used for
1.	Pulping (Isolation of cellulosic fibres from raw materials)	Mechanical	Grinding of raw materials such as fibre, lignin etc.
		Chemical	Removal of lignin and fibrous raw materials from mechanical pulp
		Semi-chemical	For pulp softening
2.	Pulp Cleaning		Removal of unwanted pulp materials. In first stage we remove knots while in next stage centrifugal screening of shives, dirt & grit.
3.	Pulp bleaching		For colouring in pulp. In single stage bleaching calcium hypochlorite ($CaOcl_2$) is used and in multi stage bleaching chlorine gas (Cl_2) and chlorine dioxide (ClO_2) is used for bleaching.

Sl No.	Process		Process used for
4.	Stock preparation	Beating	Mechanical treatment given to fibres. This increased bursting strength, tensile strength, folding endurances, smoothness and rattle of paper while decreases tearing strength, bulk, porosity, opacity and dimensional stability. Thus physical characteristics of paper depends on beating.
		Sizing	Sizing is done for making paper more or less impervious ti ink penetration. Rosin, soap along with aluminium sulphate is used.
		Loading	Addition of non-fibrous mineral matters called fillers. These improve the dimensional stability as well as brightness of paper. Common fillers used by industry are china clay, talk powder.
		Colouring	For dye the paper to correct shade
		Other additives	Starch, vegetables gums, resin also used to give special characteristics to paper
5.	Sheet formation		After proper drying and calendring it is rolled over spools finally cut, sorted & packed

67. Elements and Function of Xylem & Phloem

Parameters	Xylem	Phloem
Term	Nageli	Nageli
Elements	Four (4) elements: Tracheids – Dead Elements (DE) Vessels- Dead Elements Xylem Fibres- Dead Elements Xylem Parenchyma- Living Elements (LE)	Four (4) elements: Sieve cells/sieve tubes- LE Companion cells- LE Phloem fibres-DE Phloem parenchyma- LE
Function	Conduct water & minerals salts	Conduction of food materials

68. Height measuring instruments

Based on similar triangle	Based on Trigonometric principle
Christen's Hypsometer Smythie's Hypsometer Improved Callipers	Brandis Hypsometer Abney's Level Hega Altimeter Topographical Abney's Level Relaskop Tele Relaskop Bar & Sound Dendrometer Blume-Leiss Hypsometer

69. Classification of Volume Table

Classification on the basis of variables	
Volume table based on one variable	Diameter
Volume table based on two variable	Diameter and Height
Volume table based on three variable	Diameter, Height and Form quotient
Classification on the basis of scope of application	
General volume table	2 variables- Diameter & Height
Regional volume table	-
Local volume table	One variables- Diameter
Classification on the basis of outturn kind	
Standard volume table	For standard round timber estimation
Commercial volume table	For commercial round timber estimation
Sawn outturn table	For sawn timber estimation
Assortment table	Round timber volume in various form
Sawn assortment outturn table	Sawn outturn in the form of standardized sized pieces

70. Classification of aerial photographs

On the basis of film used			
Photographs name	**Film used**	**Remarks**	**Used for**
Panchromatic Black & White photograph	Panchromatic film	This film has low sensitivity for green region	Not suitable for plant species identification
Infra-red (IR) black & white photograph	Infra-red (IR) film with proper filter	This film is used with red filter for identification	Identification of broad leaved & coniferous tree

<table>
<tr><th colspan="4">On the basis of film used</th></tr>
<tr><th>Photographs name</th><th>Film used</th><th>Remarks</th><th>Used for</th></tr>
<tr><td>Colour photograph</td><td>Colour films</td><td>Colour films has three emulsion layers (blue, red and green) Green produces burnt magenta while red produced cyan and blue produces yellow colour</td><td>Identification of species and diseases</td></tr>
<tr><td>Infrared colour photograph</td><td>Kodak Aero-chrome Infra-red</td><td>Also called false colour photograph because object shows different colour from their actual colour</td><td>Green vegetation is seen red to magenta colour in this film</td></tr>
<tr><th colspan="4">On the basis of the devised used</th></tr>
<tr><th colspan="2">Photographs Name</th><th colspan="2">Devised used</th></tr>
<tr><td colspan="2">Single lens photograph</td><td colspan="2">Single lens camera</td></tr>
<tr><td colspan="2">Multi-spectral photograph</td><td colspan="2">Single camera with more than one lens</td></tr>
<tr><td colspan="2">Multi-spectral imagery</td><td colspan="2">Optical mechanical scanner</td></tr>
<tr><th colspan="4">On the basis of scale of photograph</th></tr>
<tr><th>Photographs</th><th>Scale</th><th colspan="2">Used for</th></tr>
<tr><td>Very large scale photograph</td><td>1:5,000 to 1:10,000</td><td colspan="2" rowspan="2">Working plan purposes, detailed management inventory, exploitation planning, insect & disease survey and small height classes classification</td></tr>
<tr><td>Large scale photograph</td><td>1:10,000 to 1: 20,000</td></tr>
<tr><td>Medium scale photograph</td><td>1:20,000 to 1:40,000</td><td colspan="2">Forest inventory preparation, detailed land use classification, forest stratification, broad height classification</td></tr>
<tr><td>Small scale photograph</td><td>1: 40,000 to 1: 70,000</td><td colspan="2" rowspan="2">Broad land use classification, forest type identification</td></tr>
<tr><td>Very small scale photograph</td><td>1:70,000</td></tr>
<tr><th colspan="4">On the basis of scale of position of optical axis of camera</th></tr>
<tr><th colspan="2">Photograph name</th><th colspan="2">Deviation</th></tr>
<tr><td colspan="2">Vertical photograph</td><td colspan="2">Upto 4^0 from perpendicular position</td></tr>
<tr><td colspan="2">Oblique photograph</td><td colspan="2">Low oblique – Upto30^0
Hogh oblique- More than 30^0</td></tr>
</table>

71. Optimum season for aerial photography in different forest

Vegetational form	Optimum season for aerial photography
Coniferous forests in Himalayan	October – November
Mixed deciduous forests	December – February
Sal forest	Mid March – Mid April
Teak forest	September – December
Tropical evergreen & moist deciduous	January – February

72. Function and deficiency symptoms of mineral nutrition in plants

Name	Absorption form	Role/Function	Deficiency symptoms
Nitrogen (N)	NO^{3-} (Nitrate)	Constituent of Plant hormones (IAA) and ATPs. Present in Porphyrins of Chlorophylls & cytochromes	Chlorosis in older leaves Anthocynin formed in stem Late flowering & plant becomes more susceptible to fungal disease due to excess nitrogen
Sulphur (S)	SO^{4-} (Suplhate)	Parts of cystinine, cysteine & methionine amino acids. Vit. Biotene, thiamine, Co-A in respiration Role in oil synthesis, chlorophyll synthesis & part of ferredoxin Root nodule formation	Chlorosis or yellowing in younger leaves with anthocyanin Tips & margins of leaves curves inwardly "tea yellow disease"
Phosphorus (P)	$H_2PO_4^-$ (Ortho-phosphate)	Very important in RNA, DNA, Phospholipid (cell membrane) NADP (Co-enzyme) ATP (Energy reaction) Very important in photosynthesis (NADP), Protein synthesis.	Premature leaf fall, necrosis and anthocyanin formation
Calcium (Ca)	Ca^{2+}	Mechanical strength because Ca is constituent of middle lamella Permeability of bio-membrane is maintained. Stability of chromosome structure & in spindle formation.	Disintegration of growing apices Irregular cell divisions and death of meristem

Name	Absorption form	Role/Function	Deficiency symptoms
Molybdenum (Mo)	MoO_4^{2-} (Molybdate)	Role as prosthetic group of nitrate reductase and nitrogenise in nitrogen metabolism	Whip tail of cauliflower
Pottasium (K)	K^+ (Monovalent cation)	Key role in stomatal movement & transpiration	Mottled chlorosis Die back disease Decrease the apical dominace
Magnesium (Mg)	Mg^{2+}	Constituent of chlorophyll Activator of many enzymes in carbohydrate metabolism	Interveinal chlorosis on large scle and formation of anthocyanin in older leaves.
Iron (Fe)	Fe^{2+} (Ferrus)	Absorption from acidic soil Iron porphyrin protein for cytochromes, Peroxidase & catalases Important for ferro=edoxin, biological N_2 fixation & ETS Essential for chlorophyll synthesis	Rapid interveinal chlorosis
Manganese (Mn)	Mn^{2+}	Essential for O_2 evolution and photolysis of water in light reaction	Marsh spot of pea and grey speak of oat Chlorosis in young & older leaves
Boron (B)	$H_3BO_3^-$ (Borate)	Boron is only micronutrient, which is not associated with enzymes. Required for uptake and utilization of Ca^{2+}, membrane functioning, cell elongation and differentiation and carbohydrate translocation. Essential in pollen tube formation. Lethal effect at carbohydrate metabolic site.	Stem & root tips dies Top rotten in tobacco, water core in turnip, brown heartrot of beets, heart rot in carrot & marigold.
Copper (Cu)	Cu^{2+}	Vit.-C (Ascorbic Acid) formation	Die back of citrus and other fruit trees exanthema in the trees Reclamation disease of cereals & legume crops

Name	Absorption form	Role/Function	Deficiency symptoms
Zinc (Zn)	Zn^{2+}	Specific role in Auxin (IAA) synthesis. Ctivator of Carbonic anhydrase, carboxylases, alcohol dehydrogenase and peptidase.	Mottle leaf disease in fruit trees "little leaf disease' Khaira disease of paddy White bud disease in maize.

73. Comparision between C_3, C_4 and CAM pathway

C_3- pathway	C_4- pathway	CAM-pathway
First stable compound is-3 carbon PGA	First stable compound is-4 carbon OAA	First stable compound is-4 carbon OAA
18 ATP & 12 $NADPH_2$ are used for 1 glucose form	30 ATP & 12 $NADPH_2$ are used for 1 glucose form	30 ATP & 12 $NADPH_2$ are used for 1 glucose form
Kranz anatomy absent	Kranz anatomy present	Kranz anatomy present
One type of carboxylase enzyme-Rubisco present	One type of carboxylase enzyme-Rubisco present & PEP case present	One type of carboxylase enzyme-Rubisco present & PEP case present
CO_2 acceptor is RUBP	Primary CO_2 acceptor are RUBP & PEP case	Primary CO_2 acceptor are RUBP & PEP case
High CO_2 compensation point (40-100PPM)	Low CO_2 compensation point (8-10PPM)	High CO_2 compensation point (40-100PPM)
Transpiration ratio is 500-1000	Transpiration ratio is 200-300	Transpiration ratio is 50-100
Examples-mostly belongs to dicot family but wheat & rice are C_3-plants.	Examples-mostly monocot belons to family-Gramineae & Cyperaceae. Such as – maize, Sugarcane, Some dicot-Euphorbia, triplex, Amaranthus, Portulaca, Rosea, Tribulus & Chenopodium	Examples-Family Crassulaceae-kalanchoe, Bryophyllum, Sedum, Kleinia, opuntia, Crassula, Agave, Aloe, Pineapple & Welwistchia etc.
Discover-Calvin	Discover-Hatch & Slack	Discover- Oleary & Rouhani
Known as-Dark reaction, Blackman Reaction, Calvin cycle, C_3-cycle, Biochemical phase, carbon assimilation, Photosynthetic carbon reduction cycle (PCR-cycle), Reductive pentose phosphates pathway etc.	Known as-CO_2 conc. mechanism, Dicarboxylic acid cycle (DCA-cycle), C_4- cycle) Hatch & Slack pathway etc.	Known as- Dark CO_2 fixatopn, Crassulacean acid metabolism (CAM) and dark acidification etc.

74. Debadghao and Shankar Narayan (1973) have recognised 5 types of grass covers in India.

Grass cover type	Found in	Covered area	Soil type
Sehima/Dichanthium	Tropical area	1.74 million Sq. km	Red, Black Yellow & Laterite soil
Dichantium-Cenchrus-Lasiurus	Sub-tropical & semi-arid region	4 lakh Sq. km	Alluvial & grey brown soil
Phragmites-Saccharum-Imperata	Gangetic plan & Brahmaputra valley	2.8 million Sq. km	lluvial soil
Themeda-Arundinella	Hilly region	2.3 lakh Sq. km	-
Temperate-Alpine	High Hilly region (Northern)	-	-

75. Important grasses with their avg. Forage yield.

Common Name	Botanical Name	Average forage Yield
Buffalo grass / Para grass	*Brachiaria mutica*	2000-2500 Qt./ha
Anjan grass	*Cenchrus ciliaris*	300-350 Qt./ha
Bermunda Grass/ Doob grass	*Cynodon dactylon*	300-350 Qt./ha
Sain grass	*Sehima nervosum*	250-300 Qt./ha
Dinanath grass	*Pennisetum pedicellatum*	750-1000 Qt./ha
Jarga grass	*Dichantium annulatum*	200-250 Qt./ha
Barware grass	*Panicum antidotale*	200-300 Qt./ha
Guinea grass	*Panicum maximum*	750-1600 Qt./ha
Sudan grass	*Sorghum sudanense*	260-300 Qt./ha
Napier grass, elephant grass	*Pennisetum perpureum*	40 ton/ha
Tiger grass	*Thysanolaena maxima*	4ton/ha

76. Duration of important census in India

Name	Duration (In years)
Livestock population census	5
Population census	10
Wildlife population census	5
Forest survey report	2
Working plan	10

77. Important forest trees and diseases

Tree	Root rot	Heart rot	Wilt
Dalbergia sisoo	*Ganoderma lucidum*	-	*Fusarium solani*
Acacia catechu	*Ganoderma lucidum*	*Fomes badius*	-
Shorea robusta	*Polyporus shoreae*	*Hymenochaete rubiginosa*	-
Tectona grandis	*Polyporus zonalis & Peniophora rhizomorpha-sulphurea*	-	*Pseudomonas solanecearum*
Eucalyptus hybrid	*Ganoderma lucidum*	-	-
Casuarina equisetifolia	-	-	*Trichosporium vesiculosum*
Abies pindrew	*Armillarea mellea*	*Fomes spp.*	-
Picea smithiana	*Armillarea mellea*	*Fomes spp.*	-
Cedrus deodara	*Fomes annosus*	-	-

78. Classification of forest based on

Age	Regeneration	Composition	Ownership	Management	Growing stock
Even-aged	High forest	Pure forest	Public forest 1. Reserved 2.Protected 3.Village	Protection forest	Normal forest
Uneven-aged	Coppice forest	Mixed forest	Private Forest	Production forest Social forest	Abnormal forest

79. Some commonly used formulae in Thinning

Formula	Used for	Given by
d=D	Deodar	Glover
d= 1.5 D	Irregular sal crops	Laurie
d=2D	Sissoo	Sagreiya
d=3D	Irregular teak crops	Howard

80. List of successful exotics in India

Species	Place of introduction	Native/Origin Place
Acacia auriculiformis	FRI, West bBengal, Bihar	Queensland, Australia
Acacia tortilis	Jodhpur (Rajsthan)	Israel
Casuarina equisetifolia	TamilNadu	Indonesia
Cinnamomum camphora	West Bengal	China
Cryptomeria japonica	DehraDun	Japan
Dendrocalamus giganteus	Rajsthan	Burma
Eucalyptus alba	Dehra Dun	Indonesia
Eucalyptus globules	Nilgiris	Australia
Fraxinus excelsa	Kulu (H.P.)	Europe
Hevea brasiliensis	Kerala	Brazil
Grevillea pteridifolia	Amarkantak (M. P.)	Australia
Leucaena laucocephala	Dehra Dun, Gujarat	Hawai, Peru
Pinus nigra	West Bengal, H. P.	Europe
Populus deltoides	Haryana, Punjab	British Forestry Commission
Prosopis juliflora	Rajsthan	Mexico
Prunus armeniaca	Kashmir, H. P.	China
Pseudotsuga taxifolia	H. P.	North merica
Salix alba	Kasmir, H. P.	China
Swietenia macrophylla	West Bengal, Karnatak	Central America
Swietenia mahogoani	Maharashtra	Cuba, Jamaica

81. Different crossing with their characteristics

Cross Name	Studied character	Phenotypic Ratio	Genotypic Ratio
Monohybrid Cross	One trait	3:1	1:2:1
Dihybrid cross	Two -pair	9:3:3:1	1:2:2:4:1:2:1:2:1
Monohybrid test cross	-	1:1	1:1
Dihybrid test cross	-	1:1:1:1	1:1:1:1

82. Name of carbohydrates with known as

Name of Carbohydrates	Known as
Glucose	Dextrose, Grape Sugar, Blood Sugar
Fructose	Fruit Sugar

Name of Carbohydrates	Known as
Galactose	Brain Sugar
Maltose	Malt Sugar
Sucrose	Invert Sugar, Cane Sugar, Table Sugar
Glycogen	Animal Starch
Pectin	Plant Cement

83. Different type of RNA with there features

Features	r-RNA	t-RNA	m-RNA
Discoverer	Kuntze	Hogland, Zemecknike	Jacob & Monad
Stability	Most stable	Stable	Least stable
Ribosome	80s	4s	-
Function	Provide attachment site to r-RNA (larger site) & m-RNA	Carrier of amino acids.	New polypeptide chain formation

84. Different type of seed with their characteristics

Seed Name	Produced by	Purity	Certification	Colour (Tag)
Nucleus seed	Nucleus seed	G & P-100%	No Need	-
Breeder Seed	Nucleus seed	G & P-100%	No Need	Yellow
Foundation Seed	Breeder Seed	G-100% & P-98%	Required	White
Registered Seed	Foundation Seed	G-100% & P-98%	Required	Purple
Certified Seed	Registered Seed	G-100% & P-98%	Required	Azar Blue

85. A vertical section of soil shows five master horizons

Horizon Name	Other Names
O	Organic Horizon (Forest Soil)
A	Washing out or Elluvial horizon or Leaching horizon
B	Washing in horizon or Illuvial horizon
C	Unconsolited materials
E/R	Unweathered rocks

86. Retained potential of water at holding stage in soil

Soil Water	Retained potential at (in Bar unit)
Gravitational water	-1/3 bar
Capillary water	-1/3 to -31 bar
Hygroscopic water	> -31 bar
Available soil water	-1/3 to -15 bar
Unavailable soil water	>-15 bar

87. Classification of soil silicates clay into 4 categories on the basis of Aluminium and silica sheet arrangement.

Silicate Groups	Examples
1:1 types	Kaolinite
2:1 types	Montmorillonite
2:1 non-expanding types	Hydrous mica or Illite
2:2 types	Chlorites

88. Soil orders (12) proposed in the 7th Approximation and their derivatives.

Soil order names	Symbols	Sub-orders
Entisol	Ent	4
Vertisol	Ert	4
Inceptisol	Ept	5
Aridosol	Id	2
Mollisol	Oll	7
Spodosol	Od	4
Alfisol	Alf	5
Ultisol	Ult	5
Oxisol	Ox	5
Histosol	Ist	4
Andisol	and	7
Gelisol	el	3

89. Types of rock with their examples

Rocks	Examples
Igneous	Granite and Basalt
Sedimentary	Limestone, Sand stone and Dolomite
Metamorphic	Gneisis, Marble, Quartz and Slate

90. Classification of soil particle size.

Classification	According to IISS diameter (in mm)
Gravel	>2.0
Coarse sand	0.2-2.0
Fine sand	0.2 – 0.02
Silt	0.02 – 0.002
Clay	< 0.002

91. Nutrient contents of some important manure.

Products	N	P	K
FYM	0.5	0.2	0.5
Town Compost	1.5	0.4	1.4
Vermicompost	3.0	1.0	1.5
Night Soil	5.5	4.0	2.0
Ground nut cakes	7.3	1.5	1.3
Safflower cakes	7.9	2.2	1.9

92. Transported soil

Transported soil by	Called as
Gravity	Colluvial
Water	Alluvial
Ice	Glacial
Wind	Eolian

93. Land capability classification (LCC) has eight classes of soils.

Soil class	Suitability	Used for	Class colour
I	Very good land	Cultivation	Light green
II	Good land	Cultivation	Yellow
III	Moderately good land	Cultivation	Red

Soil class	Suitability	Used for	Class colour
IV	Fairly good land	Restrict cultivation	Blue
V	Not good land	Pasture & Grass land	Dark green
VI	Not good land	Pasture & Grass land	Orange
VII	Not good land	Pasture & Grass land	Brown
VIII	Not good land	Wildlife & Watershed	Purple

94. Important characteristics of the adopters categories

Categories	Per cent	Characteristics
Innovators	2.5	First to adopt, very few in number
Early adopters	13.5	Localite and doesn't test untried ideas
Early majority	34	Adopt after ideas just before avg. members of community
Late majority	34	Adopt new ides just after avg. Members of community
laggards	16	Traditional, belongs to backword classes

95. Different Experimental Design

Experimental design Name	Local Control	Applied to	Error	Remarks
Copletely randomised design (CRD)	Not used	Homogenous materials	(n-1)	One way classification
Randomised block design (RBD)	One way control	Upto 20 treatments	(n-1) x (r-1)	two way classification, block=treatments
Latin square design (LSD)	Used	5-8 or 5-12 treatments	(n-1) x (r-2)	rows=columns Replication=treatment
Split plot design	Used	sub plots	-	-

96. Test of Significance

Test Name	Given by	Applied to
Chi-square	Karl Pearson	Sample size ≥ 50
z-test	R. A. Fisher	Sample size > 30
Student t-test	W.S. Gosset	Sample size < 30

97. Range of important test or distribution

Tests/ Distribution	Range
t-test	$-\infty$ to $+\infty$
Normal distribution	$-\infty$ to $+\infty$
Poisson distribution	$0 \leq x \leq \infty$
f-test	$0 \leq x \leq \infty$
Correlation coefficient	-1 to +1
Regression coefficient	$-\infty$ to $+\infty$
Probability	0 to 1

98. International Years

Year	Called as/ Details
1965	International Cooperation Year
1967	International Tourism Year
1968	International Year for Human Rights
1970	International Education Year
1971	International Year for Action to Combat Racism & Racical Discrimination
1974	World Population Year
1975	International Women's Year
1978	International Anti-Apartheid Year
1979	International Year of the Child
1981	International Year of Disabled persons
1982	International Year of Mobilisation for Sanctions Against South Africa
1983	World Communication Year
1985	International Youth Year
1986	International Year of Peace
1987	International Year of Ahelter for homeless
1990	International Literacy Year
1992	International Space Year
1993	International Year of Indegenous population
1994	International Year of Family
1995	International Year of Tolerance
1998	International Year of Ocean
1999	International Year of Older people
2000	International Year of Empowerment
2001	International Year for the Eradication of Mental disease

Year	Called as/ Details
2002	International Mountain year; International Year of Ecotourism
2003	International Fresh Water Year
2004	International Rice Year
2005	International Year of Physics; International year of Microcredit; International Year for Sport and Physical Education
2006	International Year of Deserts and Desertification
2007-08	International Polar Year; Year of good Governance
2009	International Year of Astronomy and Year of Natural Fibres
2010	International Year of Biodiversity
2011	International Year of Forests; International Year of Chemistry
2012	International Year of Sustainable energy for all
2013	International Year of Water Cooperation
2014	International Year of Family Farming; International year of Small Island developing states
2015	International Year of Light and Light based Technologies; International year of soils
2016	International Year of Pulses and International year of Camelids
2017	International Year of Sustainable Tourism for Development
2019	International Year of Indegeneous Language
2020	International Year of Plant Health
2021	International Year of Peace and Trust; International Year of Creative Economy for Sustainable Development; International Year of Fruits and Vegetables; International Year for the Elimination of Child Labor.

99. List of Botanical Garden in India

Sl. No.	Name	Location
1.	Indian Botanic Garden	Howrah
2.	Botanic Garden of Indian Republic	Noida
3.	Experimental Botanic Garden Andaman & Nicobar Circle	Dhanikheri
4.	Experimental Botanical Garden Arunachal Field station	Sankie View
5.	Experimental Botanical Garden Arid Zone Circle	Jodhpur
6.	Experimental Botanical Garden Central Circle	Allahabad
7.	Experimental Botanical Garden Eastern Circle (Shillong)	Barapani
8.	Experimental Botanical Garden Northern Circle	Pauri
9.	Experimental Botanical Garden Northern Circle	Khirsu

Sl. No.	Name	Location
10.	Experimental Botanical Garden Northern Circle	Dehradun
11.	Experimental Botanical Garden Sikkim Himalayan Circle	Gangtok
12.	Experimental Botanical Garden Southern Circle	Yercaud
13.	Experimental Botanical Garden Western Circle	Mundhwa (Pune)

100. Choice of silvicultural system in different conditions

Conditions	Silvicultural systems
Light Demandr Species	Clear felling/ Clear strip system
Shade Bearer Species	Shelter wood/ Selection system
Rocky & Precipitous Slope	Selection system
Catchment of Rivers	Selection/ Indian Irregular shelter wood system
Snow prone Areas	Selection/ Indian Irregular shelter wood system
Fuel wood/small timber	Coppice systems
Large sized timber	High Forests system
Canal plantation	Clear felling system
Road side plantations	Clear felling with reserves/ simple coppice system